ENCYCLOPEDIA OF MATHEMATICS AND ITS APPLICATIONS

FOUNDED BY G.-C. ROTA

Editorial Board
R. S. Doran, P. Flajolet, M. Ismail, T.-Y. Lam, E. Lutwak

Volume 88

Solving Polynomial Equation Systems I

ENCYCLOPEDIA OF MATHEMATICS AND ITS APPLICATIONS

4   W. Miller, Jr. *Symmetry and separation of variables*
6   H. Minc *Permanents*
11 W. B. Jones and W. J. Thron *Continued fractions*
12 N. F. G. Martin and J. W. England *Mathematical theory of entropy*
18 H. O. Fattorini *The Cauchy problem*
19 G. G. Lorentz, K. Jetter and S. D. Riemenschneider *Birkhoff interpolation*
21 W. T. Tutte *Graph theory*
22 J. R. Bastida *Field extensions and Galois theory*
23 J. R. Cannon *The one-dimensional heat equation*
25 A. Salomaa *Computation and automata*
26 N. White (ed.) *Theory of matroids*
27 N. H. Bingham, C. M. Goldie and J. L. Teugels *Regular variation*
28 P. P. Petrushev and V. A. Popov *Rational approximation of real functions*
29 N. White (ed.) *Combinatorial geometrics*
30 M. Pohst and H. Zassenhaus *Algorithmic algebraic number theory*
31 J. Aczel and J. Dhombres *Functional equations containing several variables*
32 M. Kuczma, B. Chozewski and R. Ger *Iterative functional equations*
33 R. V. Ambartzumian *Factorization calculus and geometric probability*
34 G. Gripenberg, S.-O. Londen and O. Staffans *Volterra integral and functional equations*
35 G. Gasper and M. Rahman *Basic hypergeometric series*
36 E. Torgersen *Comparison of statistical experiments*
37 A Neumaier *Intervals methods for systems of equations*
38 N. Korneichuk *Exact constants in approximation theory*
39 R. A. Brualdi and H. J. Ryser *Combinatorial matrix theory*
40 N. White (ed.) *Matroid applications*
41 S. Sakai *Operator algebras in dynamical systems*
42 W. Hodges *Model theory*
43 H. Stahl and V. Totik *General orthogonal polynomials*
44 R. Schneider *Convex bodies*
45 G. Da Prato and J. Zabczyk *Stochastic equations in infinite dimensions*
46 A Bjorner, M. Las Vergnas, B. Sturmfels, N. White and G. Ziegler *Oriented matroids*
47 E. A. Edgar and L. Sucheston *Stopping times and directed processes*
48 C. Sims *Computation with finitely presented groups*
49 T. Palmer *Banach algebras and the general theory of *-algebras*
50 F. Borceux *Handbook of categorical algebra I*
51 F. Borceux *Handbook of categorical algebra II*
52 F. Borceux *Handbook of categorical algebra III*
54 A. Katok and B. Hassleblatt *Introduction to the modern theory of dynamical systems*
55 V. N. Sachkov *Combinatorial methods in discrete mathematics*
56 V. N. Sachkov *Probabilistic methods in discrete mathematics*
57 P. M. Cohn *Skew fields*
58 Richard J. Gardner *Geometric tomography*
59 George A. Baker, Jr. and Peter Graves-Morris *Padé approximants*
60 Jan Krajicek *Bounded arithmetic, propositional logic, and complex theory*
61 H. Gromer *Geometric applications of Fourier series and spherical harmonics*
62 H. O. Fattorini *Infinite dimensional optimization and control theory*
63 A. C. Thompson *Minkowski geometry*
64 R. B. Bapat and T. E. S. Raghavan *Nonnegative matrices and applications*
65 K. Engel *Sperner theory*
66 D. Cvetkovic, P. Rowlinson and S. Simic *Eigenspaces of graphs*
67 F. Bergeron, G. Labelle and P. Leroux *Combinatorial species and tree-like structures*
68 R. Goodman and N. Wallach *Representations of the classical groups*
69 T. Beth, D. Jungnickel and H. Lenz *Design theory I 2 ed.*
70 A Pietsch and J. Wenzel *Orthonormal systems and Banach space geometry*
71 George E. Andrews, Richard Askey and Ranjan Roy *Special Functions*
72 R. Ticciati *Quantum field theory for mathematicians*
76 A. A. Ivanov *Geometry of sporadic groups I*
78 T. Beth, D. Jungnickel and H. Lenz *Design theory II 2 ed.*
80 O. Stormark *Lie's structural approach to PDE systems*
88 T. Mora *Solving polynomial equation systems I*

ENCYCLOPEDIA OF MATHEMATICS AND ITS APPLICATIONS

Solving Polynomial Equation Systems I

The Kronecker–Duval Philosophy

TEO MORA

University of Genoa

PUBLISHED BY THE PRESS SYNDICATE OF THE UNIVERSITY OF CAMBRIDGE
The Pitt Building, Trumpington Street, Cambridge, United Kingdom

CAMBRIDGE UNIVERSITY PRESS
The Edinburgh Building, Cambridge CB2 2RU, UK
40 West 20th Street, New York, NY 10011-4211, USA
477 Williamstown Road, Port Melbourne, VIC 3207, Australia
Ruiz de Alarcón 13, 28014 Madrid, Spain
Dock House, The Waterfront, Cape Town 8001, South Africa

http://www.cambridge.org

© Cambridge University Press 2003

This book is in copyright. Subject to statutory exception
and to the provisions of relevant collective licensing agreements,
no reproduction of any part may take place without
the written permission of Cambridge University Press.

First published 2003

Printed in the United Kingdom at the University Press, Cambridge

Typeface Times 10/13 pt   *System* LATEX 2_ε [TB]

A catalogue record for this book is available from the British Library

Library of Congress Cataloguing in Publication data

Mora, Teo.
Solving polynomial equation systems : the Kronecker-Duval philosophy / Teo Mora.
p. cm. – (Encyclopedia of mathematics and its applications; v. 88)
Includes bibliographical references and index.
ISBN 0 521 81154 6
1. Equations–Numerical solutions. 2. Polynomials. 3. Iterative methods (Mathematics)
I. Title. II. Series.
QA218 .M64   2002
512.9′4–dc21   2001043132

ISBN 0 521 81154 6 hardback

In the beginning God created the heaven and the earth.
Genesis

Quapropter bono christiano, sive mathematici, sive quilibet impie divinatium, maxime dicentes vera, cavendi sunt.
St Augustine, *De genesis ad literam*

The most effective way for solving polynomial equation systems is just to interpret such a system as a tool for solving itself, by building programs which use this tool to manipulate its own roots.

Therefore, the best way for solving is to return the equations (well, perhaps after some massaging) shouting sufficiently loudly that *that* is the solution.

This really means that instead of working hard to build programs which *compute* the solutions, one should work hard to build programs which use the given equations in order to *manipulate* the solutions, without even computing them.

That is the Kronecker–Duval Philosophy.
R.F. Ree, *The foundational crisis, a crisis of computability?*

Since the desperate cry of Galois, 'I have no time', but even since the scribbled note of Fermat, 'I have no space', Mathematics has been forced to investigate Complexity.
E.B. Gebstadter, *Copper, Silver, Gold: an Indestructible Metallic Alloy*

Contents

Preface		*page* xi
	Part one: The Kronecker – Duval Philosophy	1
1	**Euclid**	3
	1.1 The Division Algorithm	4
	1.2 Euclidean Algorithm	6
	1.3 Bezout's Identity and Extended Euclidean Algorithm	8
	1.4 Roots of Polynomials	9
	1.5 Factorization of Polynomials	10
	1.6* Computing a gcd	12
	1.6.1* *Coefficient explosion*	12
	1.6.2* *Modular Algorithm*	16
	1.6.3* *Hensel Lifting Algorithm*	16
	1.6.4* *Heuristic gcd*	18
2	**Intermezzo: Chinese Remainder Theorems**	23
	2.1 Chinese Remainder Theorems	24
	2.2 Chinese Remainder Theorem for a Principal Ideal Domain	26
	2.3 A Structure Theorem (1)	29
	2.4 Nilpotents	32
	2.5 Idempotents	35
	2.6 A Structure Theorem (2)	39
	2.7 Lagrange Formula	41
3	**Cardano**	47
	3.1 A Tautology?	47
	3.2 The Imaginary Number	48
	3.3 An Impasse	51
	3.4 A Tautology!	52

4 Intermezzo: Multiplicity of Roots — 53
- 4.1 Characteristic of a Field — 54
- 4.2 Finite Fields — 55
- 4.3 Derivatives — 57
- 4.4 Multiplicity — 58
- 4.5 Separability — 62
- 4.6 Perfect Fields — 64
- 4.7 Squarefree Decomposition — 68

5 Kronecker I: Kronecker's Philosophy — 74
- 5.1 Quotients of Polynomial Rings — 75
- 5.2 The Invention of the Roots — 76
- 5.3 Transcendental and Algebraic Field Extensions — 81
- 5.4 Finite Algebraic Extensions — 84
- 5.5 Splitting Fields — 86

6 Intermezzo: Sylvester — 91
- 6.1 Gauss Lemma — 92
- 6.2 Symmetric Functions — 96
- 6.3* Newton's Theorem — 100
- 6.4 The Method of Indeterminate Coefficients — 106
- 6.5 Discriminant — 108
- 6.6 Resultants — 112
- 6.7 Resultants and Roots — 115

7 Galois I: Finite Fields — 119
- 7.1 Galois Fields — 120
- 7.2 Roots of Polynomials over Finite Fields — 123
- 7.3 Distinct Degree Factorization — 125
- 7.4 Roots of Unity and Primitive Roots — 127
- 7.5 Representation and Arithmetics of Finite Fields — 133
- 7.6* Cyclotomic Polynomials — 135
- 7.7* Cycles, Roots and Idempotents — 141
- 7.8 Deterministic Polynomial-time Primality Test — 148

8 Kronecker II: Kronecker's Model — 156
- 8.1 Kronecker's Philosophy — 156
- 8.2 Explicitly Given Fields — 159
- 8.3 Representation and Arithmetics — 164
  - 8.3.1 *Representation* — 164
  - 8.3.2 *Vector space arithmetics* — 165
  - 8.3.3 *Canonical representation* — 165
  - 8.3.4 *Multiplication* — 167
  - 8.3.5 *Inverse and division* — 167

		8.3.6 *Polynomial factorization*	168
		8.3.7 *Solving polynomial equations*	169
		8.3.8 *Monic polynomials*	169
	8.4	Primitive Element Theorems	170
9	**Steinitz**		**175**
	9.1	Algebraic Closure	176
	9.2	Algebraic Dependence and Transcendency Degree	180
	9.3	The Structure of Field Extensions	184
	9.4	Universal Field	186
	9.5*	Lüroth's Theorem	187
10	**Lagrange**		**191**
	10.1	Conjugates	192
	10.2	Normal Extension Fields	193
	10.3	Isomorphisms	196
	10.4	Splitting Fields	203
	10.5	Trace and Norm	206
	10.6	Discriminant	212
	10.7*	Normal Bases	216
11	**Duval**		**221**
	11.1	Explicit Representation of Rings	221
	11.2	Ring Operations in a Non-unique Representation	223
	11.3	Duval Representation	224
	11.4	Duval's Model	228
12	**Gauss**		**232**
	12.1	The Fundamental Theorem of Algebra	232
	12.2	Cyclotomic Equations	237
13	**Sturm**		**263**
	13.1*	Real Closed Fields	264
	13.2	Definitions	272
	13.3	Sturm	275
	13.4	Sturm Representation of Algebraic Reals	280
	13.5	Hermite's Method	284
	13.6	Thom Codification of Algebraic Reals (1)	288
	13.7	Ben-Or, Kozen and Reif Algorithm	290
	13.8	Thom Codification of Algebraic Reals (2)	294
14	**Galois II**		**297**
	14.1	Galois Extension	298
	14.2	Galois Correspondence	300
	14.3	Solvability by Radicals	305
	14.4	Abel–Ruffini Theorem	314

14.5*	Constructions with Ruler and Compass	318

Part two: Factorization — 327

15 Prelude — 329
- 15.1 A Computation — 329
- 15.2 An Exercise — 338

16 Kronecker III: factorization — 346
- 16.1 Von Schubert Factorization Algorithm over the Integers — 347
- 16.2 Factorization of Multivariate Polynomials — 350
- 16.3 Factorization over a Simple Algebraic Extension — 352

17 Berlekamp — 361
- 17.1 Berlekamp's Algorithm — 361
- 17.2 The Cantor–Zassenhaus Algorithm — 369

18 Zassenhaus — 380
- 18.1 Hensel's Lemma — 381
- 18.2 The Zassenhaus Algorithm — 389
- 18.3 Factorization Over a Simple Transcendental Extension — 391
- 18.4 Cauchy Bounds — 395
- 18.5 Factorization over the Rationals — 398
- 18.6 Swinnerton-Dyer Polynomials — 402
- 18.7 L^3 Algorithm — 405

19 Finale — 415
- 19.1 Kronecker's Dream — 415
- 19.2 Van der Waerden's Example — 415

Bibliography — 420

Index — 422

Preface

If you HOPE too much from this SPES book, you will probably be disappointed: in fact not only is it nothing more than an extension of some notes of my undergraduate course, but also my *horror vacui* compelled me to fill it with irrelevant information.

If, notwithstanding this *incipit*, you are not yet disinterested by SPES, I will now provide a quick résumé of this volume.

In the first part, *The Kronecker–Duval Philosophy*, my aim is to discuss recent approaches to **S**olving **P**olynomial **E**quation **S**ystems endorsed by the Project **PoSSo**[1] through the most elementary case: the solution of a single univariate polynomial $f(X) \in \mathbb{Q}[X]$.

It requires an introduction to Kronecker's theory (finitely generated field extensions, algebraic extensions, splitting fields) and allows us to stress the importance of the revolutionary approach introduced by Kronecker: before him, the notion of 'solving' meant producing techniques for computing the roots of the equation $f(X) = 0$; Kronecker interpreted 'solving' as producing techniques for computing **with** the roots: this in a nutshell is the rôle of algebraic extensions.

This change of perspective about 'solving', stressing more the manipulation than the computations of roots, is now central to the approaches for solving polynomial equation systems: in this volume I will sketch the significantly of the Duval model and of the Thom codification of real algebraic numbers.

Such an introduction of Kronecker's theory forced me to orient the volume toward a presentation of the theory of algebraic field extensions: in this task my *livre de chevet* was naturally van der Waerden's *Algebra*[2].

---

[1] **Po**lynomial **S**ystem **So**lving, ESPRIT-BRA 6846.
[2] B. L. van der Waerden, *Algebra*, vol. I, Ungar, New York.
  I mainly used the 1950 translation more than the 1970 one; the choice is 'political'.

A discussion of Kronecker's theory requires a discussion of polynomial factorization which is the content of the second part, *Factorization*[3], which is devoted to a discussion of factorization over extension fields of prime fields, in particular the Berlekamp–Hensel–Zassenhaus factorization algorithm – but I have also included a sketch of the L^3 one.

This volume should be seen as a part of a more general survey of solving polynomial equation systems; while I already have a plan of the structure[4] and of the content of that survey, I would prefer not to bind myself too much discussing it here.

As any writer knows, the number of hidden mistakes in a draft is always larger than the number of the found ones; this text is no different. I am very grateful to Mariemi Alonso, Domenico Arezzo, Miguel Anger Borges Trenard and Maria Grazia Marinari who saved me from making some mathematical mistakes and, at the same time, detected many misspellings. I want to apologize to the reader for any errors (both misspellings and mathematical) which may still lurk.

I am grateful to the ISSAC'96 Conference in Zurich, where I was an invited tutorial speaker. This book grew out of the notes of my talks there. I am also grateful to Mika Seppala and the Mathematics Department of Florida State University in Tallahassee for inviting me to be a visiting professor in 1999. That semester in Tallahassee gave me an opportunity to test these notes in the course I offered there.

It is my firm belief that the best way of understanding a theory and an algorithm is to verify it through computation; therefore the book contains many examples which have been mainly developed via paper-and-pencil computations[5] – an approach which naturally is strongly prone to further mistakes; the readers are encouraged to follow them and, better, to test their own examples.

In order to help the readers to plan their journey through this book, some sections, containing only some interesting digressions, are indicated by asterisks in the table of contents.

---

[3] In the preparation of this part, I was mainly dependent on E. Kaltofen, Factorization of polynomials in B. Buchberger, G.E. Collins, R. Loos (Eds.) *Computer Algebra, System and Algebraic Computation*, Springer, 1982 and J.H. Davenport, Y. Siret, E. Tournier, *Computer Algebra*, Academic Press, (1988) where the reader can also find a vast bibliography.

[4] The reader can guess it from the quotations.

[5] I used computer algebra systems only to perform the four operations with polynomials.

A possible short cut which allows the readers to appreciate the discussion, without being too bored by the details (and which I usually employ in my lessons) is Chapters 1–5, 8, 11 to which I strongly suggest adding, according to the reader's interest, one of the two short tours devoted to real numbers (Section 12.1 and Chapter 13) and to Galois theory (Section 12.2, Chapter 14).

Well, good reading!

Part one
The Kronecker – Duval Philosophy

And I saw when the Lamb opened one of the seals, and I heard, as it were the noise of thunder, one of the four beasts saying, Come and see.
    And I saw, and behold a white horse: and he that sat on him had a bow; and a crown was given unto him: and he went forth conquering, and to conquer.
Revelations

The things depending from Saturn: bile, lead, onyx, asphodel, mole, hoopoe, eel.
E.C. Agrippa, *De occulta phylosophia*

Soon we will drink blood for wine.
Revolutionary of the Upper Rhine, *Book of a hundred chapters*

1
Euclid

This preliminary chapter is just devoted to recalling the Euclidean Algorithms over a univariate polynomial ring and its elementary applications: roughly speaking they are essentially the obvious generalization of those over integers.

The fundamental tool related to the Euclidean Algorithms and to solving univariate polynomials is nothing more than the elementary Division Algorithm (Section 1.1), whose iterative application produces the Euclidean Algorithm (Section 1.2), which can be extended to prove and compute Bezout's Identity (Section 1.3).

The Division- and Euclidean Algorithms and theorems have many important consequences for solving polynomial equations: they relate roots and linear factors of a polynomial (Section 1.4) allowing them, at least, to be counted, and are the basis for the theory (not the practice) of polynomial factorization (Section 1.5).

They also have another, more important, consequence which is a crucial tool in solving: they allow a computational system to be developed within quotients of polynomial rings; the discussion of this is postponed to Section 5.1.

A direct implementation of the Euclidean Algorithm provides an unexpected phenomenon, the 'coefficient explosion': during the application of the Euclidean Algorithm to two polynomials whose coefficients have small size, polynomials are produced with huge coefficients, even if the final output is simply 1. Finding efficient implementations of the Euclidean Algorithm was a crucial subject of research in the early days of Computer Algebra; in Section 1.6 I will briefly discuss this phenomenon and present efficient solutions to this problem.

1.1 The Division Algorithm

Throughout this chapter k will be a field and $\mathcal{P} := k[X]$ the univariate polynomial ring over k.

If $f = \sum_{i=0}^{n} a_i X^i \in \mathcal{P}$ with $a_n \neq 0$, denote by $\mathrm{lc}(f) := a_n$ the *leading coefficient* of f.

Theorem 1.1.1 (Division Theorem). *Given $A(X), B(X) \in \mathcal{P}$, $B \neq 0$, there are unique $Q(X), R(X) \in \mathcal{P}$ such that*

(1) $A(X) = Q(X)B(X) + R(X)$;
(2) $R \neq 0 \implies \deg(R) < \deg(B)$.

We call Q the quotient and R the remainder of A modulo B in $\mathcal{P}$.

Proof <u>Existence</u>: The proof is by induction on $\deg(A)$.

If $A = 0$ or $\deg(A) < \deg(B)$, then $Q := 0$ and $R := A$ obviously satisfy the thesis.

If $\deg(A) = n \geq m = \deg(B)$, we inductively assume that the theorem is true for each polynomial A_0 such that $A_0 = 0$ or $\deg(A_0) < n$. We then have

$$A(X) = a_n X^n + A_1(X), \quad B(X) = b_m X^m + B_1(X),$$

with $a_n \neq 0$, $b_m \neq 0$, $A_1 = 0$ or $\deg(A_1) < n$, $B_1 = 0$ or $\deg(B_1) < m$.
Let

$$A_0(X) := A(X) - a_n b_m^{-1} X^{n-m} B(X),$$

which, if non-zero, has degree less than n; by the inductive assumption there are then Q_0, R_0 such that

(1) $A_0(X) = Q_0(X)B(X) + R_0(X)$,
(2) $R_0 \neq 0 \implies \deg(R_0) < \deg(B)$,

so that

$$A(X) = (a_n b_m^{-1} X^{n-m} + Q_0(X))B(X) + R_0(X)$$

and therefore

$$Q(X) := a_n b_m^{-1} X^{n-m} + Q_0(X), R(X) := R_0(X)$$

satisfy the requirement.

<u>Uniqueness</u>: Assume that

(1) $A(X) = Q_1(X)B(X) + R_1(X)$,
(2) $A(X) = Q_2(X)B(X) + R_2(X)$,

(3) $R_i \neq 0 \implies \deg(R_i) < \deg(B), 1 \leq i \leq 2$,

so that

$$R_1(X) - R_2(X) = (Q_2(X) - Q_1(X)) B(X).$$

If $R_1 \neq R_2$ then

$$\deg(R_1 - R_2) < \deg(B) \leq \deg(Q_2 - Q_1) + \deg(B) = \deg(R_1 - R_2)$$

giving a contradiction.
Therefore $R_1 - R_2 = 0$ and (since $B \neq 0$) also $Q_2 - Q_1 = 0$. ♄

Corollary 1.1.2. *The ring $\mathcal{P}$ is a euclidean domain.* ♄

In further applications, denote

$$Q := \mathbf{Quot}(A, B), R := \mathbf{Rem}(A, B).$$

Because of their uniqueness in $\mathcal{P}$, if K is a field such that $K \supseteq k$, the quotient and the remainder of A modulo B in $K[X]$ are still Q and R.

Algorithm 1.1.3. An inductive proof can be transformed into a recursive algorithm: If we assume k to be effective[1] then the iterative algorithm in Figure 1.1 performs polynomial division.

---

[1] The concept of effectiveness was first introduced as the notion of *endlichvielen Schritten* (finite number of steps) by Grete Hermann in 1926 for polynomial ideals in the fundamental paper

G. Hermann, *Die Frage der endlich vielen Schritte in der Theorie der Polynomideale, Math. Ann.* **95** (1926) 736–788,

where she wrote:

Die Behauptung, eine Berechnung kann mit endlich vielen Schritten durchgeführt werden, soll dabei bedeuten, es kann eine *obere Schranke für die Anzahl der zur Berechnung notwendigen Operationen* angegeben werden. Es genügt also z. B. nicht, ein Verfahren anzugeben, von dem man theoretisch nachweisen kann, daß es mit endlich vielen Operationen zum Ziele führt, wenn für die Anzahl dieser Operationen keine obere Schranke bekannt ist.
The assertion that a computation can be carried through in a finite number of steps shall mean that an upper bound for the number of operations needed for the computation *can be given. Thus it is not sufficient, for example, to give a procedure for which one can theoretically verify that it leads to the desired result in a finite number of operations, so long as no upper bound is known for the number of operations,*

To this, van der Waerden in

B.L. van der Waerden, Eine Bemerkung über die Unzelegbarkeit von Polynomen, *Math. Ann.* **102** (1930), 738–739,

Fig. 1.1. Polynomial Division Algorithm

(Q,R) := **PolynomialDivision**(A,B)
where
    $A, B \in k[X], B \neq 0$
    $Q, R \in k[X]$ are such that
    − $A = QB + R$
    − $R \neq 0 \implies \deg(R) < \deg(B)$
$b := \mathrm{lc}(B), m := \deg(B)$
$A_0 := A, Q := 0$
While $A_0 \neq 0$ **and** $\deg(A_0) \geq \deg(B)$ **do**
    $a := \mathrm{lc}(A_0), n := \deg(A_0)$
    $Q := Q + ab^{-1}X^{n-m}$
    $A_0 := A_0 - ab^{-1}X^{n-m}B$
$R := A_0$

1.2 Euclidean Algorithm

Let $P_0, P_1 \in \mathcal{P}$, with $P_1 \neq 0$ (and, to dispose of the trivial cases, assume also that $P_0 \neq 0$). Let $P_2 := \mathbf{Rem}(P_0, P_1)$ and inductively, define

$$P_{i+1} := \mathbf{Rem}(P_{i-1}, P_i)$$

while $P_i \neq 0$. It is clear that the sequence $P_0, P_1, \ldots, P_i, \ldots$ (which is called the *polynomial remainder sequence* (PRS) of P_0, P_1) is finite since, otherwise,

---

added the note

Ein Körper *K* soll *explizite-bekann* heißen, wenn seine Elemente Symbole aus einem bekannten abzählbaren Vorrat von unterscheidbaren Symbolen sind, deren Addition, Multiplikation, Subtraktion und Division sich in endlichvielen Schritten ausführen lassen.

A field K is called explicitly given *when its elements are symbols from a known numerable set of distinguishable symbols, whose addition, multiplication, subtraction and division can be performed in* a finite number of steps.

In this book I will happily drop Hermann's requirement that an algorithm must be provided with its complexity evaluation, and will mainly follow Macaulay's opinion in

F.S. Macaulay, *The Algebraic Theory of Modular Systems*, Cambridge University Press (1916).

Macaulay considered the practical feasibility of an algorithm to be more crucial:

[The theory of polynomial ideals] might be regarded as in some measure complete if it were admitted that a problem is solved when its solution has been reduced to a finite number of feasible operations. If, however, the operations are too numerous or too involved to be carried out in practice the solution is only a theoretical one.

1.2 Euclidean Algorithm

each P_i must be non-zero which would give an infinite decreasing sequence of natural numbers:

$$\deg(P_1) > \deg(P_2) > \cdots > \deg(P_i) > \cdots.$$

Let $D(X)$ denote the last non-zero element P_r of the sequence, and note that $r \leq \min(\deg(P_0), \deg(P_1))$. Also denote $Q_i := \mathbf{Quot}(P_{i-1}, P_i)$.

Proposition 1.2.1. $D(X) = \gcd(P_0, P_1)$.

Proof Since $P_{r-1} = Q_r P_r$, then P_r divides P_{r-1}. So let us assume that P_r divides P_i for $i > k$ and prove that it divides P_k: this is obvious from the identity

$$P_k = Q_{k+1} P_{k+1} + P_{k+2}.$$

Therefore $D = P_r$ is a common divisor of P_0 and P_1.
If $S(X)$ divides both P_0 and P_1, then since

$$P_2 = P_0 - Q_1 P_1,$$

it divides P_2. Assuming that S divides P_i, for $i < k$, then by the identity

$$P_k = P_{k-2} - Q_{k-1} P_{k-1},$$

it also divides P_k, therefore it divides P_r. ♄

Greatest common divisors in $\mathcal{P}$ are obviously not unique, but they are associate (cf. Definition 1.5.1).

Again if K is a field such that $K \supseteq k$, $\gcd(A, B)$ and the PRS of A and B are the same in $K[X]$ as in $\mathcal{P}$.

Algorithm 1.2.2. If k is effective, the algorithm in Figure 1.2 computes the gcd of two polynomials; it actually computes the PRS of the two polynomials and also computes all the intermediate quotients Q_j.

Fig. 1.2. Euclidean Algorithm

$D := \mathbf{GCD}(A, B)$
where
    $A, B \in \mathcal{P}, A \neq 0, B \neq 0$
    D is a $\gcd(A, B)$
$D := A, U := B$
While $U \neq 0$ **do**
    $(Q, V) := \mathbf{PolynomialDivision}(D, U)$
    $D := U, U := V$

1.3 Bezout's Identity and Extended Euclidean Algorithm

Proposition 1.3.1 (Bezout's Identity). *Let $P_0, P_1 \in \mathcal{P} \setminus k$, and let us denote $D := \gcd(P_0, P_1)$. Then there are $S, T \in \mathcal{P} \setminus \{0\}$ such that*

(i) $P_0 S + P_1 T = D$
(ii) $\deg(S) < \deg(P_1), \deg(T) < \deg(P_0)$

Proof Let $P_0, P_1, \ldots, P_i, \ldots, P_r = D$ be the PRS of P_0 and P_1. Also, for $i = 0, \ldots, r-1$, let $Q_i := \mathbf{Quot}(P_{i-1}, P_i)$. Inductively define:

$$\begin{aligned} S_0 &:= 1, & T_0 &:= 0; \\ S_1 &:= 0, & T_1 &:= 1; \\ S_i' &:= S_{i-2} - Q_{i-1} S_{i-1}, & T_i' &:= T_{i-2} - Q_{i-1} T_{i-1}, & 2 \leq i \leq r; \\ S_i &:= \mathbf{Rem}(S_i', P_1), & T_i &:= T_i' + \mathbf{Quot}(S_i', P_1) P_0, & 2 \leq i \leq r. \end{aligned}$$

We claim that for $i = 0, \ldots, r$:

(i) $P_0 S_i + P_1 T_i = P_i$;
(ii) $\deg(S_i) < \deg(P_1), \deg(T_i) < \deg(P_0)$.

In fact the claims are trivial for $i = 0, 1$, and so, inductively assuming them to be true for $i < k$, and denoting $U_k := \mathbf{Quot}(S_k', P_1)$, so that

$$S_k' = U_k P_1 + S_k, \quad T_k = T_k' + U_k P_0,$$

we have

$$\begin{aligned} P_k &= P_{k-2} - Q_{k-1} P_{k-1} \\ &= P_0 S_{k-2} + P_1 T_{k-2} - Q_{k-1} P_0 S_{k-1} - Q_{k-1} P_1 T_{k-1} \\ &= P_0 (S_{k-2} - Q_{k-1} S_{k-1}) + P_1 (T_{k-2} - Q_{k-1} T_{k-1}) \\ &= P_0 S_k' + P_1 T_k' \\ &= P_0 U_k P_1 + P_0 S_k + P_1 T_k - P_1 U_k P_0 \\ &= P_0 S_k + P_1 T_k. \end{aligned}$$

Clearly $\deg(S_k) < \deg(P_1)$ and therefore also $\deg(T_k) < \deg(P_0)$, otherwise

$$\deg(P_1 T_k) \geq \deg(P_1 P_0) > \deg(S_k P_0)$$

and $\deg(P_1 T_k) > \deg(P_1) \geq \deg(P_k)$ would lead to an obvious contradiction. ♄

Corollary 1.3.2. *The ring $\mathcal{P}$ is a principal ideal domain.* ♄

Fig. 1.3. Extended Euclidean Algorithm

$(D, S, T) := \textbf{ExtGCD}(A, B)$
where
   $A, B \in \mathcal{P}, A \neq 0, B \neq 0$
   D is a $\gcd(A, B)$
   $SA + BT = D$
   $\deg(S) < \deg(B), \deg(T) < \deg(A)$
$D := A, U := B$
$S_0 := 1, S_1 := 0$
$\rightarrow T_0 := 0, T_1 := 1$
While $U \neq 0$ **do**
   $(Q, V) := \textbf{PolynomialDivision}(D, U)$
   $D := U, U := V$
   $S := S_0 - QS_1,$
   $\rightarrow T := T_0 - QT_1$
   $(Q, S) := \textbf{PolynomialDivision}(S, B)$
   $\rightarrow T := T + QA$
   $S_0 := S_1, S_1 := S$
   $\rightarrow T_0 := T_1, T_1 := T$
$S := S_0,$
$\rightarrow T := T_0$

Algorithm 1.3.3. Again, on an effective field, S and T can be computed by the algorithm in Figure 1.3.

Algorithm 1.3.4. The so-called Half-extended Euclidean Algorithm allows us to compute S, without having to compute T; it simply involves removing the lines marked by $\rightarrow$ in the algorithm in Figure 1.3. It is useful to compute inverses of field elements (see Remark 5.1.4).

1.4 Roots of Polynomials

The Division Theorem also has an obvious but important consequence on the solving of polynomial equations:

Corollary 1.4.1. *For $f(X) \in \mathcal{P}$, and $\alpha \in k$ we have:*

$$f(\alpha) = 0 \iff (X - \alpha) \text{ divides } P(X).$$

Proof Let

$$Q(X) := \textbf{Quot}(f(X), X - \alpha), \quad R(X) := \textbf{Rem}(f(X), X - \alpha);$$

since $(X - \alpha)$ is linear, either $R(X) = 0$ or $\deg(R) = 0$, i.e. $R(X)$ is a constant $r \in k$.

Therefore,
$$f(X) = Q(X)(X - \alpha) + r,$$
and evaluating in α obtains $f(\alpha) = r$, from which the proof follows. ♄

As a consequence a polynomial cannot have more roots than its degree.

1.5 Factorization of Polynomials

Definition 1.5.1. *In a domain D:*

(i) *two elements a and b are called* associate *if there exists $c \in D$, with c invertible, such that $a = bc$;*
(ii) *a non-zero and non-invertible element a is called* irreducible *if it is divisible only by invertible elements and by its associates, i.e.*

$a = bc$, *and b non-invertible* $\implies$ *c is invertible and so b is associate to a.*

Definition 1.5.2. *A domain D is a* unique factorization domain *if for each non-invertible $a \in D \setminus \{0\}$*

(i) *there is a factorization $a = p_1 \ldots p_r$ where each p_i is irreducible;*
(ii) *the factorization is unique in the following sense:*

if $a = q_1 \ldots q_s$ is another factorization with q_i irreducible, then

- *$r = s$,*
- *each p_i is associate to some q_j,*
- *each q_j is associate to some p_i.*

Lemma 1.5.3. *If $p(X) \in k[X]$ is irreducible, p divides $q_1 q_2$ and p does not divide q_2, then p divides q_1.*

Proof Since $\gcd(p, q_2)$ divides p, it either is associate to p or is a unit; since p does not divide q_2, we can then conclude that $\gcd(p, q_2) = 1$.
By Bezout's Identity, there are $s, t \in k[X]$, such that $sp + tq_2 = 1$ and therefore $spq_1 + tq_1 q_2 = q_1$, so that p divides q_1. ♄

Lemma 1.5.4. *Let $f \in k[X]$; Let $f = p_1 \ldots p_r$, $f = q_1 \ldots q_s$ be two factorizations in irreducible factors. Then*

(i) *$r = s$,*
(ii) *each p_i is associate to some q_j,*
(iii) *each q_j is associate to some p_i.*

1.5 Factorization of Polynomials

Proof The proof is by induction on r. If $r = 1$, then $p_1 = f = q_1 \ldots q_s$, so that $s = 1$ and $p_1 = q_1$ because p_1 is irreducible.

Assume therefore that each polynomial that has a factorization with less than r irreducible factors, has a unique factorization and let $f = p_1 \ldots p_r$, $f = q_1 \ldots q_s$ be two factorizations of f in irreducible factors. Then p_1 divides $q_1 \ldots q_s$ and therefore, by Lemma 1.5.3, it must divide one among the q_is, say q_j.

Since q_j is irreducible, we have $p_1 = uq_j$ for some $u \in k \setminus \{0\}$. We then have

$$f = uq_j p_2 \ldots p_r = q_1 \ldots q_s,$$

and, dividing out q_j,

$$(up_2)p_3 \ldots p_r = q_1 \ldots q_{j-1}q_{j+1} \ldots q_s.$$

The proof can then be completed using the inductive assumption. ♄

Lemma 1.5.5. *Each non-constant polynomial $f \in k[X]$ has a factorization into irreducible factors.*

Proof The proof is by induction on $\deg(f)$.

Since linear polynomials are obviously irreducible, the result is true for polynomials of degree 1.

Assume next that it is true for polynomials $g \in k[X]$, $\deg(g) < n$, and let $f \in k[X]$ be such that $\deg(f) = n$. Either f is irreducible, so that f satisfies the lemma, or f is not irreducible, so that $f = f_1 f_2$ where neither f_1 nor f_2 is a constant and each has degree less than n; therefore there are factorizations $f_1 = p_1 \ldots p_r$ and $f_2 = q_1 \ldots q_s$ in irreducible factors, and

$$f = p_1 \ldots p_r q_1 \ldots q_s$$

is then a factorization of f. ♄

Theorem 1.5.6. $k[X]$ *is a unique factorization domain.*

Proof Existence of a factorization is guaranteed by Lemma 1.5.5, uniqueness by Lemma 1.5.4. ♄

Remark 1.5.7. It is important to note that, unlike the other results of this chapter, Theorem 1.5.6 does not give any way of computing a factorization. In fact the argument of Lemma 1.5.5, that either f is irreducible or it has a proper factorization, does not give any hint of how to decide which is the case, nor how to find proper divisors. We will show in Part II that there are factorization algorithms for polynomials over all fields which are important for our theory (namely all finite fields and all finite extensions of the rationals).

However, there exist effective fields k such that it is undecidable whether the polynomial $X^2 + 1 \in k[X]$ is irreducible or not, the reason being that it is undecidable whether the imaginary number i is in k (see Section 19.2).

1.6 Computing a gcd

1.6.1 Coefficient explosion

Example 1.6.1. Let us assume that we need to compute the gcd of the two polynomials

$$P_0 := X^8 + X^6 - 3X^4 - 3X^3 + 8X^2 + 2X - 5,$$
$$P_1 := 3X^6 + 5X^4 - 4X^2 - 9X + 21,$$

in $\mathbb{Z}[X]$; we need of course to apply the Euclidean Algorithm; let us even assume that we have available nothing more than a pocket calculator, so that we can compute only in $\mathbb{Z}$ but not in $\mathbb{Q}$.

Well, that is not a serious problem: in fact, since the gcd is stable under associate elements, it is clear that by substituting the line of the algorithm of Figure 1.1

$$A_0 := A_0 - ab^{-1}X^{n-m}B$$

by

$$A_0 := bA_0 - aX^{n-m}B,$$

the answer is correct.

In this way we obtain the following PRS:

$$P_2 := -15X^4 + 3X^2 - 9,$$
$$P_3 := -15795X^2 - 30375X + 59535,$$
$$P_4 := 1254542875143750X - 1654608338437500,$$
$$P_5 := 12593338795500743100931141992187500,$$

from which, provided we are able to complete this computation, we deduce that

$$\gcd(P_0, P_1) = 1.$$

Clearly, we can perform rational arithmetic, even if it is not available on our pocket calculator, using simply the Euclidean Algorithm for the integers; the computation is of course more complex and the answer is

$$P_2 := -\frac{5}{9}X^4 + \frac{1}{9}X^2 - \frac{1}{3},$$
$$P_3 := -\frac{117}{25}X^2 - 9X + \frac{441}{25},$$
$$P_4 := \frac{233150}{6591}X - \frac{102500}{2197},$$
$$P_5 := \frac{1288744821}{543589225}.$$

Having already used stability under associate elements, we could, at each step, force each P_i to become monic; this requires more integer Euclidean Algorithms, but we could hope to do it with small size elements; in fact we get:

$$P_2 := X^4 - \frac{1}{5}X^2 + \frac{3}{5},$$
$$P_3 := X^2 + \frac{25}{13}X - \frac{49}{13},$$
$$P_4 := X - \frac{6150}{4663},$$
$$P_5 := 1.$$

Historical Remark 1.6.2. The amusing assumption of having just a pocket calculator, while not realistic, has a meaning. In fact, the above example is taken from the second volume of Knuth's book *The Art of Computer Programming*.

That book was published in 1969, when programs were input via punched cards... and computer algebra was being born. In fact, an analysis of the unexpected phenomenon of *coefficient growth explosion*, and the first tentative steps taken for solving it, marked the beginning of the unexpected phenomenon of computer algebra's rapid growth.

Independently Collins and Brown[2], applying subresultant theory, showed that in computing the PRS over $\mathbb{Z}$ it was possible at each step, while producing an element P_i, to predict an integer c_i dividing each coefficient of P_i, and thereby, performing the substitution $P_i \leftarrow P_i/c_i$, get smaller size coefficients;

---

[2] See

G.E. Collins, Subresultants and Polynomial Remainder Sequence, *J. ACM* **14** (1967), 128–142; W.S. Brown, On Euclid's Algorithm and the Computation of Polynomial and Greatest Common Divisors, *J. ACM* **18** (1971), 478–504.

The discussion (and the computations) of the example are taken from Brown's paper.

for instance, in the example above we get:

$$P_2 := 15X^4 - 3X^2 + 9,$$
$$P_3 := 65X^2 + 125X - 245,$$
$$P_4 := 9326X - 12300,$$
$$P_5 := 260708.$$

Research on how to compute the polynomial gcd continues; on the basis of general knowledge, there are three competing approaches[3]:

modular algorithm based on the Chinese Remainder Theorem (Brown, 1971);
the *Hensel Lifting Algorithm* (Moses–Yun, 1973; Wang, 1980) based on Hensel's Lemma (cf. Section 18.1);
the *Heuristic GCD* (Char–Geddes–Gonnet, 1984; Davenport–Padget, 1985).

In the following sections we will briefly discuss these three algorithms[4], using freely some facts that will be proved later:

Fact 1.6.3. *Let $f \in \mathbb{Z}[X]$ be a polynomial. Then:*

(1) *there is a computable integer $\mathfrak{B} \in \mathbb{N}$ such that for each factor $\sum a_i X^i$ of f, we have $-\mathfrak{B} < a_i \leq \mathfrak{B}$;*
(2) *there is a computable integer $\mathfrak{r} \in \mathbb{N}$ such that for each root $\rho \in \mathbb{C}$ of f, we have $|\rho| < \mathfrak{r}$.*

Proof cf. Section 18.4. ♄

For each $p \in \mathbb{N}$ let us denote the canonical projection morphism as $-_p : \mathbb{Z}[X] \mapsto \mathbb{Z}_p[X]$; conversely, we can consider the (implicit) immersion $\mathbb{Z}_p[X] \subset \mathbb{Z}[X]$, where each polynomial $f(X) \in \mathbb{Z}_p[X]$ can be interpreted,

---

[3] See

W.S. Brown, On Euclid's Algorithm and the Computation of Polynomial and Greatest Common Divisors, *J. ACM* **18** (1971), 478–504;
J. Moses, D.Y.Y. Yun, The EZ GCD Algorithm, in *Proc. of the ACM Annual Conference* (1973), 159–166;
P. Wang, The EZZ-GCD Algorithm, *SIGSAM Bulletin* **14** (1980), 50–60;
B.W. Char, K.O. Geddes, G. H. Gonnet, GCDHEU: Heuristic Polynomial GCD Algorithm Based On Integral GCD Computation, *L. N. Comp. Sci.* **174** (1984), Springer, 285–296;
J. Davenport, J. Padget, HEUGCD: How Elementary Upperbounds Generate Cheaper Data, *L. N. Comp. Sci.* **204** (1985), Springer, 18–28.

[4] The presentation of modular algorithm depends freely on the results discussed in Section 2.1 and the presentation of the Hensel Lifting Algorithm in Section 18.1. It is suggested that the interested reader go to those sections first.

with a slight abuse of notation, as a polynomial $\mathsf{f}(X) := \sum_{i=0}^{n} a_i X^i \in \mathbb{Z}[X]$ such that

$$f(X) = \mathsf{f}_p(X),$$
$$-p/2 < a_i \leq p/2,$$

from which we can readily identify f and f.

Let $f, g \in \mathbb{Z}[X]$, $h := \gcd(f, g)$ and let $p \in \mathbb{N}$ be a prime. Then:

Lemma 1.6.4. *With the above notation:*

(1) h_p *divides* $\gcd(f_p, g_p)$;
(2) if $\mathrm{lc}(f) \not\equiv 0 \not\equiv \mathrm{lc}(g) \pmod{p}$, *then*

$$\deg(\gcd(f_p, g_p)) \geq \deg(h_p) = \deg(h).$$

Proof Part 1 is obvious and implies $\deg(h_p) \leq \deg(\gcd(f_p, g_p))$. The assumption of Part 2 implies that $\mathrm{lc}(h) \not\equiv 0 \pmod{p}$ so that

$$\deg(h) = \deg(h_p) \leq \deg(\gcd(f_p, g_p)).$$

$\boxed{\hbar}$

Fact 1.6.5. *If* $\mathrm{lc}(f) \not\equiv 0 \not\equiv \mathrm{lc}(g) \pmod{p}$, *then there exists* $\mathfrak{R} \in \mathbb{Z}$ *such that*

$$p \text{ does not divide } \mathfrak{R} \implies h_p = \gcd(f_p, g_p).$$

Proof (sketch) Corollary 6.6.6 will show that, given $f', g' \in \mathbb{Z}[X]$, there is $\mathfrak{R} \in \mathbb{Z}$ such that the following are equivalent

$\mathfrak{R} \not\equiv 0 \pmod{p}$;
$\gcd(f'_p, g'_p) = 1.$

Therefore we only have to apply this result to $f' := f/h$ and $g' := g/h$ since

$$\gcd(f_p, g_p) = h_p \gcd(f'_p, g'_p).$$

$\boxed{\hbar}$

Corollary 1.6.6. *There are only finitely many primes* $p \in \mathbb{N}$ *for which*

$$\gcd(f_p, g_p) = h_p$$

does not hold.

Proof We only need to discard those primes which divide either $\mathrm{lc}(f)$, $\mathrm{lc}(g)$ or $\mathfrak{R}$.

$\boxed{\hbar}$

1.6.2 Modular Algorithm

On the basis of the above result, denoting by P the set of integer primes, the modular algorithm consists of computing

$$\mathbf{h}^{(p)} := \gcd(f_p, g_p)$$

for several primes $p \in P \subset \mathbb{N}$ until we obtain a subset $\mathsf{P} \subset P$ such that

p does not divide $\mathrm{lc}(f)\,\mathrm{lc}(g)$, for all $p \in \mathsf{P}$;
$\deg(\mathbf{h}^{(p)}) \leq \deg(\mathbf{h}^{(q)})$, for all $p \in \mathsf{P}$, for all $q \in P$;
$\prod_{\mathsf{P}} p \geq \mathfrak{B}$,

where $\mathfrak{B}$ satisfies Fact 1.6.3.1, for both f and g.

Then,

either for all $p \in \mathsf{P}$, $\deg(\mathbf{h}^{(p)}) = \deg(h)$ and so $\mathbf{h}^{(p)} = h_p$, in which case we can apply the Chinese Remainder Theorem (Corollary 2.1.5) in order to compute the single element $\mathbf{h} = \sum a_i X^i \in \mathbb{Z}[X]$ such that

$-\mathfrak{B} < a_i \leq \mathfrak{B}$, for all i;
$\mathbf{h}_p = \mathbf{h}^{(p)} = h_p$,

from which

$$\mathbf{h} = h = \gcd(f, g);$$

or for all $p \in \mathsf{P}$, we have $\deg(\mathbf{h}^{(p)}) > \deg(h)$, which happens with low probability; in this case the above computation gives a wrong answer, but this can be detected by checking whether $\mathbf{h}$ divides f and g: in fact, if the answer is positive then we can deduce that $\mathbf{h}$ divides $h = \gcd(f, g)$ and since $\deg(\mathbf{h}) \geq \deg(h)$ we can deduce that $\mathbf{h} = h = \gcd(f, g)$.

Algorithm 1.6.7. This approach leads to the algorithm presented in Figure 1.4.

1.6.3 Hensel Lifting Algorithm

The algorithm is based on the following

Fact 1.6.8. *Let $p \in \mathbb{N}$ be a prime and let $f(X) \in \mathbb{Z}[X]$ satisfy*

$$\mathrm{lc}(f) \not\equiv 0 \,(\mathrm{mod}\ p).$$

Let $\mathsf{f}, \mathsf{h} \in \mathbb{Z}[X]$ *satisfy*

(1) $f \equiv \mathsf{fh} \,(\mathrm{mod}\ p)$,
(2) $\deg(f) = \deg(\mathsf{f}) + \deg(\mathsf{h})$,
(3) $\gcd(\mathsf{f}_p, \mathsf{h}_p) = 1$.

1.6 Computing a gcd

Fig. 1.4. Modular GCD

---

h := **GCD**(f,g)
where
    $f, g \in \mathbb{Z}[X]$,
    $h := \gcd(f, g)$
Repeat
    choose a prime $p \in \mathbb{N}$ such that p does not divide $\operatorname{lc}(f)\operatorname{lc}(g)$
    $\mathbf{h}^{(p)} := \gcd(f_p, g_p)$
    $\mathbf{p} := p, h := \mathbf{h}^{(p)}, d := \deg(h)$
    Repeat
        If $\deg(\mathbf{h}^{(p)}) < d$ **then**
            $\mathbf{p} := p, h := \mathbf{h}^{(p)}, d := \deg(h)$
        else
            If $d = 0$ **then**
                $h := 1$
            else
                choose a prime $p \in \mathbb{N}$ such that p does not divide $\mathbf{p}\operatorname{lc}(f)\operatorname{lc}(g)$
                $\mathbf{h}^{(p)} := \gcd(f_p, g_p)$
                If $\deg(\mathbf{h}^{(p)}) = \deg(h)$ **then**
                      Compute by the Chinese Remainder Theorem h' such that

$$h' \equiv \begin{cases} h \pmod{\mathbf{p}} \\ \mathbf{h}^{(p)} \pmod{p} \end{cases}$$

                $h := h', \mathbf{p} := \mathbf{p}p$
    until $\mathbf{p} \geq \mathfrak{B}$
until h divides f and g

Then for each $n \in \mathbb{N}$, denoting $q := p^n$, it is possible to compute

$$\mathsf{f}', \mathsf{h}' \in \mathbb{Z}[X]$$

such that

(1) $f \equiv \mathsf{f}'\mathsf{h}' \pmod{q}$,
(2) $\mathsf{f}' \equiv \mathsf{f} \pmod{p}$, $\mathsf{h}' \equiv \mathsf{h} \pmod{p}$,
(3) $\deg(\mathsf{f}') = \deg(\mathsf{f})$, $\deg(\mathsf{h}') = \deg(\mathsf{h})$.

Moreover there is an algorithm (the Hensel Lifting Algorithm) for computing them.

Proof Compare with Theorem 18.1.2.   ♃

Let $f, g \in \mathbb{Z}[X]$, and $h := \gcd(f, g)$. After computing $\gcd(f_p, g_p)$ for several primes $p \in \mathbb{N}$, we will probabilistically obtain an element $\mathsf{h} := \gcd(f_p, g_p)$

for a suitable prime $p \in \mathbb{N}$, such that $\deg(\mathsf{h}) = \deg(h)$, choosing only the one for which $\deg(\mathsf{h})$ is minimal.

Denoting $\mathsf{f} := f/\mathsf{h}$, then f and h satisfy the assumptions of the above Fact. Therefore choosing $n \in \mathbb{N}$ such that $q := p^n \geq \mathfrak{B}$, we can obtain the polynomials $\mathsf{f}', \mathsf{h}' = \sum a_i X^i$ satisfying the above condition.

Therefore

$\deg(\mathsf{h}') = \deg(\mathsf{h}) \geq \deg(h)$, and
$-\mathfrak{B} < a_i \leq \mathfrak{B}$, for all i, so that

if h' divides f and g then $\mathsf{h}' = \gcd(f, g)$.

1.6.4 Heuristic gcd

As both the modular and the Hensel lifting gcds are based on restricting the mapping

$$-_p : \mathbb{Z}[X] \mapsto \mathbb{Z}_p[X]$$

to the suitable subset

$$S := \left\{ \sum_{i=0}^{n} a_i X^i : -\frac{p}{2} < a_i \leq \frac{p}{2}, \text{ for all } i \right\} \subset \mathbb{Z}[X]$$

so that the restriction of $-_p$ to S is an isomorphism, the heuristic gcd is based on the restriction of a different projection to a subset in order to make it invertible.

Let us just consider, for each $\xi \in \mathbb{Z}$, the evaluation map $\mathrm{ev}_\xi : \mathbb{Z}[X] \mapsto \mathbb{Z}$ defined by $\mathrm{ev}_\xi(h) := h(\xi)$, for all $h(X) \in \mathbb{Z}[X]$.

Lemma 1.6.9. *Let*

$$S := \left\{ h(X) = \sum_{i=0}^{n} a_i X^i \in \mathbb{Z}[X] : -\frac{\xi}{2} < a_i \leq \frac{\xi}{2}, \text{ for all } i \right\} \subset \mathbb{Z}[X].$$

Then the restriction of ev_ξ to S is an isomorphism between it and $\mathbb{Z}$. ♄

It is clear how to compute $\mathrm{ev}_\xi^{-1}(\gamma)$ for each integer γ (cf. Fig. 1.5).

Theorem 1.6.10. *Let $f, g \in \mathbb{Z}[X]$ and let $\mathfrak{r} \in \mathbb{N}$ be a bound for all the roots of both f and g (cf. Fact 1.6.3).*

Let $\xi \in \mathbb{Z}$ be such that $|\xi| > 1 + \mathfrak{r}$; let $m := f(\xi), n := g(\xi), \gamma := \gcd(m, n)$ and $h(X) := \mathrm{ev}_\xi^{-1}(\gamma)$.

1.6 Computing a gcd

Fig. 1.5. Computation of ev_ξ^{-1}

$h := \text{ev}_\xi^{-1}(\gamma)$
where
  $\xi \in \mathbb{Z}$,
  $\gamma \in \mathbb{Z}$,
  $h(X) \in S \subset \mathbb{Z}[X]$,
  $h(\xi) = \gamma$
$\mathsf{h} := \gamma, h := 0, i := 0,$
While $\mathsf{h} \neq 0$ **do**
  Let $a \in \mathbb{Z}$ be the unique element such that
    $a \equiv \mathsf{h} \pmod{\xi}$,
    $-\xi/2 < a \leq \xi/2$
  $\mathsf{h} := (\mathsf{h} - a)/\xi, h := h + aX^i, i := i + 1$

Then the following conditions are equivalent:

$h(X)$ *divides both* $f(X)$ *and* $g(X)$;
$h(X) = \gcd(f, g)$.

Proof If h divides both $f(X)$ and $g(X)$ and therefore $\gcd(f, g)$, then there exists $H \in \mathbb{Z}[X]$ such that $\gcd(f, g) = hH$; then we have

$$h(\xi) = \gamma = \gcd(m, n) = \gcd(f(\xi), g(\xi)) \geq \gcd(f, g)(\xi) = h(\xi)H(\xi),$$

so that $H(\xi) = \pm 1$.
Since, by the Fundamental Theorem of Algebra, $H(X) = \prod_i (X - \alpha_i)$ for suitable $\alpha_i \in \mathbb{C}$, we can deduce that $\prod_i (\xi - \alpha_i) = H(\xi) = \pm 1$ and that there is an α such that $|\xi - \alpha| \leq 1$, giving the contradiction

$$|\xi| > 1 + \mathfrak{r} \geq 1 + |\alpha| \geq |\xi|.$$

$\square$

This leads to the probabilistic algorithm presented in Figure 1.6.

Example 1.6.11. An example is

$$\begin{array}{rcll}
f(X) & := & X^3 - 3X^2 - X + 3 & = (X-1)(X+1)(X-3) \\
g(X) & := & X^3 + X^2 - 9X - 9 & = (X+1)(X-3)(X+3) \\
\xi & := & 10 & \\
m & := & 693 & = 9 \cdot 11 \cdot 7 \\
n & := & 1001 & = 11 \cdot 7 \cdot 13 \\
\gamma & := & 77 & = 11 \cdot 7 \\
h(X) & := & X^2 - 2X - 3 & = (X+1)(X-3)
\end{array}$$

Fig. 1.6. Heuristic GCD

$h := \mathbf{HEUGCD}(f, g)$
where
    $f, g \in \mathbb{Z}[X]$,
    $h(X) = \gcd(f, g)$.
Choose $e \in \mathbb{R}, e > 1$
Choose $\xi \in \mathbb{Z}$
Repeat
    $\xi := \lfloor \xi e \rfloor$
    $m := f(\xi), n := g(\xi)$,
    $\gamma := \gcd(m, n)$
    $h(X) := \mathrm{ev}_\xi^{-1}(\gamma)$
⋆ $h(X) := \mathrm{Prim}(h)$
until h divides both f and g

Example 1.6.12. However, if you consider

$$f(X) := (X+1)(X+2)(X+3) \text{ and } g(X) := (X-2)(X-1)X$$

it is clear that $\gcd(f, g) = 1$, and $m \equiv 0 \equiv n \pmod 6$, for all $\xi \in \mathbb{Z}$, so that $h(X) \neq \gcd(f, g)$, for all $\xi \in \mathbb{Z}$,

and the algorithm cannot terminate.

However, when $\xi > 12$ and $\gcd(m, n) = 6$, the algorithm returns $h(X) = 6$ which is associate to $\gcd(f, g)$.

This suggests that we remove the *content* of h[5] by adding the line marked by ⋆ in Figure 1.6.

The correctness of this amended algorithm is given by

Theorem 1.6.13. *Let $f, g \in \mathbb{Z}[X]$ and let $\mathfrak{r} \in \mathbb{N}$ be a bound for all the roots of f and g.*

Let $\xi \in \mathbb{Z}$ be such that

$$|\xi| \geq 1 + 2\mathfrak{r},$$

and let $m := f(\xi)$, $n := g(\xi)$, $\gamma := \gcd(m, n)$, $h'(X) := \mathrm{ev}_\xi^{-1}(\gamma)$, $c := \mathrm{cont}(h')$, and $h := \mathrm{Prim}(h') = c^{-1}h'$.

---

[5] We recall that for a polynomial $h(X) := \sum a_i X_i$, the content of h is

$$\mathrm{cont}(h) := c := \gcd_i(a_i)$$

and we will denote $\mathrm{Prim}(h) := c^{-1}h(X)$ (cf. Section 6.1).

1.6 Computing a gcd

Then the following conditions are equivalent:

$h(X)$ *divides both* $f(X)$ *and* $g(X)$;
$h(X) = \gcd(f, g)$.

Proof If h divides both $f(X)$ and $g(X)$ and therefore $\gcd(f, g)$, then there exists $H \in \mathbb{Z}[X]$ such that $\gcd(f, g) = hH$; thus we have

$$ch(\xi) = h'(\xi) = \gamma = \gcd(m, n) = \gcd(f(\xi), g(\xi))$$
$$\geq \gcd(f, g)(\xi) = h(\xi)H(\xi)$$

so that $H(\xi) \leq \pm c$. Since each coefficient of h' is bounded by $\xi/2$, we have $c < \xi/2$.
therefore, by the same argument as in Theorem 1.6.10, there is an α such that

$$|\xi - \alpha| \leq c < \frac{\xi}{2},$$

so that $|\alpha| \geq \xi/2 > \mathfrak{r}$, which is a contradiction. ♮

Lemma 1.6.14. *Let* $f, g \in \mathbb{Z}[X]$ *be such that* $\gcd(f, g) = 1$. *Then there is* $M \in \mathbb{N}$ *such that*

$$\forall \xi \in \mathbb{Z}, \gcd(f(\xi), g(\xi)) \leq M.$$

Proof By assumption there are $a'(X), b'(X) \in \mathbb{Q}[X]$ such that $a'f + b'g = 1$; eliminating denominators, we obtain polynomials $a(X), b(X) \in \mathbb{Z}[X]$ and an integer $M \in \mathbb{N}$ such that

$$a(X)f(X) + b(X)g(X) = M.$$

Therefore for all $\xi \in \mathbb{Z}$, $a(\xi)f(\xi) + b(\xi)g(\xi) = M$, from which the proof follows. ♮

Corollary 1.6.15. *Let* $f, g \in \mathbb{Z}[X]$, $h(X) := \gcd(f, g)$. *Then there is* $M \in \mathbb{N}$ *such that*

$$\forall \xi \in \mathbb{Z}, \gcd(f(\xi), g(\xi)) \leq Mh(\xi).$$

Proof Apply the above lemma to the polynomials f/h and g/h. ♮

Corollary 1.6.16. *Let* $f, g \in \mathbb{Z}[X]$ *and* $\xi > 2M\mathfrak{B}$.
Let $m := f(\xi)$, $n := g(\xi)$, $\gamma := \gcd(m, n)$, $h'(X) := \text{ev}_\xi^{-1}(\gamma)$, $c := \text{cont}(h')$, *and* $h := \text{Prim}(h') = c^{-1}h'$.

Then $h(X) = \gcd(f, g)$.

Proof Denoting $h(X) := \sum_i a_i X^i$, we have
$$2M|a_i| \leq 2M\mathfrak{B} < \xi, \text{ for all } i,$$
so that $\text{ev}_\xi^{-1}(\gamma) = Mh(X)$. ♄

Corollary 1.6.17. *The algorithm of Figure 1.6 terminates.* ♄

2
Intermezzo: Chinese Remainder Theorems

The Chinese Remainder Theorem is the tool which allows us to compute the integer satisfying some given congruences modulo primes (Section 2.1).

Since the techniques needed to develop it are those of a principal ideal domain, namely Bezout's Identity and factorization, the Chinese Remainder Theorem and Algorithm can be generalized as a tool for studying congruences modulo elements in that setting (Section 2.2).

More precisely (Section 2.3), given a principal ideal domain D, Chinese Remaindering allows us to relate the factorization of elements $m \in D$ and the structure of the residue rings of D; in particular, given an element $m \in D$ we can consider its factorization into powers of irreducible elements $m = \sum_{i=1}^{n} m_i$ and its residue ring $R = D/(m)$: Chinese Remaindering lets us study the structure of R as a *direct sum decomposition* of the residue rings $R_i = D/(m_i)$:

$$D/(m) = D/(m_1) \oplus \cdots \oplus D/(m_i) \oplus \cdots \oplus D/(m_n).$$

In this analysis I must introduce and study useful notions and tools such as *nilpotent rings* (Section 2.4) and *primitive idempotents* (Section 2.5), before I can describe the structure of residue rings of a principal ideal domain (Section 2.6).

Chinese Remaindering can be computed using one of two methods:

by an iterative approach (due to Newton) which is the one discussed in Section 2.1, or

by a global evaluation represented by a formula (due to Lagrange) from which Chinese Remaindering can be interpreted as an interpolation problem, and whose specialization returns the Lagrange Interpolation formula.

The last section is devoted to Lagrangian results (Section 2.7).

2.1 Chinese Remainder Theorems

Theorem 2.1.1 (Chinese Remainder Theorem for Integers).
Let $m_1, \ldots, m_n \in \mathbb{N}$ be such that $\gcd(m_i, m_j) = 1$, for all i, j, and let $m = \prod_i m_i$. Let $a_1, \ldots, a_n \in \mathbb{Z}$.
There is an $a \in \mathbb{Z}$ such that

(1) $a \equiv a_i \pmod{m_i}$, for all i;
(2) if $b \in \mathbb{Z}$ is such that $b \equiv a_i \pmod{m_i}$, for all i, then $b \equiv a \pmod{m}$.

Proof Let us first prove that if both a and b are such that

$$a \equiv a_i, b \equiv a_i \pmod{m_i}, \text{ for all } i,$$

then $b \equiv a \pmod{m}$.
This follows immediately because then $a \equiv b \pmod{m_i}$, for all i, and so

$$a \equiv b \pmod{m}$$

since $m = \text{lcm}(m_1, \ldots, m_n)$.
Now let us prove that there exists an a satisfying (1). It is clearly sufficient to prove this when $n = 2$, because then one can use induction.
Since $\gcd(m_1, m_2) = 1$, by the Bezout Identity for the integers, there are $c_1, c_2 \in \mathbb{Z}$ such that

$$c_1 m_1 + c_2 m_2 = 1.$$

Let

$$u := c_1(a_2 - a_1)$$

and

$$a := a_1 + u m_1.$$

Then $a \equiv a_1 \pmod{m_1}$; Let us now show that $a \equiv a_2 \pmod{m_2}$ too.
Denoting

$$v := c_2(a_2 - a_1)$$

we have

$$a_2 - a_1 = c_1(a_2 - a_1)m_1 + c_2(a_2 - a_1)m_2 = u m_1 + v m_2,$$

and so

$$a_2 = a_1 + u m_1 + v m_2 = a + v m_2 \equiv a \pmod{m_2}.$$

♄

2.1 Chinese Remainder Theorems

Corollary 2.1.2. *Let $m_1, \ldots, m_n \in \mathbb{N}$ be such that*

$$\gcd(m_i, m_j) = 1, \text{ for all } i, j,$$

and let $m = \prod_i m_i$. Let $a_1, \ldots, a_n \in \mathbb{Z}$.
Then there is a unique $a \in \mathbb{Z}$ such that

$$a \equiv a_i \pmod{m_i}, \text{ for all } i; \text{ and } -m/2 < a \leq m/2.$$

Proof In each congruence class mod m, there is a unique element a such that $-m/2 < a \leq m/2$. ♄

Historical Remark 2.1.3. Is the Chinese Remainder Theorem a Chinese Theorem on remainders or a theorem on Chinese Remainders?

While the Chinese Remainder Theorem is usually and correctly related to astronomy and calendars, there is a folklore story which relates it to military technology.

According to this folklore story, by ordering an army to arrange itself in rows of 3, 5, 7, ... and counting the remainders, it was possible to compute how many soldiers started the war and how many came back, and therefore the number of losses.

In this sense, the theorem is really related to Chinese Remainders!

Corollary 2.1.4. *Let $m_1, \ldots, m_n \in \mathbb{N}$ be such that $\gcd(m_i, m_j) = 1$, for all i, j, and let $m = \prod_i m_i$. Let $a_1, \ldots, a_n$ be such that $a_i \in \mathbb{Z}_{m_i}$.*
Then there is a unique $a \in \mathbb{Z}_m$ such that

$$a \equiv a_i \pmod{m_i}, \text{ for all } i.$$

♄

Corollary 2.1.5. *Let $m_1, \ldots, m_n \in \mathbb{N}$ be such that $\gcd(m_i, m_j) = 1$, for all i, j, and let $m = \prod_i m_i$. Let $f_1(X), \ldots, f_n(X) \in \mathbb{Z}[X]$.*
Then there is a unique $f(X) := \sum_{i=0}^{d} c_i X^i \in \mathbb{Z}[X]$ such that

$f(X) \equiv f_i(X) \pmod{m_i}$, *for all i;*
$\deg(f) = d = \max(\deg(f_i))$;
$-m/2 < c_i \leq m/2$, *for all i.*

Proof We have only to apply the Chinese Remainder Theorem to each of the coefficients of f_i at each degree $\delta \leq d$, to obtain the coefficients of f at that same degree. ♄

Theorem 2.1.6 (Chinese Remainder Theorem for Polynomials).
Let $g_1, \ldots, g_n \in k[X]$ be such that $\gcd(g_i, g_j) = 1$, for all i, j, and let $g = \prod_i g_i$. Let $f_1, \ldots, f_n \in k[X]$.
Then there is a unique $f \in k[X]$ such that

$$f \equiv f_i \ (\mathrm{mod} g_i), \text{ for all } i \text{ and } \deg(f) < \deg(g).$$

Proof Uniqueness is proved by noting that if f and h are such that

$$f \equiv f_i, h \equiv f_i (\mathrm{mod}\ g_i), \text{ for all } i,$$

then $f \equiv h (\mathrm{mod}\ g)$. Hence

$$\deg(f) < \deg(g), \deg(h) < \deg(g) \implies f = h.$$

Existence is proved by rephrasing the proof of the Chinese Remainder Theorem for Integers. Again it is sufficient to prove this when $n = 2$.
Since $\gcd(g_1, g_2) = 1$, by the Bezout Identity, there are $s, t \in k[X]$ such that

$$sg_1 + tg_2 = 1.$$

Let

$$u := s(f_2 - f_1)$$

and

$$f := f_1 + ug_1,$$

so that $f \equiv f_1 (\mathrm{mod}\ g_1)$. Then, denoting

$$v := t(f_2 - f_1),$$

we have

$$f_2 - f_1 = s(f_2 - f_1)g_1 + t(f_2 - f_1)g_2 = ug_1 + vg_2,$$

which implies that

$$f_2 = f_1 + ug_1 + vg_2 = f + vg_2 \equiv f(\mathrm{mod}\ g_2).$$

♄

2.2 Chinese Remainder Theorem for a Principal Ideal Domain

Recall that

Fact 2.2.1. *In a principal ideal domain D we have:*

(Bezout's Identity) for all $a, b \in D, \exists s, t \in D : as + bt = \gcd(a, b);$

2.2 Chinese Remainder Theorem for a Principal Ideal Domain

(Factorization) for each non-invertible $a \in D$ there is a unique factorization into irreducible factors p_i:

$$m = p_1^{e_1} \cdots p_n^{e_n}.$$

Since in the proof of the Chinese Remainder Theorem we only used Bezout's Identity, both for integers and for polynomials over a field, it is easy to modify the statement and the proof of the Chinese Remainder Theorem for it to hold in a principal ideal domain[1]:

Theorem 2.2.2 (Chinese Remainder Theorem for a PID).
Let D be a principal ideal domain (PID).
Let $m_1, \ldots, m_n \in D$ be such that $\gcd(m_i, m_j) = 1$, for all i, j, and let $m = \prod_i m_i$. Let $a_1, \ldots, a_n \in D$.
Then there is an $a \in D$ such that

(1) $a \equiv a_i \pmod{m_i}$, for all i,
(2) if $b \in D$ is such that

$$b \equiv a_i \pmod{m_i}, \text{ for all } i,$$

then $a \equiv b \pmod{m}$.

Proof First let us prove that if both a and b are such that

$$a \equiv a_i, \ b \equiv a_i \pmod{m_i}, \forall i,$$

then $b \equiv a \pmod{m}$.
This follows immediately because then $a \equiv b \pmod{m_i}$, for all i and so

$$a \equiv b \bmod \mathrm{lcm}(m_1, \ldots, m_n) = m.$$

Now let us prove that there exists an a satisfying (1). It is clearly sufficient to prove this when $n = 2$, because then one can use induction.
Since $\gcd(m_1, m_2) = 1$, by the Bezout Identity, there are $c_1, c_2 \in D$ such that

$$c_1 m_1 + c_2 m_2 = 1.$$

Let

$$u := c_1(a_2 - a_1)$$

and

$$a := a_1 + u m_1.$$

---

[1] Note, however, that there is a still more general version of the Chinese Remainder Theorem, which holds under more relaxed assumptions and is applicable to ideals in multivariate polynomial rings.

Then $a \equiv a_1 \pmod{m_1}$ and we can show that $a \equiv a_2 \pmod{m_2}$ too. Denoting
$$v := c_2(a_2 - a_1)$$
we have
$$a_2 - a_1 = c_1(a_2 - a_1)m_1 + c_2(a_2 - a_1)m_2 = um_1 + vm_2,$$
and so
$$a_2 = a_1 + um_1 + vm_2 = a + vm_2 \equiv a \pmod{m_2}.$$

♄

Corollary 2.2.3. *Let D be a principal ideal domain. Let $m_1, \ldots, m_n \in D$ be such that $\gcd(m_i, m_j) = 1$, for all i, j, and let $m = \prod_i m_i$.*
Let $a_1, \ldots, a_n$ be such that $a_i \in D/(m_i)$, for all i.
Then there is a unique $a \in D/(m)$ such that $a \equiv a_i \pmod{m_i}$, for all i.

♄

Proposition 2.2.4 (Newton–Garner). *Let D be a principal ideal domain. Let $m_1, \ldots, m_n \in D$ be such that $\gcd(m_i, m_j) = 1$, for all i, j, and let*
$$q_0 := 1, \quad q_j := \prod_{i=1}^{j-1} m_i, \ j = 1, \ldots, n, \ m := \prod_i m_i.$$
Let $a_1, \ldots, a_n \in D$.
Then there are b_i, $0 \leq i < n$, such that
$$a := \sum_{i=0}^{n-1} b_i q_i$$
satisfies
$$a \equiv a_i \pmod{m_i}, \forall i.$$

Proof By the Bezout Identities, for all i, j, there are elements s_{ij}, s_{ji} such that
$$s_{ij} m_i + s_{ji} m_j = 1.$$
Now define
$$t_1 := 1, \ t_{k+1} := t_k s_{kj} = \prod_{i=1}^{k} s_{ij}, \ 1 \leq k < j, \ s_j := t_{j-1}$$
so that
$$s_j q_j \equiv 1 \pmod{m_j}.$$

Then put $b_0 := a_1$ and assume inductively that we have already computed elements $b_i : i = 0, \ldots, j - 1$ such that

$$\sum_{i=0}^{j-1} b_i q_i \equiv a_k (\bmod\ m_k), \text{ for all } k < j.$$

Let us then compute

$$v_1 := b_{j-1}(\bmod\ m_j),\ v_i := v_{i-1}m_{j-i} + b_i(\bmod\ m_j),\ i = 2, \ldots, k-1$$

so that

$$v_j = \sum_{i=0}^{j-1} b_i q_i (\bmod\ m_j);$$

therefore setting

$$b_j := (a_j - v_j)s_j$$

we have

$$\sum_{i=0}^{j} b_i q_i \equiv \begin{cases} v_j + (a_j - v_j)s_j q_j \equiv a_j (\bmod\ m_j) \\ \sum_{i=0}^{j-1} b_i q_i \equiv a_k (\bmod\ m_k),\ k < j. \end{cases}$$

♄

Historical Remark 2.2.5. The corresponding algorithm, the Garner Algorithm, is essentially the generalization from $k[X]$ to the PIDs of the Newton Interpolation Formula which, given roots $\alpha_1, \ldots, \alpha_n \in k$ and values $a_1, \ldots, a_n \in K$, allow us to interpolate a polynomial $f(X)$, $\deg(f) < n$, such that for all i : $f(\alpha_i) = a_i$. In other words, it lets us apply the Chinese Remainder Theorem to the case

$$m_i := X - \alpha_i,\ m := \prod_i m_i.$$

2.3 A Structure Theorem (1)

The Chinese Remainder Theorem for principal ideal domains can be interpreted as a structural result for residues of a PID. The rest of this chapter is devoted to describing this.

Throughout, we will fix a principal ideal domain D, an element $m \in D$, and the residue ring $R := D/(m)$.

Let us factor m in D as

$$m = p_1^{e_1} \cdots p_n^{e_n}$$

so that, denoting $m_i := p_i^{e_i}$, we have

$$m = \prod_i m_i \text{ and } \gcd(m_i, m_j) = 1, \text{ for all } i, j,$$

i.e. the exact assumptions of Corollary 2.2.3.

Moreover we will denote $R_i := D/(m_i)$; also, for each $g \in D$, we will denote by g the canonical projection g $: D \mapsto D/(g)$.

Finally all rings considered in the rest of this chapter are implicitly commutative.

First we recall the notion of *direct sum decomposition*: if R_1, R_2 are two rings, we can consider the set

$$R_1 \oplus R_2 := \{(a_1, a_2) : a_1 \in R_1, a_2 \in R_2\}.$$

It is easy to verify that $R_1 \oplus R_2$ is a ring under the sum

$$(a_1, a_2) + (b_1, b_2) := (a_1 + b_1, a_2 + b_2)$$

and the product

$$(a_1, a_2)(b_1, b_2) := (a_1 b_1, a_2 b_2),$$

the null element being $(0, 0)$ and the identity being $(1, 1)$.

Note that since

$$(1, 0)(0, 1) = (0, 0),$$

even if both R_1 and R_2 are domains, $R_1 \oplus R_2$ has zero divisors.

We call $R_1 \oplus R_2$ the *direct sum* of R_1 and R_2.

In the same way, given n rings $R_1, \ldots, R_n$ we define the direct sum $R_1 \oplus \cdots \oplus R_n$ to be the set of n-tuples $(r_1, \ldots, r_n)$ with $r_i \in R_i$, for all i, and the sum and product defined componentwise.

Definition 2.3.1. *If $R, R_1, \ldots, R_n$ are rings such that*

$$R \cong R_1 \oplus \cdots \oplus R_n,$$

then $R_1 \oplus \cdots \oplus R_n$ is called a direct sum decomposition *of R.*

Note that if

$$\phi : R \mapsto R_1 \oplus \cdots \oplus R_n$$

is an isomorphism, then there are canonical projections

$$\pi_i : R \mapsto R_i$$

and canonical immersions

$$\eta_i : R_i \mapsto R$$

2.3 A Structure Theorem (1)

defined by

for all $a \in R$, $\pi_i(a) := a_i$, for all i, where $a_i \in R_i$ are defined by the relation $\phi(a) = (a_1, \ldots, a_n)$; and

$\eta_i(a_i)$ is the unique $a \in R$ such that $\pi_j(a) = \begin{cases} a_i & \text{if } j = i \\ 0 & \text{otherwise}; \end{cases}$

note that they satisfy the properties

for all $a_i \in R_i$, $\pi_i \eta_i(a_i) = a_i$; and for all $a \in R$, $a = \sum_i \eta_i \pi_i(a)$.

Let us extend our notation by defining the maps

$\pi_i : R \mapsto R_i$ to be the canonical projection such that

for all $a \in D$, $\pi_i \mathsf{m}(a) = \mathsf{m}_i(a)$,

i.e. for all $a \in R$, $\pi_i(a)$ is the congruence class of $a \pmod{m_i}$;

$\phi : R \mapsto R_1 \oplus \cdots \oplus R_n$ to be the morphism such that

$\phi(a) = (\pi_1(a), \ldots, \pi_n(a))$, for all $a \in R$;

$\psi : R_1 \oplus \cdots \oplus R_n \mapsto R$ to be the morphism such that $\psi(a_1, \ldots, a_n)$ is the only element $a \in R$ satisfying $a \equiv a_i \pmod{m_i}$ for all i – its existence and uniqueness being guaranteed by Corollary 2.2.3;

$\eta_i : R_i \mapsto R$ to be the morphism such that $\eta_i(a_i) = \phi(0, \ldots, 0, a_i, 0, \ldots, 0)$, i.e. $\eta_i(a_i)$ is the only element $a \in R$ such that

$$a \equiv \begin{cases} a_i \pmod{m_i} \\ 0 \pmod{m_j} & j \neq i. \end{cases}$$

Using the notion of direct sum decomposition and the notation we have introduced, we can now interpret the Chinese Remainder Theorem as follows:

Theorem 2.3.2. *With the notation above, we have that:*

(1) ϕ and ψ are inverse isomorphisms;
(2) R is isomorphic to $R_1 \oplus \cdots \oplus R_n$;
(3) π_i and η_i are respectively the canonical projections and the canonical immersions.

Proof (2) and (3) are obvious consequences of (1).
To prove (1) we only have to prove that ϕ and ψ are inverse applications, which is just an obvious restatement of Corollary 2.2.3. ♄

Example 2.3.3. Let us consider a trivial example. We choose $D := \mathbb{Q}[X]$, $m_1 := X, m_2 := X - 1, m_3 := X + 1$, so that $m = X^3 - X$,

$$R = D/(m) = \{a + bX + cX^2 : a, b, c \in \mathbb{Q}\}$$

and $R_i = D/(m_i) = \mathbb{Q}$, for all i.

We then have:

$$\pi_1(a + bX + cX^2) = a;$$
$$\pi_2(a + bX + cX^2) = a + b + c;$$
$$\pi_3(a + bX + cX^2) = a - b + c;$$
$$\phi(a + bX + cX^2) = (a, a + b + c, a - b + c);$$
$$\psi(\alpha, \beta, \gamma) = \alpha + ((\beta - \gamma)/2)X + ((\beta + \gamma)/2 - \alpha)X^2;$$
$$\eta_1(\alpha) = \alpha - \alpha X^2 = \alpha(1 - X^2);$$
$$\eta_2(\beta) = (\beta/2)X + (\beta/2)X^2 = (\beta/2)(X + X^2);$$
$$\eta_3(\gamma) = -(\gamma/2)X + (\gamma/2)X^2 = (\gamma/2)(-X + X^2).$$

2.4 Nilpotents

Definition 2.4.1. *Let R be a ring. An element $r \in R$ is called* nilpotent *if there is $n \in \mathbb{N}$ such that $r^n = 0$.*

A ring R is called a nilpotent ring *if each non-invertible element of R is nilpotent.*

Note that a field is an obvious case of a nilpotent ring, since the only non-invertible element is zero, which is obviously nilpotent.

A more general case is that of any ring $R_i := D/(m_i) = D/(p_i^{e_i})$, where D is a PID. In fact, the following results hold:

Proposition 2.4.2. *Let D be a principal ideal domain, and $m \in D$ be a non-invertible element. Let $R := D/(m)$, the ring of congruence classes modulo m, and let $\mathsf{m} : D \mapsto R$ be the canonical projection.*

Then:

(1) for all $h \in D$, $\mathsf{m}(h)$ is invertible $\iff \gcd(m, h) = 1$;
(2) m is irreducible if and only if R is a field;
(3) m is associate to the power of an irreducible element if and only if R is nilpotent.

2.4 Nilpotents

Proof

(1) Let $h \in D$; then

 $\mathsf{m}(h)$ is invertible in R
$\iff$ there is $s \in D$ such that $\mathsf{m}(h)\mathsf{m}(s) = 1$
$\iff$ there are $s, t \in D$ such that $sh + tm = 1$
$\iff$ $\gcd(m, h) = 1$.

The assumption that D is a principal ideal domain has been used in the application of the Bezout Identity.

(2) We have:

 m is irreducible
$\iff$ $\forall h \in D$, $\gcd(m, h)$ is associate to either 1 or m
$\iff$ $\forall h \in D$, either $\mathsf{m}(h) \in R$ is invertible or $\mathsf{m}(h) = 0$
$\iff$ each non-zero element of R is invertible
$\iff$ R is a field.

(3) Let us assume R is nilpotent and let $m := p_1^{e_1} \ldots p_n^{e_n}$ be a factorization of m. We want to prove that $n = 1$.

Note that, since R is nilpotent, for all $h \in D$, either $\mathsf{m}(h)$ is invertible in R or there is a t such that $\mathsf{m}(h)^t = 0$. Therefore, for all $i \leq n$, either:

$\mathsf{m}(p_i^{e_i})$ is invertible in R, from which we deduce that

$$1 = \gcd(m, p_i^{e_i}) = p_i^{e_i};$$

or there is a t such that $\mathsf{m}(p_i^{e_i})^t = 0$, i.e. $p_i^{e_i t}$ is a multiple of m.

In conclusion, there is a single value i such that

$$m = p_i^{e_i} \text{ and } 1 = \gcd(m, p_j^{e_j}) = p_j^{e_j}, \text{ for all } j \neq i.$$

Conversely let us assume that there is an irreducible $p \in D$ and $e \in \mathbb{N}$ such that $m = p^e$ and let us prove that R is nilpotent. Let $h \in D$; then

(i) either $\gcd(h, m) = 1$ and so $\mathsf{m}(h)$ is invertible in R,
(ii) or $\gcd(h, m) = p^{\varepsilon}$ for some $\varepsilon \leq e$, in which case $h = p^{\varepsilon} s$ for a suitable $s \in D$. Then $h^e = p^{\varepsilon e} s^e = m^{\varepsilon} s^e$ so that $\mathsf{m}(h)^e = 0$.

Thus we conclude that for each $h \in D$, $\mathsf{m}(h)$ is either irreducible or nilpotent. $\boxed{\natural}$

Definition 2.4.3. *Let D be a unique factorization domain and let $m \in D$; m is called* **squarefree** *if it has no multiple factors, or, equivalently, if it is a product of irreducible factors.*

Theorem 2.4.4. *Let D be a principal ideal domain, $m \in D$ be a non-invertible element and $R := D/(m)$. Then:*

(1) R is a direct sum of nilpotent rings;
(2) R is a direct sum of fields iff m is squarefree.

Proof Because D is a principal ideal domain, it is factorial; let

$$m = p_1^{e_1} \cdots p_n^{e_n}$$

be its factorization into powers of distinct non associate irreducible elements p_i; denote $m_i := p_i^{e_i}$, $R_i := R/(m_i)$.
By Theorem 2.3.2 we have that R is isomorphic to $R_1 \oplus \cdots \oplus R_n$. By Proposition 2.4.2 each R_i is nilpotent; moreover each R_i is a field iff $e_i = 1$, $\forall i$, i.e. iff m is a product of irreducible distinct factors. $\boxed{\natural}$

Theorem 2.4.4 gives us another step towards the structure theorem for residues of a PID. Before going on let us note

Lemma 2.4.5. *Let R be a ring; then the set of nilpotent elements of R is an ideal.*

Proof Denote by $\mathbf{N} := \{r \in R : r \text{ is nilpotent}\}$ the set which we need to prove is an ideal.
Let $r_1, r_2 \in \mathbf{N}$ and let $n, m \in \mathbb{N}$ be such that $r_1^n = r_2^m = 0$. Setting $N = n + m - 1$, we have

$$(r_1 + r_2)^N = \sum_{i=0}^{N} c_i r_1^i r_2^{N-i}$$

for some $c_i \in R$. In each term of this expansion either $i \geq n$, so $r_1^i = 0$, or $N - i \geq N - n + 1 = m$, so $r_2^{N-i} = 0$. Therefore $(r_1 + r_2)^N = 0$ and $r_1 + r_2 \in \mathbf{N}$.
Let $r \in R$, $r_1 \in \mathbf{N}$, $n \in \mathbb{N}$ be such that $r_1^n = 0$. Then $(rr_1)^n = r^n r_1^n = 0$ so that $rr_1 \in \mathbf{N}$. $\boxed{\natural}$

Remark 2.4.6. With the notation we are using throughout this chapter, let us consider a nilpotent ring $R = D/(m)$, so that $m = m_1 = p_1^{e_1}$ for an irreducible element $p := p_1 \in D$.

Then for each $h \in D$ either:

$\mathsf{m}(h)$ is invertible, in which case there is $s \in D$ such that $\mathsf{m}(s)\mathsf{m}(h) = 1$, and therefore $\mathsf{p}(s)\mathsf{p}(h) = 1$, and $h \notin (p)$;

or $\mathsf{m}(h)$ is nilpotent, in which case there is $g \in D$ such that $h^e = gm = gp^{e_1}$, i.e., since p is irreducible, $h \in (p)$.

In conclusion, the set of the nilpotent elements of R is the ideal generated by $\mathsf{m}(p)$.

2.5 Idempotents

To complete our analysis of the structure of residue R of a PID D, we need to consider the following definition

Definition 2.5.1. *Let R be a ring. An element $r \in R$ is called* idempotent *if $r = r^2$.*

We then have

Lemma 2.5.2. *Let R be a ring. Then:*

(1) $r \in R$ is idempotent $\iff 1 - r$ is idempotent.
(2) If r_1, r_2 are idempotent, then $r_1 + r_2 - r_1 r_2$, $r_1 r_2$, $r_1 - r_1 r_2$, $r_2 - r_1 r_2$ are idempotent.

Proof

(1) Assume r is idempotent. Then
$$(1-r)^2 = 1 - 2r + r^2 = 1 - 2r + r = 1 - r$$
so that $1 - r$ is idempotent.

Since $r = 1 - (1 - r)$ the converse is obvious.

(2) We have
$$\begin{aligned}(r_1 + r_2 - r_1 r_2)^2 &= r_1^2 + r_2^2 + r_1^2 r_2^2 + 2r_1 r_2 - 2r_1^2 r_2 - 2r_1 r_2^2 \\ &= r_1 + r_2 + r_1 r_2 + 2r_1 r_2 - 2r_1 r_2 - 2r_1 r_2 \\ &= r_1 + r_2 - r_1 r_2;\end{aligned}$$

$$(r_1 r_2)^2 = r_1^2 r_2^2 = r_1 r_2;$$

$$(r_1 - r_1r_2)^2 = r_1^2 + r_1^2r_2^2 - 2r_1^2r_2 = r_1 + r_1r_2 - 2r_1r_2$$
$$= r_1 - r_1r_2. \qquad ♮$$

Lemma 2.5.3. *Let R be a nilpotent ring. Then its only idempotents are its identity and its zero.*

Proof Let c be an idempotent of a nilpotent ring R which is neither 1 nor 0. Since c and $1-c$ are both non-zero, and since $c = c^2$ and $c(1-c) = 0$, both are zero-divisors and so they are not invertible in R. Therefore there is an n such that $c^n = 0$. However, $c = c^2 = c^3 = \ldots = c^n = 0$, which is absurd. ♮

By Theorem 2.3.2 and Theorem 2.4.4 we know that R is the direct sum of the nilpotent rings

$$R \cong R_1 \oplus \cdots \oplus R_n;$$

therefore, in order to study the idempotents of R, on the basis of Lemma 2.5.3 let us denote, $\forall i$, $e_i := \eta_i(1_{R_i}) \in R$ so that

$$\pi_j(e_i) = \begin{cases} 1 & \text{if } i = j \\ 0 & \text{otherwise,} \end{cases}$$

which, as we will soon see are the building blocks for all the idempotents of R.

Lemma 2.5.4. *Under this notation:*

(1) $e_i e_j = 0$ *if* $i \neq j$;
(2) $e_i^2 = e_i$, $\forall i$;
(3) $1_R = e_1 + \cdots + e_n$;
(4) $\forall a \in R$, *let* $a_i = ae_i$; *then*

$$a = \sum_k ae_k = \sum_k a_k;$$

(5) $\forall a \in R$,

$$a_i = \pi_i \eta_i(a_i) = \eta_i \pi_i(a_i) = \eta_i \pi_i(a);$$

(6) the ideal

$$(e_i) = \{ae_i : a \in R\} \subset R$$

is isomorphic to R_i, under the restriction of π_i to it.

2.5 Idempotents

Proof (1), (2), (3) are obvious.
For (4) we have
$$a = a1_R = \sum_k ae_k = \sum_k a_k.$$

Therefore

$$\pi_i(a_i) = \pi_i(a)\pi_i(e_i) = \sum_k \pi_i(a)\pi_i(e_k) = \pi_i(\sum_k ae_k) = \pi_i(a);$$
$$\pi_j(a_i) = \pi_j(ae_i) = \pi_j(a)\pi_j(e_i) \qquad\qquad\qquad = 0, \forall j \neq i,$$

so that a_i is the only element in R such that

$$\pi_j(a_i) = \begin{cases} \pi_i(a) & \text{if } i = j \\ 0 & \text{otherwise,} \end{cases}$$

i.e. $a_i = \eta_i \pi_i(a_i)$.
Moreover,

$$\eta_i \pi_i(a) = \eta_i \pi_i \left(\sum_k a_k \right) = \eta_i \left(\sum_k \pi_i(a_k) \right) = \eta_i \pi_i(a_i) = a_i;$$

proving (5).
Because of (5)

$$\text{Im}(\eta_i) = \{ae_i : a \in R\}$$

and $\eta_i \pi_i(ae_i) = ae_i$, from which (6) follows immediately. ♃

Definition 2.5.5. *A set of idempotents* $\{e_1, \ldots, e_n\}$ *satisfying conditions* (1), (2), (3) *of Lemma 2.5.4 will be called a* primitive set of idempotents *for R and each e_i will be called a* primitive idempotent *for R.*

Corollary 2.5.6. *Let* $R_1 \oplus \cdots \oplus R_n$ *be a direct sum decomposition of R into nilpotent rings* R_i.
For each $J \subseteq \{1, \ldots, n\}$ *denote* $e_J := \sum_{j \in J} e_j$, *so that*

$$\pi_j(e_J) = \begin{cases} 1 & \text{if } j \in J \\ 0 & j \notin J. \end{cases}$$

Then the set of idempotents of R is $\{e_J : J \subseteq \{1, \ldots, n\}\}$.

Proof Let $J = \{i_1, \ldots, i_k\}$; then

$$\begin{aligned} e_J^2 &= (e_{i_1} + \ldots + e_{i_k})^2 \\ &= \sum_j e_{i_j}^2 + \sum_{j \neq l} e_{i_j} e_{i_l} \\ &= e_{i_1} + \cdots + e_{i_k} \\ &= e_J. \end{aligned}$$

Conversely if $\sum_i a_i e_i$ is idempotent, then

$$\sum_i a_i e_i = \left(\sum_i a_i e_i\right)^2 = \sum_i a_i^2 e_i$$

so that $a_i = a_i^2$, for all i, and a_i is an idempotent of R_i, for all i.
Since R_i is nilpotent this means that a_i is either 1 or 0.
Then, setting $J := \{i : a_i = 1\}$, we have $\sum_i a_i e_i = e_J$. ♮

Theorem 2.5.7. *Let $R_1 \oplus \cdots \oplus R_n$ be a direct sum decomposition of R into nilpotent rings R_i.*

Then this decomposition is unique in the following sense: if $S_1 \oplus \cdots \oplus S_m$ is another direct sum decomposition of R into nilpotent rings S_i, then $n = m$ and there is a permutation ρ of $\{1, \ldots, n\}$ such that S_j is isomorphic to $R_{\rho(j)}$.

Proof Denote by $\pi_j : R \mapsto R_j, \eta_j : R_j \mapsto R, \sigma_j : R \mapsto S_j, \zeta_j : S_j \mapsto R$ the canonical projections and immersions.

Denote $e_i := \eta_i(1_{R_i})$, $\epsilon_i := \zeta_i(1_{S_i})$, $e_I := \sum_{i \in I} e_i$, $\epsilon_I := \sum_{i \in I} \epsilon_i$.
Because of Proposition 2.5.6 the set of idempotents of R is both

$$\{e_I : I \subseteq \{1, \ldots, n\}\}$$

and

$$\{\epsilon_I : I \subseteq \{1, \ldots, m\}\}.$$

A cardinality count is then sufficient to prove that $n = m$.
Moreover, $\forall j$ there is $I \subseteq \{1, \ldots, n\}$ such that $\epsilon_j = e_I$. Let $i \in I$; there is $J \subseteq \{1, \ldots, n\}$ such that $e_i = \epsilon_J$. Therefore we have

$$e_i = e_i e_I = \epsilon_J \epsilon_j = \epsilon_j.$$

In this way we prove the existence of a permutation ρ such that $\forall j, \epsilon_j = e_{\rho(j)}$.

Let $\mathcal{I} \subseteq R$ be the ideal generated by $e_{\rho(j)} = \epsilon_j$; by Lemma 2.5.4, both S_j and $R_{\rho(j)}$ are isomorphic to $\mathcal{I}$, from which the proof follows. ♮

In this context, it is worthwhile noting that the operations between idempotents described in Lemma 2.5.2 have an easy set-theoretical interpretation:

Proposition 2.5.8. *Let $I, J \subseteq \{1, \ldots, n\}$. Then:*

(1) $1 - e_J = e_{\mathcal{C}(J)}$ *where* $\mathcal{C}(J) = \{i : 1 \le i \le n, i \notin J\}$;
(2) $e_I e_J = e_{I \cap J}$;
(3) $e_I + e_J - e_I e_J = e_{I \cup J}$;
(4) $e_I - e_I e_J = e_{I \setminus J}$ *where* $I \setminus J = \{i : i \in I, i \notin J\}$;
(5) $1_R = e_1 + \cdots + e_n = 1_{R_1} + \cdots + 1_{R_n}$;
(6) $e_I e_J = e_J \iff J \subseteq I$.

♮

Example 2.5.9. With the same notation as Example 2.3.3 we can list the idempotents of R, which are

$e_1 = 1 - X^2$,
$e_2 = (X + X^2)/2$,
$e_3 = (-X + X^2)/2$,
$e_{\{1,2\}} = 1 + \frac{1}{2}X - \frac{1}{2}X^2$,
$e_{\{1,3\}} = 1 - \frac{1}{2}X - \frac{1}{2}X^2$,
$e_{\{2,3\}} = X^2$.

2.6 A Structure Theorem (2)

We are now able to summarize the results obtained in this analysis of the structure of the residue ring of the principal ideal domain in

Theorem 2.6.1. *Let D be a principal ideal domain. Then:*

(1) Let $m \in D$, $R := D/(m)$, and let $m = \prod_{i=1}^n p_i^{e_i}$ be its factorization in D so that, denoting $m_i := p_i^{e_i}$, we have

$$m = \prod_i m_i \text{ and } \gcd(m_i, m_j) = 1, \forall i, j.$$

Moreover we will set $R_i := D/(m_i)$ and the maps

$\pi_i : R \mapsto R_i$ *to be the canonical projection;*

$\phi : R \mapsto R_1 \oplus \cdots \oplus R_n$ to be the morphism such that

$$\phi(a) = (\pi_1(a), \ldots, \pi_n(a)), \forall a \in R;$$

$\psi : R_1 \oplus \cdots \oplus R_n \mapsto R$ to be the morphism such that $\psi(a_1, \ldots, a_n)$ is the only element $a \in R$ satisfying $a \equiv a_i \pmod{m_i} \forall i$;
$\eta_i : R_i \mapsto R$ to be the morphism such that

$$\eta_i(a_i) = \phi(0, \ldots, 0, a_i, 0, \ldots, 0).$$

Then R is a direct sum of nilpotent rings

$$R \cong R_1 \oplus \cdots \oplus R_n$$

via the inverse isomorphisms ϕ and ψ; moreover this decomposition is unique.

(2) Conversely let R be a residue ring of D which is a direct sum of nilpotents

$$R \cong R_1 \oplus \cdots \oplus R_n$$

so that for each i,

R_i is a residue ring of D, $R_i = D/(m_i)$, and
m_i is a power of an irreducible element, $m_i := p_i^{e_i}$ (cf. Proposition 2.4.2).

Then denoting $m = \prod_{i=1}^n p_i^{e_i}$, we have that $R = D/(m)$.

Proof The assertions follow from Theorem 2.3.2, Theorem 2.4.4 and Theorem 2.5.7, except the last one, which follows easily by noting that, setting $\mathsf{m} : D \mapsto R$ and $\mathsf{m_i} : D \mapsto R_i$ to be the canonical projections, for each $a \in D$ we have

$$\begin{aligned} \mathsf{m}(a) = 0 &\iff \mathsf{m_i}(a) = 0, \text{ for all } i \\ &\iff a \equiv 0 \pmod{m_i} \text{ for all } i \\ &\iff a \equiv 0 \pmod{m} \end{aligned}$$

so that $R = D/\ker(\mathsf{m}) = D/(m)$. ♄

Corollary 2.6.2. *With the notation above, R is also a direct sum of fields* iff m *is squarefree.* ♄

2.7 Lagrange Formula

In Theorem 2.2.2 we have presented the formula and algorithm for applying the Chinese Remainder Theorem proposed by Newton; we intend here to present those proposed by Lagrange; we assume the circumstances discussed in Theorem 2.6.1 still hold and we denote for each $I \subset \{1, \ldots n\}$ the elements

$$g_I := \prod_{i \in I} m_i, \quad h_I := \prod_{i \notin I} m_i$$

so that

$m = g_I h_I;$
$\exists s_I, t_I \in D$ such that $1 = s_I g_I + t_I h_I.$

To simplify the notation, for each i, $1 \leq i \leq n$, we write g_i, h_i, s_i, t_i for g_I, h_I, s_I, t_I where $I = \{i\}$.

Theorem 2.7.1 (Lagrange). *There are γ_i, $1 \leq i \leq n$, such that*

$$1 = \sum_{i=1}^{n} \gamma_i h_i.$$

Proof An elementary proof can easily be derived by iteration from the Bezout Identity.
If $n = 2$, we have $h_2 = g_1$, and therefore the result holds setting

$$\gamma_1 := t_1, \quad \gamma_2 := s_1.$$

Thus, let $H_i := h_i / g_n$ for $i < n$; by induction there are χ_i, $1 \leq i \leq n-1$ such that

$$1 = \sum_{i=1}^{n-1} \chi_i H_i.$$

Therefore, setting

$$\gamma_i := \begin{cases} s_n \chi_i & \text{if } i < n \\ t_n & \text{otherwise} \end{cases}$$

we have

$$\begin{aligned}
1 &= s_n g_n + t_n h_n = s_n \left(\sum_{i=1}^{n-1} \chi_i H_i \right) g_n + t_n h_n \\
&= \sum_{i=1}^{n-1} (s_n \chi_i)(H_i g_n) + t_n h_n = \sum_{i=1}^{n} \gamma_i h_i.
\end{aligned}$$

♄

Let us denote

$$c_i := \gamma_i h_i \pmod{m}, \quad C_i := \sum_{j \neq i} c_j, \forall i,$$

and

$$c_I := \sum_{i \in I} c_i, \quad C_I := \sum_{i \notin I} c_i, \forall I \subset \{1, \ldots, n\},$$

so that

$$c_I + C_I = 1, \forall I.$$

Lemma 2.7.2. *We have*

$$c_j \equiv \begin{cases} 1 \pmod{m_i} & \text{if } i = j \\ 0 \pmod{m_i} & \text{otherwise} \end{cases}$$

Proof Since $h_j \equiv 0 \pmod{m_i}$ if $j \neq i$, we have $c_j \equiv 0 \pmod{m_i}$ and

$$c_i = 1 - \sum_{j \neq i} \gamma_j h_j \equiv 1 \pmod{m_i}.$$

♄

This lemma gives a different formula for the Chinese Remainder Theorem

Corollary 2.7.3. *Let $a_i \in D$ and let $a = \sum_{i=1}^n a_i c_i$.*
Then $a \equiv a_i \pmod{m_i}, \forall i$.

♄

Evidently the c_is are just a different way of computing the idempotents of R; in fact:

Proposition 2.7.4.

(1) $1 = \sum_{j=1}^n c_j$;
(2) $c_i c_j = 0$ if $i \neq j$;
(3) c_i is an idempotent;
(4) c_i and h_i generate the same ideal $\mathcal{I}_i$;
(5) $\{f \in R : f = f c_i\} = \mathcal{I}_i$;
(6) R_i and $\mathcal{I}_i$ are isomorphic;
(7) under this isomorphism e_i and c_i are identified.

Proof (1) is obvious and (2) follows immediately from the fact that $h_i h_j = 0$ if $i \neq j$.

As a consequence we have

$$c_i = c_i \sum_{j=1}^{n} c_j = \sum_{j=1}^{n} c_i c_j = c_i^2$$

and (3) follows.

For (4), $c_i \in (h_i)$ since $c_i = \gamma_i h_i$; conversely $h_i \in (c_i)$ because

$$h_i = \sum_{j=1}^{n} h_i c_j = h_i c_i.$$

Obviously $\{f \in R : f = fc_i\} \subset \mathcal{I}_i$, therefore (5) follows if we prove that for each $f \in \mathcal{I}_i$, we have $f = fc_i$: in fact, for each $f \in \mathcal{I}_i$, there is an $F \in R$ such that $f = Fc_i$ and (5) follows as

$$f = Fc_i = Fc_i^2 = fc_i.$$

Since $1 = s_i g_i + t_i h_i$, we have $f = fs_i g_i + ft_i h_i$ for each $f \in R$. The mapping $\Phi : R \mapsto \mathcal{I}_i$ defined by $\Phi(f) = ft_i h_i$ is surjective and its kernel is (g_i), which proves (6).

Since $\Phi(e_i)$ and c_i are idempotents of the field $\mathcal{I}_i$, (7) follows from Lemma 2.5.3. □

Remark 2.7.5. It is worthwhile and interesting to apply the Lagrange formula to the case of a squarefree polynomial which splits into linear factors

$$m(X) = \prod_{i=1}^{n}(X - \alpha_i)$$

in $K[X]$, where K is a field containing all the roots α_i of m.

For each j the polynomial $c_j(X)$ is a polynomial of degree $< n$; moreover $c_j(\alpha_i) = 0$, for all $i \neq j$, so that

$$c_j(X) = a_j \prod_{i \neq j}(X - \alpha_i),$$

for a suitable $a_j \in K$; furthermore

$$1 = c_j(\alpha_j) = a_j \prod_{i \neq j}(\alpha_j - \alpha_i)$$

so that

$$c_j(X) = \prod_{i \neq j} \frac{(X - \alpha_i)}{(\alpha_j - \alpha_i)};$$

in other words the Lagrange version of the Chinese Remainder Theorem gives the Lagrange Interpolation formula.

Example 2.7.6. Completing the computations of Example 2.3.3 and Example 2.5.9, we have

$$\begin{aligned}
g_1 &= X = h_{\{2,3\}}, \\
h_1 &= -1 + X^2 = g_{\{2,3\}}, \\
g_2 &= X - 1 = h_{\{1,3\}}, \\
h_2 &= X + X^2 = g_{\{1,3\}}, \\
g_3 &= X + 1 = h_{\{1,2\}}, \\
h_3 &= -X + X^2 = g_{\{1,2\}}.
\end{aligned}$$

Moreover

$$1g_1 - 1g_2 = 1, \quad \frac{1}{2}h_3 + \left(1 - \frac{1}{2}X\right)g_3 = 1$$

so that we obtain

$$1 = -\left(1 - \frac{1}{2}X\right)h_1 + \left(1 - \frac{1}{2}X\right)h_2 + \frac{1}{2}h_3$$

which gives

$$\begin{aligned}
\gamma_1 &= -1 + \frac{1}{2}X, \\
\gamma_2 &= 1 - \frac{1}{2}X, \\
\gamma_3 &= \frac{1}{2}, \\
c_1 &= 1 - X^2, \\
c_2 &= \frac{1}{2}X + \frac{1}{2}X^2, \\
c_3 &= -\frac{1}{2}X + \frac{1}{2}X^2, \\
c_{\{1,2\}} &= 1 + \frac{1}{2}X - \frac{1}{2}X^2, \\
c_{\{1,3\}} &= 1 - \frac{1}{2}X - \frac{1}{2}X^2, \\
c_{\{2,3\}} &= X^2.
\end{aligned}$$

Furthermore

$$a := \alpha c_1 + \beta c_2 + \gamma c_3 = \alpha + \left(\frac{\beta - \gamma}{2}\right)X + \left(\frac{\beta + \gamma}{2} - \alpha\right)X^2$$

thus confirming Corollary 2.7.3.

2.7 Lagrange Formula

Also the Lagrange Interpolation Formula can immediately be verified.

As a consequence of Theorem 2.6.1 and Proposition 2.7.4, we have:

Theorem 2.7.7. *With the above notation, there are obvious bijections between*

subsets $I \subset \{1, \ldots, n\}$;
factors g of m;
cofactors h of m;
subrings $\mathcal{R}$ of R;
idempotents c of R;
ideals $\mathcal{I}$ of R,

which are given by associating with each other:

the subset $I \subset \{1, \ldots, n\}$,
the factor $g_I = \prod_{j \in I} m_j$,
the cofactor $h_I = \prod_{j \notin I} m_j$,
the subring $\mathcal{R}_I = \oplus_{j \in I} R_j$,
the idempotent c_I,
the ideal $\mathcal{J}_I := (h_I) = (c_I) \cong \mathcal{R}_I = D/(g_I)$.

Proof The correspondence between factors, cofactors and subsets is obvious, as is the existence of that between cofactors and ideals, and this associates to any ideal its generator of minimal degree; also, we obtain from Theorem 2.6.1 the relation between the ring $\mathcal{R}_I = D/(g_I)$ and the factor g_I.
For $I = \{i\}$ we have from Proposition 2.7.4 the relation

$$\mathcal{J}_I = (h_I) = (c_I) \cong \mathcal{R}_I = D/(g_I). \tag{2.1}$$

To extend it let us consider $J \subset \{1, \ldots, n\}, i \notin J, I := J \cup \{i\}$; we need to prove that, if Equation 2.1 holds for J, then it also holds for I. In fact we have

$c_I = c_J + c_i \in (h_J, h_i) = (\gcd(h_J, h_i)) = (h_I)$
$h_I = \sum_{i \in I} h_I c_i + \sum_{j \notin I} h_I c_j = \sum_{i \in I} h_I c_i = h_I c_I$, so that
$\mathcal{J}_I := (h_I) = (c_I)$; also
$\mathcal{R}_I = \oplus_{j \in I} R_j = D/(g_I)$, and
$\mathcal{R}_I = \oplus_{j \in I} R_j = \oplus_{j \in J} R_j \bigoplus R_i \cong (h_J) + (h_i) = (\gcd(h_J, h_i)) = (h_I) = \mathcal{J}_I$.

♄

Remark 2.7.8. It is obvious that, with the notation of this section, the single primitive set of idempotents for R is $\{e_1, \ldots, e_n\}$, and by Theorem 2.7.7 we get bijections between

(i) the element $i \in \{1, \ldots, n\}$;
(ii) the factor m_i;
(iii) the cofactor $h_i = \prod_{j \neq i} m_j$;
(iv) the subring $R_i = D/(m_i)$;
(v) the idempotent e_i;
(vi) the ideal $\mathcal{J}_i := (h_i) = (e_i) \cong R_i = D/(m_i)$.

Remark 2.7.9. Let us now consider the case $D = K[X]$. Given a rational function $\frac{n(X)}{m(X)}$, $\deg(n) < \deg(m)$, let us assume – with the notation used throughout this chapter – that

$$m(X) = \prod_{i \in I} m_i(X), \quad m_i(X) := p_i(X)^{e_i},$$

with $p_i(X)$ irreducible.

We then define $\forall i$, $r_i(X) := \mathbf{Rem}(nc_i, m)$ so that

$$r_i(X) \equiv n(X)\gamma_i(X)h_i(X) (\mathrm{mod}\, m),$$

and $\deg(r_i) < \deg(m)$, so that $n(X) = \Sigma_i r_i(X)$.

Since

$$r_i(X) \equiv n(X)\gamma_i(X)h_i(X) \equiv 0 (\mathrm{mod}\, m_j), \text{ for all } j \neq i,$$

we know that $\forall i$, $r_i(X) = s_{i0}(X)h_i(X)$ and

$$n(X) = \sum_i s_{i0}(X)h_i(X)$$

for suitable $s_{i0}(X) \in K[X] : \deg(s_{i0}) < \deg(m_i)$.

For each i let us now compute iteratively, for $j = 0, \ldots, e_i - 1$,

$$t_{ij} := \mathbf{Rem}(s_{ij}, p_i), \quad s_{i\,j+1} := \mathbf{Quot}(s_{ij}, p_i)$$

so that

$$s_{i0}(X) = \sum_{j=0}^{e_i - 1} t_{ij} p_i(X)^j, \ \deg(t_{ij}) < \deg(p_i).$$

Therefore we obtain

$$\frac{n(X)}{m(X)} = \frac{\Sigma_i s_{i0}(X)h_i(X)}{\prod_{i \in I} m_i(X)} = \sum_i \frac{s_{i0}(X)}{p_i(X)^{e_i}} = \sum_i \sum_{j=0}^{e_i - 1} \frac{t_{ij}}{p_i(X)^{e_i - j}},$$

i.e. the classic decomposition of rational functions in the case $K = \mathbb{R}$, where the p_is are linear or quadratic factors.

3
Cardano

> Among those who attacked me, there was none whose knowledge went further than a schoolmaster and I do not know how they dared to insert themselves among the erudites; in any case here are their names: ...Tartaglia...
> G. Cardano, *De propria vita liber*

3.1 A Tautology?

Before discussing Solving Polynomial Equation Systems, it seems natural to ask:

'What does it mean to solve a polynomial equation system?'

For a univariate polynomial $f(X) \in \mathcal{P}$ apparently the answer is obvious: e.g. it is clear that the solutions of $f(X) := X^2 - X$ are 0 and 1. Analogously, we could then say that *'the solutions of $f(X) := X^2 - 2$ are $\sqrt{2}$ and $-\sqrt{2}$'*.

Yes, yes, ... unless somebody asks you for a definition of $\sqrt{2}$.... Well, whatever approach you use, your only possible answer is: '$\sqrt{2}$ *and* $-\sqrt{2}$ *are the solutions of* $X^2 - 2$'. Apparently, we have a strange tautology: *the solutions of $X^2 - 2$ are the solutions of $X^2 - 2$!*

If you are not really convinced by this, let me try a stronger example: you will agree that the solutions of the polynomial $X^2 + 1$ are $\pm i$ and that the imaginary number i can be defined only as that number whose square is -1, i.e. to be a solution of the polynomial $X^2 + 1$. So we truly have a tautology:

The solutions of the polynomial equation $X^2 + 1 = 0$ are the two solutions of the polynomial equation $X^2 + 1 = 0$.

To base a Solving Polynomial Equation Systems theory on a tautology is not a clever idea. So it is probably a good idea to understand better the rôle of the imaginary number.

3.2 The Imaginary Number

My story goes back to the first half of the sixteenth century. At that time it was known that a quadratic equation

$$X^2 + bX + c = 0$$

has solutions only when $b^2 - 4c$ is not negative, in which case the solutions are

$$\frac{-b}{2} \pm \sqrt{\frac{b^2}{4} - c}.$$

They also knew that up to a linear transformation[1], cubic equations could be easily reduced to the form

$$X^3 + pX + q = 0.$$

The formula giving the solutions of this equations was discovered by Tartaglia and later divulged by Cardano. The formula is:

$$\sqrt[3]{\frac{-q}{2} + \sqrt{\left(\frac{q}{2}\right)^2 + \left(\frac{p}{3}\right)^3}} + \sqrt[3]{\frac{-q}{2} - \sqrt{\left(\frac{q}{2}\right)^2 + \left(\frac{p}{3}\right)^3}}. \quad (3.1)$$

Example 3.2.1. Let us consider the equation $X^3 + 3X - 14 = 0$. It is easy to verify that it has a single (real[2]) root – the function is increasing, since

---

[1] In general, if $f(X) \in k[X]$ is a polynomial, solving the equation $f(x) = 0$ and solving the equation $\mathrm{lc}(f)^{-1} f(X) = 0$ are the same, so we can assume that we have been given a monic polynomial to solve, let us say:

$$f(X) = \sum_{i=0}^{n} a_{n-i} X^i = X^n + a_1 X^{n-1} + \cdots + a_{n-1} X + a_n.$$

Let $c \in K \setminus \{0\}$ and let us consider the polynomial

$$\sum_{i=0}^{n} b_{n-i} X^i = g(X) = f(X - c) = X^n + (a_1 - nc) X^{n-1} + \cdots$$

It is then obvious that

α is a root of $g \iff \alpha - c$ is a root of f;
if we choose $c := a_1/n$ then $b_1 = 0$.

[2] Before the imaginary number was invented, all the numbers were, of course, real!

3.2 The Imaginary Number

its derivative is positive for each real number – and that such root is 2. Let us apply Equation 3.1 with $p = 3, q = -14$, obtaining

$$\sqrt[3]{7 + \sqrt{49 + 1}} + \sqrt[3]{7 - \sqrt{49 + 1}} = \sqrt[3]{7 + 5\sqrt{2}} + \sqrt[3]{7 - 5\sqrt{2}}$$

which requires us to find numbers α such that

$$\alpha^3 = 7 \pm 5\sqrt{2}.$$

Although they couldn't prove so, it was obvious to Tartaglia and Cardano that such numbers should be of the same kind, i.e. $\alpha = a + b\sqrt{2}$ for integer numbers a, b. We thus obtain

$$\begin{aligned} 7 \pm 5\sqrt{2} &= \alpha^3 \\ &= a^3 + 3a^2b\sqrt{2} + 3ab^2(\sqrt{2})^2 + b^3(\sqrt{2})^3 \\ &= (a^3 + 6ab^2) + (3a^2b + 2b^3)\sqrt{2} \end{aligned}$$

and to compute a and b, we need to solve the system

$$\begin{cases} a^3 + 6ab^2 &= 7 \\ 3a^2b + 2b^3 &= \pm 5 \end{cases}$$

which gives the solution $\begin{cases} a &= 1 \\ b &= \pm 1 \end{cases}$ so that

$$\sqrt[3]{7 + 5\sqrt{2}} + \sqrt[3]{7 - 5\sqrt{2}} = (1 + \sqrt{2}) + (1 - \sqrt{2}) = 2.$$

Where is the rôle of the imaginary number? It appears in the following

Example 3.2.2. where we consider the equation

$$(X - 1)(X - 4)(X + 5) = X^3 - 21X + 20 = 0$$

for which we would expect that (3.1) will give us the three solutions. Let us then apply Equation (3.1) with $p = -21, q = 20$, obtaining

$$\sqrt[3]{-10 + \sqrt{100 - 343}} + \sqrt[3]{-10 - \sqrt{100 - 343}}$$
$$= \sqrt[3]{-10 + 9\sqrt{-3}} + \sqrt[3]{-10 - 9\sqrt{-3}}.$$

Unlike in the above example where we had to compute with the well-known number $\sqrt{2}$, we have to deal with the number $\sqrt{-3}$ which is well known to be non-existent. However, let us try to compute with this non-existent number as we did with the well-known number $\sqrt{2}$, looking for integer numbers a, b such that

$$\begin{aligned}-10 \pm 9\sqrt{-3} &= (a+b\sqrt{-3})^3 \\ &= a^3 + 3a^2b\sqrt{-3} + 3ab^2(\sqrt{-3})^2 + b^3(\sqrt{-3})^3 \\ &= (a^3 - 9ab^2) + (3a^2b - 3b^3)\sqrt{-3}\end{aligned}$$

where we reasonably assume that the non-existent number $\sqrt{-3}$ behaves like $\sqrt{2}$, i.e. we assume that $(\sqrt{-3})^2 = -3$ and $(\sqrt{-3})^3 = -3\sqrt{-3}$.

As before, to compute a and b, we solve the system

$$\begin{cases} a^3 - 9ab^2 &= -10 \\ 3a^2b - 3b^3 &= \pm 9 \end{cases}$$

which gives us three solutions

$$\begin{cases} a = 2 \\ b = \pm 1 \end{cases} \quad \begin{cases} a = -5/2 \\ b = \pm 1/2 \end{cases} \quad \begin{cases} a = 1/2 \\ b = \mp 3/2 \end{cases}$$

from which we obtain

$$\sqrt[3]{-10 + 9\sqrt{-3}} + \sqrt[3]{-10 - 9\sqrt{-3}}$$

$$= \begin{cases} (2 + \sqrt{-3}) + (2 - \sqrt{-3}) &= 4 \\ (-\tfrac{5}{2} + \tfrac{1}{2}\sqrt{-3}) + (-\tfrac{5}{2} - \tfrac{1}{2}\sqrt{-3}) &= -5 \\ (\tfrac{1}{2} - \tfrac{3}{2}\sqrt{-3}) + (\tfrac{1}{2} + \tfrac{3}{2}\sqrt{-3}) &= 1 \end{cases}$$

which, *mirabile dictu*, gives exactly the three roots we were expecting.

So what? The first comment to make is that we were able to find the three expected roots of our equation just by manipulating the non-existent[3] number $\sqrt{-3}$ as we did for a 'reasonable' square such as $\sqrt{2}$; as Cardano put it

Putting aside the mental tortures involved, multiply $5 + \sqrt{-15}$ by $5 - \sqrt{-15}$, making $25 - (-15)$, which is $+15$. Hence the product is $40. \ldots$ This is truly sophisticated[4].

But the moral of this computation is more significant: the computation we did with $\sqrt{-3}$ – as that with $\sqrt{2}$, – just used the same four operations and the

---

[3] Remember, we are still pretending to live in Tartaglia's time, when how to deal with integers was known – even better than we know: are you able to solve the equations relating a and b? I can not! – and when it was well known that *negative numbers have no square root*!

[4] *Ars Magna*, Chap. 37 (Translated by T.R. Wither): *The Great Art or The Rule of Algebra by Girolamo Cardano*, M.I.T. Press, Cambridge, Mass. 1986.
I took this quotation from B.L. van der Waerden, *A History of Algebra*, Springer, 1985.

'obvious' fact that $(\sqrt{-3})^2 = -3$, i.e. the simple fact that *the root which satisfies the relation $X^2 = -3$ satisfies the relation $X^2 = -3$*; another formulation of our tautology....

3.3 An Impasse

Cardano's student Ferrari solved biquadratic equations with a similar, while more complicated, approach, and research to solve the quintic equation started.

Note that a polynomial has at most as many roots as its degree, as a consequence of the Division Algorithm (cf. Corollary 1.4.1). Girard was the first to conjecture that

toutes les équations d'algèbre reçoivent autant de solutions, que la dénomination de la plus haute quantité le démonstre...[not only real roots but also] autres enveloppées, comme celles qui ont des $\sqrt{-}$, comme $\sqrt{-3}$, ou autres nombres semblables[5]

i.e. a polynomial has exactly as many roots as its degree, provided we invent other non-existent numbers. It was Euler who stated a stronger conjecture than Girard's; namely that

Theorem 3.3.1 (Euler's Conjecture). *A polynomial with real coefficients has exactly as many roots (real or complex) as its degree,* ♄

which was later (1799) proved by Gauss for even complex polynomials (Fundamental Theorem of Algebra), a Gaussian proof of which will be discussed in Section 12.1.

Comparing Euler's Conjecture with Girard's, it is important to stress that it assumed that the invention of a *single* non-existing root – the imaginary number i solving $X^2 + 1 = 0$ – is all one needs to give *any* real polynomial the proper number of roots.

The problem of solving quintic (or higher degree) equations was still baffling the mathematicians. At that time, 'solving' polynomial equations was intended to allow computation of roots by applying *five* operations to the coefficients of equations; the fifth operation to which we refer is, of course, root extraction.

It is even a temptation to translate their notion of 'solving' in to a more modern language, by saying that solving an equation is writing a program (even a straight-line-program) whose input is the coefficients of the equation and whose output is the roots, the operations allowed being the five operations, testing equalities and branching.

---

[5] I took this quotation from R. Remmert, The Fundamental Theorem of Algebra, in H.D. Ebbinghaus *et al.*, *Numbers*, Springer (1991).

The problem of solving a quintic polynomial equation was thoroughly investigated by Lagrange (1772) in *Sur la forme des racines imaginaires des équations*, where he gave a wonderful, in-depth survey on the state of the art, stressing the importance of root permutations. A careful analysis of the subgroups of the group of the permutations over a set of five elements, following his suggestions and ideas, allowed it to be proven that

Theorem 3.3.2 (Abel–Ruffini). *The generic equation of degree ≥ 5 cannot be 'solved', i.e. its roots cannot be expressed in terms of the five operations on its coefficients.* $\boxed{\hbar}$

3.4 A Tautology!

What was, therefore, the then (i.e. nineteenth century) state of the art on Solving Polynomial Equation Systems?

On one side, we have, through Gauss' proof, the information that the – now familiar – field of the complex numbers contains *all* the solutions of *each* polynomial equation. On the other side, Abel–Ruffini informed us that our century-long dream and hope of solving such equations had definitely been lost.

To move out of this impasse, Kronecker proposed modifying the meaning of 'solving' and to do that by reinterpreting the tautology we have discussed in this chapter.

4
Intermezzo: Multiplicity of Roots

Before discussing Kronecker's approach to 'solving' a polynomial equation $f(X) = 0$ and how to find all its roots α such that $f(\alpha) = 0$, it is useful to discuss some properties of the roots, which will allow us to have both a feel for how they behave and some algorithmic tools.

Let us assume we are given a polynomial $f(X) \in k[X]$ and a field $K \supseteq k$ which contains all the roots of $f(X)$ – the Fundamental Theorem of Algebra guarantees us that if $k \subset \mathbb{C}$, $K := \mathbb{C}$ is such a field; in the general case the existence of such a field will be proved in Theorem 5.5.6.

Then, as a consequence of Corollary 1.4.1, we can conclude that, in $K[X]$, $f(X)$ has a factorization into linear factors

$$f(X) = \prod_{i=1}^{s}(X - \alpha_i)^{e_i}$$

where $\alpha_1, \ldots, \alpha_s \in K$ are the roots of f, and e_i is the 'multiplicity' of α_i and, of course, the relation $\sum_{i=1}^{s} e_i = \deg(f)$ holds.

In other words we can conclude that a polynomial has as many roots as its degree, provided they are counted with their multiplicity, and this leads us to reflect on the notion of multiplicity and to look for a technique for computing it, at least one that is more efficient than repeatedly dividing f by its linear factors (Section 4.4).

It is well known that in the case of polynomials over the reals, the notion of multiplicity is given in terms of the number of consecutive derivatives of the polynomial which vanish at the root; in order to show that the same result holds in a more general situation – but not in all possible fields K – I introduce a formal notion of a derivative, which does not require analytical notions as the limit (Section 4.3).

To do so, I first need to introduce the notion of 'characteristic' (Section 4.1) and to analyse the basic properties of finite fields (Section 4.2).

However, once the notion of 'multiplicity' is introduced, we will find out that, unlike the case of polynomials over the algebraic, real and complex numbers, where the only polynomials which have null derivatives are the constants, in the general case there are fields k such that there are polynomials $f(X) \in k[X]$ which are non-constant and irreducible, but with null derivatives (Section 4.5).

Such 'monster' polynomials will be labelled as 'inseparable' and I need to analyse their existence and their behaviour. By doing that I show that such polynomials cannot exist over *perfect* fields, i.e. those which are either finite or containing $\mathbb{Q}$ (Section 4.6).

After this investigation the purpose of studying the multiplicity of the irreducible factors of a polynomial f becomes clear. With this in mind, I discuss the notions of *squarefree associates*, *distinct power factorization* and the algorithms used to compute them (Section 4.7).

4.1 Characteristic of a Field

Let k be a field. There is a ring morphism $\chi : \mathbb{Z} \mapsto k$, the *characteristic morphism*, which is defined by:

$$\chi(0) = 0,$$
$$\chi(1) = 1,$$
$$\chi(n) = \chi(n-1) + 1, \quad n \geq 2,$$
$$\chi(-m) = -\chi(m), \quad m < 0.$$

Two cases can occur:

the above morphism is injective, i.e. $\chi(n) \neq 0$ if $n \neq 0$. Then χ can be extended to a field morphism $\chi : \mathbb{Q} \mapsto k$ by

$$\chi\left(\frac{m}{n}\right) := \frac{\chi(m)}{\chi(n)},$$

which satisfies the relations $\ker(\chi) = 0$, $\text{Im}(\chi) \cong \mathbb{Q}$. So identifying $\mathbb{Q}$ with its isomorphic image we can assume k contains $\mathbb{Q}$. In this case we say that k is of *characteristic* 0, that is $\text{char}(k) = 0$.

There is $p \neq 0$ such that $\chi(p) = 0$. Therefore we can conclude that

there is a *minimal* such $p \in \mathbb{N} \setminus \{0\}$, which is prime, since, otherwise, from a factorization $p = ab$ we obtain $\chi(a)\chi(b) = \chi(p) = 0$, and, k being a field, for one of the factors, say a, we conclude that $\chi(a) = 0$, contradicting the minimality of p.

For $g \in \mathbb{N}$

$$\chi(g) = 0 \iff g \text{ is a multiple of } p;$$

by division we obtain $g = qp + r$ and because

$$\chi(r) = \chi(g) - \chi(q)\chi(p) = 0,$$

so that by the minimality of p we have $r = 0$.

As a consequence we deduce that $\ker(\chi) = (p) \subset \mathbb{Z}$ and $\operatorname{Im}(\chi) \cong \mathbb{Z}_p$.

So again, after identification, we can assume k contains $\mathbb{Z}_p$. In this case we say k is of *characteristic* p [1] – $\operatorname{char}(k) = p$ – or, if p is not specified, of *prime characteristic*.

And $\mathbb{Q}$ (respectively $\mathbb{Z}_p$) is called the *prime field* of k.

4.2 Finite Fields

Further developments require a short discussion of the main properties of finite fields[2]. We will therefore assume that F is a finite field such that $\operatorname{card}(F) = q$ and k is its prime field.

Proposition 4.2.1. *Let F be a finite field, $\operatorname{card}(F) = q$, $p := \operatorname{char}(F)$. Then $p \neq 0$ and $q = p^n$ for some n.*

Proof Since F is finite, $\mathbb{Q} \not\subseteq F$, so $\operatorname{char}(F) =: p \neq 0$, and then $F \supseteq k \cong \mathbb{Z}_p$. Therefore F is a $\mathbb{Z}_p$-vector space, by necessity of finite dimension n (because it is finite). Then $F \cong (\mathbb{Z}_p)^n$ and so $\operatorname{card}(F) = p^n$. ♄

Lemma 4.2.2. *Let F be a finite field such that $\operatorname{card}(F) = q$ and $\operatorname{char}(F) = p$. Let $a, b \in F$, $f, g \in F[X]$.*
Then:

(1) $(a+b)^p = a^p + b^p$.
(2) $(a+b)^q = a^q + b^q$.

---

[1] Apparently the notation is inconsistent; it becomes consistent if we read it in the ideal-theoretical language: the kernel of χ is then respectively generated by 0 or p.
[2] The argument will be discussed in more depth in Chapter 7.

(3) $(f+g)^p = f^p + g^p$.
(4) $(f+g)^q = f^q + g^q$.

Proof (1) and (3) follow from the binomial theorem

$$(\alpha + \beta)^p = \sum_{i=0}^{p} \binom{p}{i} a^{p-i} b^i$$

since p divides $\binom{p}{i}$ for $0 < i < p$.
(2) and (4) follow from induction:

$$(a+b)^{p^n} = \left((a+b)^{p^{n-1}}\right)^p = \left(a^{p^{n-1}} + b^{p^{n-1}}\right)^p = a^{p^n} + b^{p^n}.$$

♄

Lemma 4.2.3 (Little Fermat Theorem).
Let F be a finite field such that $\text{card}(F) = q$. Then

$$a^q = a, \ \forall a \in F.$$

Proof The statement is equivalent to

$$a^{q-1} = 1, \ \forall a \in F \setminus \{0\}$$

for which an elementary proof is as follows: since F is a field, the mapping

$$\phi_a : F \setminus \{0\} \mapsto F \setminus \{0\}$$
$$\phi_a(x) = ax, \ \forall x$$

is injective and, being F finite, also a bijection; therefore

$$\prod_{x \in F \setminus \{0\}} x = \prod_{x \in F \setminus \{0\}} ax = a^{q-1} \prod_{x \in F \setminus \{0\}} x,$$

whence the thesis after dividing out $\prod_{x \in F \setminus \{0\}} x$. ♄

Corollary 4.2.4. Let F be a finite field, $\text{char}(F) = p$, $\text{card}(F) = q = p^n$.
Let $a \in F$, $b := a^{p^{n-1}}$.
Then $b^p = a$.
Therefore, for each $a \in F$, there is a unique $b \in F$ such that $b^p = a$. As a consequence, each element of F is a pth power.

Proof The only statement which needs a proof is the uniqueness of the pth root of a: assume $b^p = a = c^p$; then $0 = b^p - c^p = (b-c)^p$ and so $b = c$. ♄

Remark 4.2.5. For each $m \in \mathbb{N}$ consider the map $\Phi_m : F \mapsto F$ defined by $\Phi_m(a) = a^{p^m}$ and let Φ denote Φ_1.

Here Φ is an automorphism – which is called the *Frobenius automorphism* –, since

it is bijective by the result above,
$(ab)^p = a^p b^p$, $\forall a, b$,
$(a+b)^p = a^p + b^p$, $\forall a, b$ by Lemma 4.2.2.

Since $\forall m$, $\Phi_m = \Phi \cdot \Phi_{m-1}$, it follows that

$\forall m$, Φ_m is an automorphism;
Φ_n is the identity, by the Little Fermat Theorem;
$\forall m_1, m_2 \in \mathbb{N}, \Phi_{m_1} = \Phi_{m_2} \iff m_1 \equiv m_2 \pmod{n}$.

4.3 Derivatives

If k is, say, $\mathbb{R}$, then the multiplicity of a root α of $f(X) \in k[X]$ can be characterized in terms of the derivatives of f at α becoming zero.

Looking at the proof of this characterization, we sees that the 'analytical' properties of the derivative are not involved, only its 'algebraic' properties. So we are going to define a formal derivative of a polynomial in $k[X]$.

Definition 4.3.1. *Let*

$$f(X) := \sum_{i=0}^{n} a_i X^i \in k[X];$$

the derivative of f, denoted f' or $D(f)$, is the polynomial

$$\sum_{i=1}^{n} i a_i X^{i-1}$$

where i represents the image of the exponent $i \in \mathbb{N}$ in the prime field of k via the characteristic morphism χ.

We will also define the ith derivative of f by the recursive definition

$$f^{(i)} := D(f^{(i-1)})$$

where $f^{(0)} := f$.

The next lemma shows that the formal properties of derivatives are satisfied by this notion; the subsequent lemma shows that something unexpected occurs in fields of prime characteristic: there are non-constant polynomials with null derivative.

Lemma 4.3.2.
$D(f + g) = D(f) + D(g);$
$D(fg) = D(f)g + fD(g).$

Proof A straightforward verification. □

Lemma 4.3.3. *(1) If* $\operatorname{char}(k) = 0$ *then*
$$D(f) = 0 \iff f \in k.$$
(2) If $\operatorname{char}(k) = p > 0$, *then*
$$D(f) = 0 \iff \exists g \in k[X] : f(X) = g(X^p).$$

Proof

(1) Requires a straightforward verification.
(2) Assume $f(X) = g(X^p)$; since
$$D(X^{ip}) = ipX^{ip-1} = 0, \forall i,$$
the thesis follows from Lemma 4.3.2.
Conversely let $f(X) = \sum_{i=0}^{n} a_i X^i$ be such that
$$D(f) = \sum_{i=1}^{n} ia_i X^{i-1} = 0;$$
then $ia_i = 0, 1 \leq i \leq n$, and $a_i = 0$ whenever i is not a multiple of p. Therefore setting $m := \lfloor n/p \rfloor$ and $g(X) = \sum_{k=0}^{m} a_{kp} X^k$,
$$f(X) = \sum_{i=0}^{n} a_i X^i = \sum_{k=0}^{m} a_{kp} X^{kp} = g(X^p).$$

□

4.4 Multiplicity

My introduction of this general notion of derivatives was aimed at generalizing, from $\mathbb{R}$ to k, its application to computing the multiplicity of roots; so let us postpone our consideration of this unexpected behaviour of derivatives in the case of prime characteristic fields, until we can see how it affects our approach to multiplicity.

4.4 Multiplicity

Therefore let us consider a field k, a polynomial $f(X) \in k[X]$, a field $K \supseteq k$ which contains all the roots of f including a root $\alpha \in K \supseteq k$; this setting and notation will be used throughout this chapter.

Definition 4.4.1. *Let* $r \in \mathbb{N} \setminus \{0\}$ *be such that in* $K[X]$,

$$f(X) = (X - \alpha)^r f_1(X), \; f_1(\alpha) \neq 0. \tag{4.1}$$

Then r *is called the* multiplicity *of the root* α *of* f.
A root is simple *if its multiplicity is* 1, multiple *otherwise*.

Lemma 4.4.2. *The definition of multiplicity, while obviously depending on* α *and* k *does not depend at all on the field* K *such that* $\alpha \in K \supseteq k$.

Proof Let L be any other field such that $\alpha \in L \supseteq k$ and let us assume that in $L[X]$

$$f(X) = (X - \alpha)^s g_1(X), \; g_1(\alpha) \neq 0. \tag{4.2}$$

Now let F be any field such that[3]

$$F \subseteq K, \quad F \subseteq L, \quad F \supseteq k, \quad \alpha \in F.$$

Clearly, being the result of successive divisions of $f(X) \in k[X] \subseteq F[X]$ by $(X - \alpha) \in F[X]$, the polynomials $f_1(X)$ and $g_1(X)$ belong to $F[X]$. Moreover $f_1(\alpha) = 0$ in F would contradict $f_1(\alpha) \neq 0$ in K. Therefore in F, $f_1(\alpha) \neq 0$ and $g_1(\alpha) \neq 0$.
From Equations 4.1 and 4.2 we obtain in $F[X]$

$$(X - \alpha)^r f_1(X) = (X - \alpha)^s g_1(X);$$

since $f_1(\alpha) \neq 0$ and $g_1(\alpha) \neq 0$, it follows immediately that $s = r$ and $f_1 = g_1$.
♁

When $K = \mathbb{R}$, we know how to use the derivatives to compute the multiplicity: the multiplicity of α is r iff the rth derivative is the first one which does not vanish at α.

In order to show that this holds over any field k such that $\mathrm{char}(k) = 0$ and to show what happens in the general case $\mathrm{char}(k) \neq 0$, let us consider a root α of f with multiplicity r, so that there is $h(X) \in K[X]$ such that

$$f(X) = (X - \alpha)^r h(X), \; h(\alpha) \neq 0, \tag{4.3}$$

---

[3] For the sake of our argument, for such F it is sufficient to take $F := K \cap L$. In the context of the discussions of Chapter 5 the natural choice is $k(\alpha)$.

and let us derive f obtaining

$$f'(X) = r(X-\alpha)^{r-1}h(X) + (X-\alpha)^r h'(X), \tag{4.4}$$

from which we can obviously deduce the following

Lemma 4.4.3. *With the notation above,*

α *is a root of* f' *of at least multiplicity* $r-1$;
if $\mathrm{char}(k) = 0$, α *is a root of* f' *of multiplicity* $r-1$;
α *is a simple root of* f *iff* $f'(\alpha) \neq 0$;
if $\mathrm{char}(k) = p \neq 0$ *and* $r \equiv 0 \bmod (p)$, *then* α *is a root of* f' *of at least multiplicity* r.

♮

Corollary 4.4.4. *Moreover if* $\mathrm{char}(k) = 0$ *then*

α *is a root of* f *with multiplicity* r *iff*

$$f^{(i)}(\alpha) = 0, \ \forall i, \ 0 \le i < r, \ \text{and} \ f^{(r)}(\alpha) \neq 0;$$

α *is a root of* f *with multiplicity* $r > 1$ *iff* α *is a root of* $\gcd(f, f')$ *of multiplicity* $r-1$.

♮

As we have just seen, the prime characteristic case complicates the study of multiplicity both because the derivative of a non-constant polynomial can vanish and because the usual count of multiplicity cannot be generalized in this setting.

For the present to avoid these problems and better understand the behaviour of derivatives, we can try an alternative approach reducing the study to *irreducible* factors since

Lemma 4.4.5. *Let* $f(X) \in k[X]$ *and let* $\alpha \in K \supseteq k$ *be a root of* f. *Then there is a unique irreducible factor* $g(X)$ *of* $f(X)$ *such that* $g(\alpha) = 0$.

Proof Let $f = \prod_i f_i^{e_i}$ be a factorization in $k[X]$. Since

$$0 = f(\alpha) = \prod_i f_i^{e_i}(\alpha)$$

in K we conclude that there exists an irreducible factor g of f such that $g(\alpha) = 0$.

If there were two of them, say g, h, then $(X - \alpha)$ would divide $\gcd(g, h)$ in $K[X]$, contradicting the fact that $\gcd(g, h) = 1$, ♮

4.4 Multiplicity

and, introducing the obvious generalization of the notion of multiplicity of a factor as

Definition 4.4.6. *Let $g(X) \in k[X]$ be an irreducible factor of $f(X)$ in $k[X]$; let $r \in \mathbb{N}$ be such that g^r divides f and g^{r+1} does not divide f.*

Then we say r is the multiplicity *of g as a factor of f.*

Lemma 4.4.7. *Let $g(X) \in k[X]$ be an irreducible factor of $f(X)$ in $k[X]$ and let $\alpha \in K \supseteq k$ be a root of g. Denote by m the multiplicity of α as a root of g, by M its multiplicity as a root of f, and by r the multiplicity of g as a factor of f; then $M = mr$.*

Proof The result follows easily since we have

$$g(X) = (X - \alpha)^m g_1(X), \; g_1(\alpha) \neq 0,$$
$$f(X) = g(X)^r h(X), \; h(\alpha) \neq 0,$$

and so

$$f(X) = (X - \alpha)^{mr} g_1(X)^r h(X), \; g_1(\alpha)^r h(\alpha) \neq 0,$$

so that $M = mr$. ♄

Therefore we need to discuss the multiplicity of the roots of an irreducible polynomial, proving

Lemma 4.4.8. *Let $f \in k[X]$. Then*

(1) If $\gcd(f, f') = 1$, f is squarefree.
(2) If $\gcd(f, f') = 1$, all roots of f are simple.
(3) If f is irreducible and $f'(X) \neq 0$, $\gcd(f, f') = 1$.

Proof

(1) Let $g \in k[X]$ be an irreducible factor of f of multiplicity r in $k[X]$ and let $h(X) \in k[X]$ be such that $f(X) = g(X)^r h(X)$. Then

$$f'(X) = rg(X)^{r-1}h(X)g'(X) + g(X)^r h'(X).$$

Therefore, if $r > 1$, $g(X)^{r-1}$ is a common factor of f and f', and so $\gcd(f, f') \neq 1$.

(2) Since $\gcd(f, f')$ is independent of the field, we only have to apply (1) over K.

(3) Since f is irreducible and $f' \neq 0$, then $\gcd(f, f')$ is either 1 or f; the latter is impossible, because it implies that f' is a multiple of f while $\deg(f') < \deg(f)$. ♄

Theorem 4.4.9. *Let f be irreducible. Then*

if $\operatorname{char}(k) = 0$, *$f$ has only simple roots;*
if $\operatorname{char}(k) = p \neq 0$, *either*

   $f'(X) \neq 0$ and f has only simple roots, or
   $f'(X) = 0$ and there exists $g \in k[X]$ such that $f(X) = g(X^p)$.

Proof Either $f' \neq 0$ and all the roots are simple by the above lemma, or $f' = 0$ – in which case $\operatorname{char}(k) = p \neq 0$ – and the results follow from Lemma 4.3.3. ♄

4.5 Separability

It is clear from these results that the behaviour of the multiplicity of the roots of a non-constant irreducible polynomial $f(X)$ depends upon whether $f'(X)$ is zero or not; therefore let us introduce

Definition 4.5.1. *Let $f(X) \in k[X]$ be a non-constant irreducible polynomial. We say that f is* separable *if $f'(X) \neq 0$,* inseparable *otherwise,*

and let us prove

Proposition 4.5.2. *Assume $\operatorname{char}(k) = p \neq 0$ and $f(X) \in k[X]$ is a non-constant irreducible polynomial. Then there is $e \in \mathbb{N}$ and a separable polynomial $g(X) \neq 0$ such that*

(1) $f(X) = g(X^{p^e})$;
(2) all the roots of f have the same multiplicity $m = p^e$.

Proof

(1) If $f' \neq 0$, we can take $e = 0$ and $g = f$.
   If $f' = 0$, as a consequence of Lemma 4.3.3, $\exists f_1(X) \in k[X]$ such that $f(X) = f_1(X^p)$. If $f_1' = 0$, then $\exists f_2(X) \in k[X]$ such that $f_1(X) = f_2(X^p)$ so that
   $$f(X) = f_1(X^p) = f_2(X^{p^2}),$$
   whence the claim follows by iteration.

(2) Since $g'(X) \neq 0$, all its roots are simple; therefore, within a field K containing all the roots of g and f – whose existence is a consequence of Theorem 5.5.6, – we have a factorization

$$g(X) = \prod_{i=1}^{n_0} (X - \beta_i)$$

from which we deduce

$$f(X) = \prod_{i=1}^{n_0} (X^{p^e} - \beta_i).$$

For each i, let $\alpha_i \in K$ be a root of f which annihilates the polynomial $(X^{p^e} - \beta_i)$; we then conclude that $\alpha_i^{p^e} = \beta_i$ and

$$(X^{p^e} - \beta_i) = (X^{p^e} - \alpha_i^{p^e}) = (X - \alpha_i)^{p^e}$$

yielding the factorization

$$f(X) = \prod_{i=1}^{n_0} (X - \alpha_i)^{p^e}$$

and the result. ♃

Remark 4.5.3. Given a field k, an irreducible polynomial $f(X) \in k[X]$ of degree n and a field $K \supset k$ which contains all the roots $\alpha := \alpha_1, \ldots, \alpha_{n_0}$ of f, we have two situations:

f is separable, in which case $n = n_0$, all the roots of f are simple, and

$$f = \prod_{i=1}^{n} (X - \alpha_i);$$

f is inseparable, in which case, setting $p := \text{char}(k)$, there is $e > 0$ and a separable polynomial $g(X) \in k[X]$ so that $f(X) = g(X^{p^e})$; therefore $n = n_0 p^e$, all the roots of f have multiplicity p^e, and

$$f = \prod_{i=1}^{n_0} (X - \alpha_i)^{p^e}.$$

With a slight misuse of notation we will identify these two cases by stating that

Proposition 4.5.4. *Let k be a field, $p := \text{char}(k)$, $f(X) \in k[X]$ a polynomial, $K \supseteq k$ a field which contains all the roots $\alpha := \alpha_1, \ldots, \alpha_{n_0}$ of f.*

Then:

if f is irreducible, denoting $n := \deg(f)$, there is $e \geq 0$ such that all the n_0 roots of f have multiplicity p^e, $n = n_0 p^e$, and

$$f = \prod_{i=1}^{n_0}(X - \alpha_i)^{p^e};$$

otherwise, for each root α of f, let $g(X) \in k[X]$ be the unique irreducible factor of f such that $g(\alpha) = 0$; then there is $e \geq 0$ such that the multiplicity of α as a root of f is rp^e where r is the multiplicity of g as a factor of f. ♃

That is, if f is separable, the statement holds for $e = 0$.

Definition 4.5.5. *With the notation of the above proposition and with the assumption that f is irreducible, $\alpha \in K \setminus k$ is called* separable (resp. inseparable) *over K iff f is separable (resp. inseparable)*[4].

The value n_0 is called the reduced degree *of α and f, e their* exponent of inseparability, *p^e their* degree of inseparability.

4.6 Perfect Fields

It is quite natural to ask which are the fields k for which a non-constant, irreducible, inseparable polynomial $f(X) \in k[X]$ exists.

The above discussion tells us that this cannot happen if $\text{char}(k) = 0$ and suggests that, when $\text{char}(k) = p \neq 0$, the property is related to the irreducibility of the polynomials $X^{p^e} - \alpha$, $\alpha \in k$, i.e. to the impossibility of extracting pth roots in k.

In fact:

Lemma 4.6.1. *Let k be a field such that $\text{char}(k) = p > 0$ and*

$$\forall \alpha \in k, \exists \beta \in k : \beta^p = \alpha \tag{4.5}$$

and let $f \in k[X]$.

Then the following are equivalent:

(1) there exists $g \in k[X]$ such that $f(X) = g(X^p)$.
(2) there exists $h \in k[X]$ such that $f = h^p$.
(3) $f'(X) = 0$.

---
[4] Using the language of Section 5.3, if α is an algebraic number and f is its minimal polynomial, α and f share the same property of being separable or not.

4.6 Perfect Fields

Proof

(1) $\implies$ (2) Let $g(X) := \sum_{k=0}^{m} \alpha_k X^k$; let $\forall k$, β_k be such that $\beta_k^p = \alpha_k$, and denote $h(X) := \sum_{k=0}^{m} \beta_k X^k$; then

$$f(X) = g(X^p) = \sum_{k=0}^{m} \alpha_k (X^p)^k = \sum_{k=0}^{m} \beta_k^p (X^k)^p$$
$$= \left(\sum_{k=0}^{m} \beta_k X^k\right)^p = h(X)^p.$$

(2) $\implies$ (1) Let $h(X) := \sum_{k=0}^{m} \beta_k X^k$ and denote $g(X) = \sum_{k=0}^{m} \beta_k^p X^k$; then

$$f(X) = h(X)^p = \sum_{k=0}^{m} \beta_k^p X^{kp} = g(X^p).$$

(1) $\iff$ (3) This holds by Lemma 4.3.3. ♮

Corollary 4.6.2. *Let k be a field, such that either*

char$(k) = 0$ *or*
char$(k) = p > 0$ *and Equation (4.5) holds.*

Then every non-constant irreducible polynomial $f(X) \in k[X]$ is separable.

Proof Clearly if char$(k) = 0$ and $f' = 0$, then f is constant.
If char$(k) = p > 0$ and $f' = 0$, then f is a pth power by Lemma 4.6.1 and so it is not irreducible. ♮

To prove the converse of this result, we have just to consider a field k such that char$(k) = p$ which contains an element $\alpha \in k$ which is not a pth power, i.e.

$$\forall \beta \in k : \alpha \neq \beta^p,$$

and analyse the polynomial $X^p - \alpha \in k[X]$:

Proposition 4.6.3. *Under the above assumption, $f(X) := X^p - \alpha$ is irreducible.*

Proof Let us assume f is not irreducible and let $g(X) \in k[X]$ be a monic irreducible factor of f with multiplicity r so that

$$f(X) = g^r(X)h(X)$$

for some $h(X) \in k[X]$ such that $\gcd(g, h) = 1$; by derivation we obtain

$$0 = f' = g^r h' + r g^{r-1} g' h$$

and, after dividing out g^{r-1},

$$gh' = -rg'h.$$

Hence h must divide gh'; since h cannot divide g – because $\gcd(g, h) = 1$ –, it divides h'. This is possible only if $h' = 0$ and so $rg' = 0$.
Since $h' = 0$, we can conclude by Lemma 4.6.1 that

$$\exists H(X) \in k(X) : h(X) = H(X^p);$$

similarly, since $D(g^r) = rg^{r-1}g' = 0$, we have that

$$\exists G(X) \in k(X) : g^r(X) = G(X^p).$$

We can then conclude that

$$X^p - \alpha = f(X) = g^r(X)h(X) = H(X^p)G(X^p),$$

i.e. $Y - \alpha = H(Y)G(Y)$ in $k[Y]$.
Since $Y - \alpha$ is linear and G is not constant, and g is not the same, then we conclude that $H = 1$ – both $Y - \alpha$ and G are monic – and so $Y - \alpha = G(Y)$, i.e.

$$f(X) = g^r(X).$$

We now have two cases:

if r is a multiple of p, then f is a power of g^p, whose coefficients are pth powers; in particular, α would be the same, contradicting the assumption; therefore r is not a multiple of p and so, from $rg' = 0$, we conclude that $g' = 0$ and

$$\exists \mathsf{G}(X) \in k(X) : g(X) = \mathsf{G}(X^p),$$

giving

$$X^p - \alpha = f(X) = g^r(X) = \mathsf{G}^r(X^p)$$

and

$$Y - \alpha = \mathsf{G}^r(Y),$$

so that $r = 1$.

As a conclusion we have established that $f(X) = g^r(X)h(X) = g(X)$ is irreducible. ♮

This proposition can be generalized as follows:

Theorem 4.6.4. *Under the assumption above, let $e \in \mathbb{N} \setminus \{0\}$ and let*

$$f(X) := X^{p^e} - \alpha \in k[X].$$

Then $f(X)$ is irreducible.

Proof (sketch) The argument is given by iteration on e.
On the one side Proposition 4.6.3 proves the result in the case $e = 1$; on the other side the same result allows us to iterate on e: in fact, assuming that $X^{p^{e-1}} - \alpha$ is irreducible, the same argument of Proposition 4.6.3 is applicable, just changing $Y^{p^{e-1}} - \alpha$ everywhere $Y - \alpha$. ♄

After this tour de force, we can now easily prove the converse of Corollary 4.6.2:

Theorem 4.6.5. *Let k be a field; the following conditions are equivalent:*

(1) Either

$\text{char}(k) = 0$ *or*
$\text{char}(k) = p > 0$ *and Equation (4.5) holds;*

(2) every non-constant irreducible polynomial $f(X) \in k[X]$ is separable.

Proof We only have to prove (2) $\implies$ (1) since (1) $\implies$ (2) is Corollary 4.6.2: let us consider any field k such that $\text{char}(k) = p$ and Equation (4.5) does not hold.
Then there is $\alpha \in k$ such that it is not a pth power and $f(X) := X^p - \alpha \in k[X]$ is irreducible, non-constant and inseparable. ♄

On the basis of that, we introduce

Definition 4.6.6. *A field k is called* perfect *if it satisfies the conditions of Theorem 4.6.5.*

We note

Corollary 4.6.7. *A finite field is perfect.*

Proof The proof follows from Corollary 4.2.4. ♄

Note also there are infinite fields which are not perfect, such as $\mathbb{Z}_p(Y)$.

Corollary 4.6.8. *If k is perfect, then:*

(1) If $g(X) \in k[X]$ is irreducible, each root of g is simple.
(2) Let $f(X) \in k[X]$, $\alpha \in K \supseteq k$ be a root of f and $g(X) \in k[X]$ be the unique irreducible factor of f such that $g(\alpha) = 0$. The multiplicity of α as a root of f is the multiplicity of g as a factor of f.
(3) Let $f(X) \in k[X]$; f is squarefree iff $\gcd(f, f') = 1$.

Proof

(1) The proof follows from Lemma 4.4.8.
(2) The proof follows from Lemma 4.4.7.
(3) Because one implication is Lemma 4.4.8, let us assume $\gcd(f, f') \neq 1$ and let h be an irreducible common factor of f and f'. Then $f = hg$ and $f' = hg' + h'g$. Since h divides f', it divides $h'g$ and, since it is irreducible, it divides either h' or g.
However, $h' \neq 0$ and $\deg(h') < \deg(h)$ implies that h does not divide h', so it divides g: $g = hg_1$ for a suitable $g_1 \in k[X]$.
Therefore $f = hg = h^2g_1$, so that h is a multiple factor of f. ♄

Note that if k is an effective perfect field, then Corollary 4.6.8 gives an algorithm to test whether f is squarefree.

4.7 Squarefree Decomposition

It is quite clear, on the basis of the above discussion, that counting the multiplicity of roots is related to counting the multiplicity of factor polynomials.

Since factorization is far from being an easy task, the notion of *squarefree decomposition* is a very useful tool:

Proposition 4.7.1. *Let k be a perfect field.*
Let $f(X) \in k[X]$. Then there are unique (up to associates) polynomials $f_1, \ldots, f_i, \ldots \in k[X]$ such that

(1) either $f_i = 1$ or f_i is squarefree;
(2) $f = \prod_i f_i^i = f_1 f_2^2 \cdots f_{s-1}^{s-1} f_s^s$;
(3) $\gcd(f_i, f_j) = 1$ if $i \neq j$;
(4) α is a root of f with multiplicity r iff α is a (simple) root of f_r.

4.7 Squarefree Decomposition

Proof We only have to define f_i to be the product of all the irreducible factors of f which has multiplicity i.

All the properties are then obvious, except (4) which is a consequence of Corollary 4.6.8. ♄

Definition 4.7.2. *We will call the polynomial* $\mathrm{SQFR}(f) := \prod_i f_i$ *the squarefree associate of* f, *and* $f = \prod_i f_i^i$ *the distinct power factorization or squarefree decomposition of* f.

Note that $\mathrm{SQFR}(f)$ is the product of all the irreducible factors of f, each taken with multiplicity 1, and that both $\mathrm{SQFR}(f)$ and the distinct power factorization of f are independent of the field[5].

Let us again restrict ourselves to the case of a field of characteristic 0, in order to take advantage of Corollary 4.4.4.

Proposition 4.7.3. *Let* $f \in k[X]$, $p := \gcd(f, f')$, $q := f/p$, $s := \gcd(p, q)$, $t := q/s$. *Then:*

$q = \mathrm{SQFR}(f);$
$s = \mathrm{SQFR}(p) = p/\gcd(p, p');$
t *is the product of the simple irreducible factors of* f.

Proof Let $f = \prod_i f_i^i$ be the distinct power factorization of f; then, for a suitable $g \in k[X]$ such that $\gcd(g, f) = 1$, we have

$$
\begin{array}{rcccccccc}
f & = & & f_1 & f_2^2 & f_3^3 & \cdots & f_i^i & \cdots, \\
f' & = & g & f_2 & f_3^2 & \cdots & f_i^{i-1} & \cdots, \\
p & = & & f_2 & f_3^2 & \cdots & f_i^{i-1} & \cdots, \\
q & = & & f_1 & f_2 & f_3 & \cdots & f_i & \cdots,
\end{array}
$$

---

[5] Note that we introduce these concepts *only* for polynomials over a *perfect* field. For an infinite field k of finite characteristic, if we introduce these concepts, we should take care that there will be two versions, a *weak* one in which the result will hold in $k[X]$ and a *strong* one in which the result will hold in $K[X]$: if we consider the situation discussed in Proposition 4.5.2 and Remark 4.5.3, the polynomial

$$f(X) = \prod_{i=1}^{n_0}(X^{p^e} - \beta_i) = \prod_{i=1}^{n_0}(X - \alpha_i)^{p^e}$$

has the strong squarefree $\prod_{i=1}^{n_0}(X - \alpha_i)$ in $K(X)$, while it is the weak squarefree of itself in $k(X)$.

Fig. 4.1. Distinct Power Factorization Algorithm

$[f_1, \ldots, f_s] := $ **DistinctPowerFactorization**(f)
where
  $f \in k[X]$
  $f = f_1 f_2^2 \cdots f_{s-1}^{s-1} f_s^s$ is the distinct power factorization of f
  $L := []$
  $p := \gcd(f, f')$
  $q := f/p$
  Repeat
    $s := \gcd(p, q)$
    $t := q/s$
    $L := [L, t]$
    $q := s, p := p/s$
  until $\deg(s) = 0$

$$\begin{aligned} s &= & f_2 \quad f_3 \quad \cdots \quad f_i \quad \cdots, \\ p/s &= & f_3 \quad \cdots \quad f_i^{i-2} \quad \cdots, \end{aligned}$$

whence the claims. ♄

Algorithm 4.7.4. In an effective field of characteristic 0, the existence of an algorithm to compute SQFR(f) is then obvious; also, by an iterative application of Proposition 4.7.3, we obtain the algorithm in Figure 4.1 computing the distinct power factorization of f.

The generalization of this algorithm for the case of a finite field F, char(F) = p, is more complex[6].

In fact, by Lemma 4.4.3, we know that if r, the multiplicity of α, is a multiple of p, either $f' = 0$ or α is a root of f' with multiplicity at least r, so that α is a root of $\gcd(f, f')$ with multiplicity exactly r. In fact, we have

$$f(X) = (X - \alpha)^{k\mathsf{p}} h(X), \quad h(\alpha) \neq 0,$$

and

$$f'(X) = k\mathsf{p}(X - \alpha)^{k\mathsf{p}-1} h(X) + (X - a)^{k\mathsf{p}} h'(X) = (X - a)^{k\mathsf{p}} h'(X).$$

---

[6] This generalization is due to Davenport in

J.H. Davenport, On the Integration of Algebraic Functions, *L.N.C.S.* **102**, Springer (1981)

which we have followed.

4.7 Squarefree Decomposition

Therefore the roots of multiplicity $k\mathsf{p}$ of $\gcd(f, f')$ are both the roots of f of multiplicity $k\mathsf{p}+1$ and those of multiplicity $k\mathsf{p}$. Therefore in this setting the results of Props. 4.7.3 become as follows:

Lemma 4.7.5. *Let F be a field such that $\operatorname{char}(F) = \mathsf{p} \neq 0$. Let $f \in F[X]$ and $f = \prod_i f_i^i$ be its distinct power factorization. Then:*

(1) *If $f' = 0$ let $h \in F[X]$ be such that $f = h^{\mathsf{p}}$ and $h = \prod_j h_j^j$ be its distinct power factorization; then*
$$f_i := \begin{cases} 0 & \text{if } i \not\equiv 0 \bmod \mathsf{p} \\ h_j & \text{if } i = j\mathsf{p}. \end{cases}$$

(2) *If $f' \neq 0$, let $p := \gcd(f, f')$, $q := f/p$, $s := \gcd(p, q)$, $t := q/s$. Then*

(a) *for a suitable $g \subset k[X]$ such that $\gcd(g, f) = 1$ we have*

f	$=$		f_1	f_2^2	$\cdots$	$f_{k\mathsf{p}-1}^{k\mathsf{p}-1}$	$f_{k\mathsf{p}}^{k\mathsf{p}}$	$f_{k\mathsf{p}+1}^{k\mathsf{p}+1}$	$f_{k\mathsf{p}+2}^{k\mathsf{p}+2}\cdots,$
f'	$=$	g		f_2	$\cdots$	$f_{k\mathsf{p}-1}^{k\mathsf{p}-2}$	$f_{k\mathsf{p}}^{k\mathsf{p}}$	$f_{k\mathsf{p}+1}^{k\mathsf{p}}$	$f_{k\mathsf{p}+2}^{k\mathsf{p}+1}\cdots,$
p	$=$			f_2	$\cdots$	$f_{k\mathsf{p}-1}^{k\mathsf{p}-2}$	$f_{k\mathsf{p}}^{k\mathsf{p}}$	$f_{k\mathsf{p}+1}^{k\mathsf{p}}$	$f_{k\mathsf{p}+2}^{k\mathsf{p}+1}\cdots,$
q	$=$		f_1	f_2	$\cdots$	$f_{k\mathsf{p}-1}$		$f_{k\mathsf{p}+1}$	$f_{k\mathsf{p}+2}\cdots,$
s	$=$			f_2	$\cdots$	$f_{k\mathsf{p}-1}$		$f_{k\mathsf{p}+1}$	$f_{k\mathsf{p}+2}\cdots;$

(b) $t = f_1$;

(c) *if $p = \prod_j p_j^j$ is its distinct power factorization, we have*
$$p_j := \begin{cases} 0 & \text{if } j \equiv -1 \bmod \mathsf{p} \\ f_j f_{j+1} & \text{if } j \equiv 0 \bmod \mathsf{p} \\ f_{j+1} & \text{otherwise;} \end{cases}$$

(d) *denoting $\epsilon : \mathbb{N} \mapsto \{0, 1\}$ the function such that*
$$\epsilon(n) = 0 \iff n \not\equiv 1 \bmod \mathsf{p},$$
and setting $P := \prod_j p_j^{j+\epsilon(j)}$, we have

$f/P = \prod_k f_{k\mathsf{p}+1}$, *and*

$\gcd(f/P, p_{k\mathsf{p}}) = f_{k\mathsf{p}+1}.$

Algorithm 4.7.6. On the basis of the above remarks, the squarefree decomposition algorithm can be generalized to the prime characteristic field case as in Figure 4.1.

Algorithm 4.7.7. To complete this survey, we should discuss an algorithm which, for a given polynomial $f(X) \in k[X]$ in an infinite field k with finite characteristic p, computes the polynomial's splitting field K and its squarefree associate and decomposition within $K[X]$.

Fig. 4.2. Distinct Power Factorization Algorithm, characteristic p

$\bigl(f(1), \ldots, f(s)\bigr) := $ **DistinctPowerFactorization**(f)

where
    k is a finite field
    p $= \mathrm{char}(k) \neq 0$
    $f \in k[X]$
    $f = f(1)f(2)^2 \ldots f(s-1)^{s-1} f(s)^s$ is the distinct power factorization of f

If $\deg(f) = 1$ **then**
    $f(1) := f, V := \bigl(f(1)\bigr)$
    $\to K := k$
else
    If $f' = 0$ **then**
        $\to$ **let** a_i such that $\sum_{i=0}^{d} a_i X^{ip} = f(X)$
        $\to K := K(\sqrt[p]{a_0}, \ldots, \sqrt[p]{a_d})$
        let h be such that $h^p = f$
        $\bigl(h(1), \ldots, h(r)\bigr) := $ **DistinctPowerFactorization**(h)
        For $i = 1, \ldots, r$ **do**
            $f(i\mathrm{p}) := h(i)$
        For $i = 1, \ldots, r\mathrm{p}, i \not\equiv 0 \bmod \mathrm{p}$ **do**
            $f(i) := 0$
        $V := \bigl(f(1), \ldots, f(s\mathrm{p})\bigr)$
    else
        $p := \gcd(f, f')$
        If $p \in k$ **then**
            $f(1) := f, V := \bigl(f(1)\bigr)$
        else
            $q := f/p, s := \gcd(p, q)$
            $f(1) := q/s$
            $f := f/f(1)$
            $\bigl(p(1), \ldots, p(r)\bigr) := $ **DistinctPowerFactorization**(p)
            For $i = 2, \ldots, r+1$ **do**
                If p divides i **then**
                    $f(i) := p(i)$
                    $f := f/f(i)^i$
                else
                    $f(i+1) := p(i)$
                    $f := f/f(i+1)^{i+1}$
            For $i = 2, \ldots, r+1$ such that p divides i **do**
                $f(i+1) := \gcd(f, f(i))$
                $f(i) := f(i)/f(i+1),$
                $f := f/f(i+1)$
            $V := \bigl(f(1), \ldots, f(r+1)\bigr)$

V

4.7 Squarefree Decomposition

To obtain this, as has been noted by G. Kemper, whenever $f' = 0$ occurs in the algorithm in Figure 4.2, we have only to compute the polynomial $h^{\#}(X) = \sum_{i=0}^{d} a_i X^i$ such that $h^{\#}(X^p) = f$, and to extend K with the pth roots of the a_is, so that $h(X) := \sum_{i=0}^{d} \sqrt[p]{a_i} X^i$ satisfies $f(X) = h^{\#}(X^p) = h^p(X)$.

The lines which have to be added in Figure 4.2 to cover the general case are marked by $\rightarrow$.

5
Kronecker I: Kronecker's Philosophy

> Die ganzen Zahlen hat der liebe Gott gemacht,
> alles andere ist Menschenwerk
> L. Kronecker

In this chapter, I now discuss Kronecker's proposal for interpreting the concept of 'solving' and how to deal with algebraic numbers[1].

This proposal, applying only the ability to compute within the residues of a polynomial ring $k[X]$ by an element g – which is guaranteed by the Euclidean Algorithm (Section 5.1) – , introduced both a technique (Section 5.2) which, given a polynomial $f(X) \in k[X]$, allows us to build a field $K \supseteq k$ that contains all the roots of f, i.e. – according to Corollary 1.4.1 – in which $f(X)$ factorizes in linear factors

$$f(X) = \prod_{i=1}^{n}(X - \alpha_i), \qquad (5.1)$$

and the notions of *finitely generated field extensions* (Section 5.3, Section 5.4) – with their classification as *algebraic* and *transcendental* extensions – and of a *splitting field* (Section 5.5) of $f(X) \in k[X]$, which is any minimal field $K \supseteq k$ that contains all the roots of $f(X)$, so that Equation 5.1 holds.

Since the Fundamental Theorem of Algebra had already guaranteed that the roots of $f(X) \in \mathbb{Q}[X]$ – and even of those in $\mathbb{C}[X]$ – exist in $\mathbb{C}$, Kronecker's proposal raised the essential question of what is the relation of the roots of $f(X)$ constructed by Kronecker with the 'true' roots in $\mathbb{C}$. Kronecker proved that all splitting fields of f are isomorphic, so that the answer is that the 'true' roots and those 'constructed' by Kronecker – and even those obtained by any other construction – are essentially the same (Section 5.5).

---

[1] L. Kronecker, Grundzüge einer Arithmetischen Theorie der Algebraischen Grössen, *Crelle's Journal*, **92** (1882).

5.1 Quotients of Polynomial Rings

The only tool which is needed by Kronecker's construction is nothing more than another free consequence of the Euclidean algorithms: the ability to compute within the rings and fields $\mathcal{R}$ which are residues of a polynomial ring $\mathcal{P} = k[X]$ by a non-constant polynomial $g(X) \in \mathcal{P}$:

$$\mathcal{R} = k[X]/g(X).$$

We will fix $n := \deg(g) > 0$ and we will denote $\pi : \mathcal{P} \to \mathcal{R}$ to be the canonical projection.

Proposition 5.1.1. *There is a k-vector space isomorphism Ψ between $\mathcal{R}$ and the subvector space*

$$\mathrm{Span}_k\left(\{1, \ldots, X^{n-1}\}\right) \subset \mathcal{P}$$

with basis $\{1, \ldots, X^{n-1}\}$, which is defined by

$$\Psi(\pi(h)) = \mathbf{Rem}(h, g).$$

If we define a product in the latter vector space by

$$a * b := \mathbf{Rem}(ab, g)$$

then it inherits a ring structure, isomorphic under Ψ to that of $\mathcal{R}$.

Proof We only have to show that

$$\pi(h_1) = \pi(h_2) \implies \mathbf{Rem}(h_1, g) = \mathbf{Rem}(h_2, g),$$

which holds since from $h_1 = q_1 g + \mathbf{Rem}(h_1, g)$ and $h_2 - h_1 = sg$, we get $h_2 = (s + q_1)g + \mathbf{Rem}(h_1, g)$ and conclude by the uniqueness of $\mathbf{Rem}(h_2, g)$. ♄

Proposition 5.1.2. *For $h \in \mathcal{P}$:*
$\pi(h)$ *is invertible in* $\mathcal{R} \iff \gcd(g, h) = 1.$

Proof Assume $\gcd(g, h) = 1$; then by the Bezout Identity there are $S, T \in \mathcal{P}$ such that

$$hS + gT = \gcd(g, h) = 1;$$

therefore

$$\pi(h)\pi(S) = \pi(h)\pi(S) + \pi(g)\pi(T) = \pi(1) = 1,$$

i.e. $\pi(S)$ is the inverse of $\pi(h)$.

Conversely, if $\pi(h)$ is invertible, there is $S \in \mathcal{P}$ such that $\pi(h)\pi(S) = 1$ and so for some $T \in \mathcal{P}, hS + Tg = 1$. A common divisor of h and g must therefore divide 1. ♄

Proposition 5.1.3. $\mathcal{R}$ *is a field if and only if g is irreducible.*

Proof If g is irreducible and $h \in \mathcal{P}$, either h is a multiple of g (and $\pi(h) = 0$) or $\gcd(g, h) = 1$ (and so $\pi(h)$ is invertible in $\mathcal{R}$).
Conversely, if $g(X) = q_1(X)q_2(X)$ is a non-trivial factorization, then $\pi(q_i) \neq 0, \forall i$, and $\pi(q_1)\pi(q_2) = 0$, so $\mathcal{R}$ is not a field. ♄

Remark 5.1.4. Assume k is effective and let us compute
$$(D, S, T) = \mathbf{ExtGCD}(h, g).$$
Then $\pi(h)$ is invertible iff D is constant in which case the inverse of $\pi(h)$ is $\pi(D^{-1}S)$. As we remarked in Algorithm 1.3.4, we can use the Half-extended Euclidean Algorithm, which avoids the computation of T. This does not change the asymptotical complexity but roughly halves the number of field computations needed.

As a conclusion, if k is effective, the residue rings $\mathcal{R}$ are effective too, via the isomorphism Ψ.

Lemma 5.1.5. $\mathcal{R}$ *contains an isomorphic copy of k.*

Proof If g is not linear, while $\mathcal{R}$ does not contain k, its isomorphic copy
$$\mathrm{Span}_k\left(\{1, \ldots, X^{n-1}\}\right)$$
does, containing the subvector space generated by $\{1\}$; therefore $\mathcal{R}$ contains as an isomorphic copy of k the subset $\Psi^{-1}(\mathrm{Span}_k(\{1\})$, i.e. the set of all the classes mod g containing a constant polynomial.
If g, instead, is linear, say $g(X) = X - a$, then $\mathcal{R}$ is isomorphic to k, since $(X - a)$ is the kernel of the morphism $\Phi : k[X] \mapsto k$ defined by $\Phi(f) = f(a)$. ♄

5.2 The Invention of the Roots

Because of Corollary 1.4.1, we have an equivalence between the roots of a polynomial in a field k and its linear factors in $k[X]$, so that the problems of

5.2 The Invention of the Roots

finding the roots of a polynomial and of computing its linear factors are the same. While the Fundamental Theorem of Algebra informed us that there is a field ($\mathbb{C}$) which contains all the roots of each polynomial $f(X) \in \mathbb{Q}[X]$, this result is at the same time useless and tantalizing. We will show here that for each field k and each polynomial $f(X) \in k[X]$, there exists a field $K \supseteq k$ such that K contains all roots of f, i.e. such that f factorizes into linear factors in $K[X]$. Moreover the proof will be constructive – provided we have a factorization algorithm in $k[X]$ – and, if k is effective, K is also effective.

Let, again, k be an effective field, $\mathcal{P} := k[X]$ and $f(X) \in \mathcal{P}$ be a polynomial. Let $g(X) \in \mathcal{P}$ be an irreducible factor of f, $\mathcal{R} := k[X]/g(X)$ which is, therefore, an effective field, $\pi : k[X] \mapsto \mathcal{R}$ be the canonical projection and $\alpha := \pi(X) \in \mathcal{R}$.

Proposition 5.2.1. *With the notation above, $g(\alpha) = 0$.*

Proof In fact $g(\alpha) = g(\pi(X)) = \pi(g(X)) = 0$. ♄

Corollary 5.2.2. *With the notation above, α is a root of f.* ♄

Clearly, Proposition 5.2.1 is nothing more than a reinterpretation of the tautology that *the roots of g are the roots of g*; but this interpretation gives us something more: Proposition 5.2.1 tells us that

the roots α of the irreducible polynomial $g(X)$ satisfy the relation $g(\alpha) = 0$ in $\mathcal{R}$,

giving us a tool to compute in $\mathcal{R}$. On the basis of that, we can see that Kronecker's proposal of reinterpreting the tautology we are discussing, consists of applying the Euclidean Algorithm to construct a field $k_1 := \mathcal{R}$ such that

k_1 is effective,
$k_1 \supseteq k$,
k_1 contains a root α_1 of g.

This allows us to compute a root α_1 of any polynomial $f(X) \in k[X]$: all we have to do is to factorize f, choose a factor $g(X)$ of f, construct $k_1 := k[X]/g(X)$ and take α_1 to be the class of X mod g.

Since Corollary 1.4.1 guarantees for us that

$$f(X) = (X - \alpha_1)f_1(X), \ f_1(X) \in k_1[X],$$

if we assume we are able to factorize $f_1(X)$ over $k_1(X)$, the same idea can be repeated, allowing us to 'solve' f by finding a field K and elements $\alpha_1, \ldots,$

$\alpha_n \in K$ – where $n = \deg(f)$ – such that

K is effective,
$K \supseteq k$,
$\alpha_1, \ldots, \alpha_n \in K$ are all the roots of f, so that

$$f(X) = \prod_{i=1}^{n}(X - \alpha_i). \tag{5.2}$$

Theorem 5.2.3. *Let $f(X) \in \mathcal{P}$. Then there is a field K, $K \supseteq k$, such that $f(X)$ splits, i.e. it factorizes into linear factors in $K[X]$.*

Proof We are going to define a tower of fields

$$k = k_0 \subseteq k_1 \subseteq \ldots \subseteq k_{n-1} =: K,$$

where $n = \deg(f)$, such that f factorizes into linear factors over K. Let g_0 be an irreducible factor of f and denote

$k_1 := k[X]/g_0$,
$\pi : k[X] \mapsto k[X]/g(X)$ the canonical projection,
$\alpha_1 := \pi(X)$.

Then, by Corollary 5.2.2, $f(\alpha_1) = 0$, so that, by Corollary 1.4.1,

$$f(X) = (X - \alpha_1)f_1(X) \text{ for some } f_1(X) \in k_1[X].$$

So let us inductively assume we have obtained

a tower of fields $k = k_0 \subseteq k_1 \subseteq \ldots \subseteq k_r$,
elements $\alpha_1, \ldots, \alpha_r$,
polynomials $f_i(X) \in k_i[X]$,

so that for all i

$\alpha_i \in k_i$,
$f(X) = (X - \alpha_1) \ldots (X - \alpha_i)f_i(X)$ in $k_i[X]$,
$k_i = k_{i-1}[X]/g_{i-1}$ where g_{i-1} is an irreducible factor of f_{i-1} in $k_{i-1}[X]$[2],

and we will show that we are able to build $k_{r+1}, \alpha_{r+1}, f_{r+1}$ satisfying the same properties.

If $r = n - 1$, then, since $\deg(f_r) = n - r$, f_r is linear and f factorizes into linear factors over k_{n-1}.

If $r < n - 1$, let g_r be an irreducible factor of f_r. If g_r is linear, $g_r = (X - a)$, let $k_{r+1} := k_r, \alpha_{r+1} := a, f_{r+1}$ such that $f_r(X) = (X - a)f_{r+1}$.

---

[2] Note that here we do not claim that $\alpha_i \notin k_{i-1}$; if $\alpha_i \in k_{i-1}$, then $k_i \cong k_{i-1}$.

5.2 The Invention of the Roots

Otherwise let $k_{r+1} := k_r[X]/g_r$, $\pi : k_r[X] \mapsto k_r[X]/g_r$ the canonical projection, $\alpha_{r+1} := \pi(X)$, so that (again by Corollary 5.2.2 and Corollary 1.4.1) $f_r(X) = (X - \alpha_{r+1})f_{r+1}(X)$ for some $f_{r+1}(X) \in k_{r+1}[X]$. ♄

Remark 5.2.4. If k is effective, since K is obtained from it by at most $n - 1$ constructions of residue rings $k_i[X]/g_i(X)$, then K is effective too.

Moreover the above construction could be translated into an algorithm provided we know how to factorize polynomials in each of the fields k_i.

Example 5.2.5. We apply the Kronecker technique to compute the roots of the polynomial

$$f(X) := X^3 - 2 \in \mathbb{Q}[X],$$

i.e. to build a field $K \supseteq \mathbb{Q}$ and to find three elements

$$\alpha_1, \alpha_2, \alpha_3 \in K : f(\alpha_i) = 0, \forall i.$$

Preliminarily let us remark that f is not factorizable in $\mathbb{Q}[X]$, otherwise it would contain a linear factor, i.e. a rational root, which of course it cannot have.

We can then consider the field

$$k_1 := \mathbb{Q}[X]/f(X)$$

and the element

$$\beta := \pi_1(X) \in k_1,$$

where $\pi_1 : \mathbb{Q}[X] \mapsto k_1$ is the canonical projection.

An elementary division gives us

$$X^3 - 2 = (X - \beta)(X^2 + \beta X + \beta^2) + \beta^3 - 2$$

and, since

$$\beta^3 - 2 = f(\beta) = 0$$

in k_1, we have found

$k_1 := \mathbb{Q}[X]/f(X),$
$\alpha_1 := \beta,$
$f_1(X) := X^2 + \beta X + \beta^2 \in k_1[X].$

Since f_1 is quadratic it is easy to check that it is not factorizable since its discriminant $-3\beta^2 \neq 0$. We could in fact even compute its roots by the quadratic formula, but I would prefer not to do that, in order to apply Kronecker's proposal up to the end.

We then build the field

$$k_2 := k_1[X]/f_1(X)$$

and we consider the element

$$\gamma := \pi_2(X) \in k_2,$$

where $\pi_2 : k_1[X] \mapsto k_2$ is the canonical projection. By division we obtain

$$X^2 + \beta X + \beta^2 = (X - \gamma)(X + \beta + \gamma) + \gamma^2 + \beta\gamma + \beta^2$$

and, since

$$\gamma^2 + \beta\gamma + \beta^2 = f_1(\gamma) = 0,$$

we have found the linear factorization

$$f(X) = (X - \beta)(X - \gamma)(X + \beta + \gamma)$$

in $k_2[X]$, and so the three roots $\beta, \gamma, -\beta - \gamma$.

Example 5.2.6. It is, however, worthwhile seeing what happens if, after having built k_1 and β and obtained the factorization

$$X^3 - 2 = (X - \beta)(X^2 + \beta X + \beta^2),$$

we apply the quadratic formula to

$$f_1(X) := X^2 + \beta X + \beta^2.$$

First of all, we compute the root of the discriminant, getting

$$\sqrt{-3\beta^2} = \beta\sqrt{-3}.$$

By the above computation we know that $\sqrt{-3} \notin k_1$ – since $\gamma \notin k_1$ – so we need to extend k_1 to a field in which we have the roots of -3, which we know how to do: we consider the polynomial

$$h(X) := X^2 - 3,$$

we build the field

$$k_2' := k_1[X]/h(X)$$

and we consider the element

$$\delta := \pi_2'(X) \in k_2',$$

where $\pi_2' : k_1[X] \mapsto k_2'$ is the projection. Then, by the quadratic formula we have that the other two roots of f are

$$\frac{\beta}{2}(-1 \pm \delta)$$

so that we obtain the field k_2', the linear factorization in $k_2'[X]$

$$f(X) = (X - \beta)\left(X + \frac{\beta}{2} - \frac{\beta\delta}{2}\right)\left(X + \frac{\beta}{2} + \frac{\beta\delta}{2}\right),$$

and so the three roots

$$\beta, \quad \frac{-\beta + \beta\delta}{2}, \quad \frac{-\beta - \beta\delta}{2}.$$

5.3 Transcendental and Algebraic Field Extensions

The field K obtained in the above section is obtained by adjoining to k the elements $\alpha_1, \ldots, \alpha_{n-1}$. Let us therefore study in detail the effect of this operation.

Let $k \subset K$ be two fields, and $x_1, \ldots, x_n \in K \setminus k$. It follows then that we consider the subfield of K which consists of all the numbers obtainable starting from those in k and the x_is and repeatedly performing the four operations. It is quite clear that this subfield of K can be interpreted as the set of the numbers which are obtained evaluating all the rational functions in $k(X_1, \ldots, X_n)$ at $x_1, \ldots, x_n$.

This leads us to consider the ring morphism

$$\Phi : k[X_1, \ldots, X_n] \mapsto K$$
$$\Phi(f) = f(x_1, \ldots, x_n)$$

whose image can alternatively be described as

the smallest subring of K containing k and $x_1, \ldots, x_n$,

the ring containing all the numbers of K which are obtained by recursively applying the *three* operations starting with $x_1, \ldots, x_n$ and the elements of k.

This ring is an integral domain, being contained in the field K, and we will denote it by $k[x_1, \ldots, x_n]$.

Further we consider the following subset of K:

$$k(x_1, \ldots, x_n) := \{\alpha\beta^{-1} \mid \alpha, \beta \in k[x_1, \ldots, x_n], \beta \neq 0\}$$

which is

the field of fractions of $k[x_1, \ldots, x_n]$,

the smallest subfield of K containing k and $x_1, \ldots, x_n$, the field containing all the elements of K that are obtained by recursively applying the four operations starting with $x_1, \ldots, x_n$ and the elements of k.

Definition 5.3.1. *A field $K \supset k$ such that*

$$\exists x_1, \ldots, x_n \in K \setminus k : K = k(x_1, \ldots, x_n)$$

is called a finitely generated extension *of k by $x_1, \ldots, x_n$.*

It is called simple *if it is generated by a single element x, i.e. $K = k(x)$.*

Two field extensions of k, K and K', are called k-isomorphic if there is an isomorphism $\Psi : K \mapsto K'$ such that $\Psi(a) = a, \forall a \in k$.

Remark 5.3.2. More generally, we could consider two fields $k \subset K$ and a set $\mathcal{S} \subset K$, not necessarily a finite one, and denote the smallest subring of K containing k and $\mathcal{S}$ by $k[\mathcal{S}]$, and the smallest subfield of K containing k and $\mathcal{S}$ by $k(\mathcal{S})$.

In this setting, we say that $k(\mathcal{S})$ is obtained from k by the *adjunction* of $\mathcal{S}$.

Note that each field $K \supset k$ can be obtained from k by the adjunction of itself: $K = k(K)$. As a consequence, each field $K \supset k$ is called a *field extension* of k.

In order to consider the effect of generating extensions of a field by adjoining elements onto it, it is reasonable to start by considering the simple extensions. In Proposition 5.2.1, we extended k by adding an indeterminate X and requiring that it satisfy an irreducible polynomial relation $g(X) \in k[X]$, obtaining the field $k[X]/g(X)$; another way to obtain a simple extension field is to add an indeterminate X and require that it does not satisfy any polynomial relation; we then obtain the rational function field $k(X)$. As we should expect, these are the only ways of getting a simple extension field:

Definition 5.3.3. *Let k, K be two fields, $k \subset K$, and let $\alpha \in K \setminus k$. We say that α is* transcendental *over k if*

$$\forall f \in k[X] \setminus \{0\}, \quad f(\alpha) \neq 0;$$

algebraic *over k if*

$$\exists f \in k[X] \setminus \{0\} : f(\alpha) = 0.$$

Remark 5.3.4. The definition above depends on α (obviously) but not on K. It depends, however, on k. In fact π is transcendental over $\mathbb{Q}$ (not an elementary result) but is algebraic over $\mathbb{Q}(\pi^2)$ since it is a root of $X^2 - \pi^2$.

5.3 Transcendental and Algebraic Field Extensions

Proposition 5.3.5. *Let $\alpha \in K \setminus k$ be transcendental over k. Let $\Phi : k(X) \mapsto K$ be the map defined by*

$$\Phi(f/g) = \frac{f(\alpha)}{g(\alpha)}.$$

Then $\mathrm{Im}(\Phi) = k(\alpha)$ is a subfield of K isomorphic to $k(X)$.

Proof Φ is well defined since $g(\alpha) \neq 0$, $\forall g \in k[X]\setminus\{0\}$; it is then easy to verify that Φ is a field morphism. The rest of the statement is obvious. ♮

Proposition 5.3.6. *Let $\alpha \in K \setminus k$ be algebraic over k. Let $\Phi : k[X] \mapsto K$ be the morphism defined by $\Phi(g) = g(\alpha)$. Then:*

(1) *there is a unique monic polynomial $f(X) \in k[X]$ of least degree, such that $f(\alpha) = 0$, which is called the* minimal polynomial *of α over k.*
(2) *f is irreducible.*
(3) *$g(\alpha) = 0 \iff g$ is a multiple of f.*
(4) *$\mathrm{Im}(\Phi)$ is isomorphic to $k[X]/f(X)$.*
(5) *$\mathrm{Im}(\Phi) = k[\alpha]$ is the smallest subfield of K containing both k and α.*

Proof

(1) If f_1 and f_2 are both monic polynomials such that $f_i(\alpha) = 0$ and of least degree, let $g := f_1 - f_2$. If $g \neq 0$, then $\deg(g) < \deg(f_i)$ and

$$g(\alpha) = f_1(\alpha) - f_2(\alpha) = 0,$$

contradicting the fact that f_i is of least degree. Therefore $f_1 = f_2$.
(2) If f were not irreducible, then one irreducible factor of f would vanish in α, again contradicting the fact that f is of least degree.
(3) Let $r(X) := \mathbf{Rem}(g, f)$, $q(X) := \mathbf{Quot}(g, f)$.
Then $r(\alpha) = g(\alpha) - q(\alpha)f(\alpha) = 0$.
We cannot have $\deg(r) < \deg(f)$ and so $r(X) = 0$.
(4) This is equivalent to saying that $\ker(\Phi)$ is the ideal generated by f, which was proved for (3).
(5) This is obvious. ♮

Definition 5.3.7. *If α is algebraic over k, the* degree *of α is the degree of its minimal polynomial over k.*

Corollary 5.3.8. *If $\alpha \in K$ is algebraic over k of degree n, then $k[\alpha]$ is a k-vector space of dimension n, a basis being $\{1, \alpha, \alpha^2, \ldots, \alpha^{n-1}\}$.* ♮

5.4 Finite Algebraic Extensions

Definition 5.4.1. *Let k, K be two fields, $k \subset K$. We call K an algebraic extension of k if each element in K is algebraic over k. It is called a* transcendental extension *of k, if there is an element in K which is transcendental over k.*

Lemma 5.4.2. *Let $k \subset K \subset L$ be fields. Assume that L is a K-vector space of dimension m, and that K is a k-vector space of dimension n.*

Then L is a k-vector space of dimension mn.

Moreover if $\{\alpha_1, \ldots, \alpha_n\}$ is a k-basis of K, and $\{\beta_1, \ldots, \beta_m\}$ a K-basis of L, then

$$\{\alpha_i \beta_j \mid 1 \leq i \leq n, 1 \leq j \leq m\}$$

is a k-basis of L.

Proof It is sufficient to prove the second statement.
If $\gamma \in L$, then $\gamma = \sum_{j=1}^{m} \zeta_j \beta_j$ for some $\zeta_j \in K$. In turn $\forall j$, $\zeta_j = \sum_{i=1}^{n} a_{ij} \alpha_i$ for some $a_{ij} \in k$, so that $\gamma = \sum_{i,j} a_{ij} \alpha_i \beta_j$. Therefore

$$\{\alpha_i \beta_j \mid 1 \leq i \leq n, 1 \leq j \leq m\}$$

generates L over k.
If $0 = \sum_{i,j} a_{ij} \alpha_i \beta_j$, let $\zeta_j := \sum_{i=1}^{n} a_{ij} \alpha_i$, then $0 = \sum_{j=1}^{m} \zeta_j \beta_j$ and so $\zeta_j = 0, \forall j$, which in turn implies $a_{ij} = 0, \forall i, j$. ♮

Proposition 5.4.3. *A finitely generated extension K of k is an algebraic extension if and only if it is a finite-dimensional k-vector space.*

Proof In fact if K is a k-vector space of dimension n, then for each $\alpha \in K$, $\{1, \alpha, \alpha^2, \ldots, \alpha^n\}$ are linearly dependent over k, so that α is algebraic over k.
Conversely assume that $K = k(\alpha_1, \ldots, \alpha_n)$ is an algebraic extension of k and let $k_0 := k$, $k_i := k_{i-1}[\alpha_i]$, $1 \leq i \leq n$, so that $K = k_n$. Since K is algebraic, each α_i is algebraic over k and *a fortiori* over k_{i-1}.
Since k_i is a simple algebraic extension of k_{i-1}, and so a finite-dimensional k_{i-1}-vector space, by Lemma 5.4.2 we can then conclude that K is a finite-dimensional k-vector space. ♮

Definition 5.4.4. *A finitely generated algebraic extension K of k is usually called a* finite extension *and its k-dimension is called the* degree *of K over k and denoted $[K : k]$.*

Remark 5.4.5. If α is algebraic over k, the degree of α, the degree of its minimial polynomial and $[k[\alpha] : k]$ are the same.

5.4 Finite Algebraic Extensions

Lemma 5.4.6. *If L is a finite extension of k and $\alpha \in L$, then the degree of α divides $[L : k]$.*

Proof Let $K := k[\alpha]$. Since $L \supseteq K$, L is a K-vector space and therefore it has to be a finite-dimensional vector space. By Lemma 5.4.2,

$$[L : K][K : k] = [L : k],$$

so $[K : k]$ divides $[L : k]$. ♄

Algebraic number and extensions can be further classified:

Definition 5.4.7. *If $K \supset k$ is an algebraic extension, K is called a* separable *extension of k if each element $\alpha \in K$ is separable over k; it is called* inseparable *otherwise.*

If $\alpha \in K \supset k$ is an inseparable element, $f(X) \in k(X)$ its minimal polynomial, and $\alpha = \alpha_1, \ldots, \alpha_{n_0} \in K$ are all its roots, recall that (Proposition 4.5.4 and Definition 4.5.5)

$$f = \prod_{i=1}^{n_0}(X - \alpha_i)^{p^e},$$

where $p = \text{char}(k)$, n_0 is the reduced degree and e the exponent of inseparability of f and α.

Remark 5.4.8. If k is perfect, $K \supset k$ is an algebraic extension and $\alpha \in K$, then K and α are separable. In fact, if α were inseparable, its minimal polynomial would be a non-constant, irreducible inseparable polynomial, contradicting the hypothesis that k is perfect.

Definition 5.4.9. *With the present notation, when $\text{char}(k) = p \neq 0$, $\alpha \in K \supset k$ is called* purely inseparable *over k if $n_0 = 1$, i.e. α is the single root of its minimal polynomial f.*

A field $K \supset k$ is called a purely inseparable extension *of k if every $\alpha \in K$ is purely inseparable over k.*

Remark 5.4.10. It is obvious that $\alpha \in K \supset k$ is purely inseparable iff its minimal polynomial over k is

$$(X - \alpha)^{p^e} = X^{p^e} - \alpha^{p^e} \in k[X],$$

iff there is $e \geq 0$ such that $\alpha^{p^e} \in k$.

Therefore $[k(\alpha) : k] = p^e$ is a power of p.

Lemma 5.4.11. *If $\alpha \in K \supset k$ is both separable and purely inseparable over k, then $\alpha \in k$.*

Proof Since α is purely inseparable its minimal polynomial over k is $X^{p^e} - \alpha^{p^e}$. Since α is separable over k, then $e = 0$ and so $\alpha \in k$. ♮

5.5 Splitting Fields

Apparently Theorem 5.2.3 gives us a solution to our problem of finding an algorithm to solve a polynomial equation and to compute the roots of a polynomial $f \in k[X]$.

However, it raises a new problem: what if somebody gives a different construction leading to a different field $K' \supset k$ in which f has linear factors? What relation, if any at all, is there among the roots of f in K and those in K'? Example 5.2.5 and Example 5.2.6 illustrate this problem: we obtained, with two different approaches, two fields k_2, and k'_2 which contain the roots of f.

Moreover, the Fundamental Theorem of Algebra asserts that a polynomial in $\mathbb{Q}[X]$ has all its roots in $\mathbb{C}$ so that we can theoretically consider the ring $K := \mathbb{Q}(\alpha_1, \alpha_2, \alpha_3) \subset \mathbb{C}$, where $\alpha_1, \alpha_2, \alpha_3 \in \mathbb{C}$ are the three roots of $f(X)$.

The question, of course, is what is the relation among the quite abstract roots we obtained in k_2, those we obtained in k'_2 and the somehow more concrete ones existing in $K \subset \mathbb{C}$.

This section is devoted to giving an answer to this new question; the final result will be that any two fields, where f factorizes into linear factors that are minimal for this property, are k-isomorphic (so to all practical purposes they can be considered to be the same).

Example 5.5.1. To illustrate and introduce this result, let us go back to Example 5.2.5 and Example 5.2.6.

There we introduced

the polynomial $f(X) := X^3 - 2 \in \mathbb{Q}[X]$,

the field $k_1 := \mathbb{Q}[X]/f(X)$,

the canonical projection $\pi_1 : \mathbb{Q}[X] \mapsto k_1$,

the root $\beta := \pi_1(X) \in k_1$;

the polynomial $f_1(X) := X^2 + \beta X + \beta^2 \in k_1[X]$,

the field $k_2 := k_1[X]/f_1(X)$,

the canonical projection $\pi_2 : k_1[X] \mapsto k_2$,

the root $\gamma := \pi_2(X) \in k_2$;

5.5 Splitting Fields

the polynomial $h(X) := X^2 - 3 \in k_1[X]$,
the field $k'_2 := k_1[X]/h(X)$,
the canonical projection $\pi'_2 : k_1[X] \mapsto k'_2$,
the root $\delta := \pi'_2(X) \in k'_2$,

so that

$k_2 = \mathbb{Q}[\beta, \gamma]$,
the minimal polynomial of β over $\mathbb{Q}$ is $f(X)$,
the minimal polynomial of γ over k_1 is $f_1(X)$,
the roots of f in k_2 are $\beta, \gamma, -\beta - \gamma$;
$k'_2 = \mathbb{Q}[\beta, \delta]$,
the minimal polynomial of δ over k_1 is $h(X)$,
the roots of f in k'_2 are $\beta, \frac{-\beta - \beta\delta}{2}, \frac{-\beta + \beta\delta}{2}$.

On the other hand we know how to 'solve' the equation $X^3 - 2$ (where we use 'solve' in the pre-Abel–Ruffini meaning). The three roots are

$$\alpha_i := \sqrt[3]{2}\epsilon_i, i = 1, 2, 3$$

where ϵ_i are the three third roots of unity, i.e.

$$\epsilon_1 = 1, \quad \epsilon_2 = \frac{-1 + \sqrt{-3}}{2}, \quad \epsilon_3 = \frac{-1 - \sqrt{-3}}{2},$$

so that the three roots are

$$\alpha_1 := \sqrt[3]{2}, \quad \alpha_2 := \sqrt[3]{2}\frac{-1 + \sqrt{-3}}{2}, \quad \alpha_3 := \sqrt[3]{2}\frac{-1 - \sqrt{-3}}{2}.$$

As a consequence

$$K = \mathbb{Q}[\alpha_1, \alpha_2, \alpha_3] = \mathbb{Q}[\sqrt[3]{2}, \sqrt{-3}]$$

and it is clear that **an** isomorphism between k'_2 and K is the obvious one:

$$\Psi : k'_2 \mapsto K$$

defined by

$$\Psi(\beta) = \sqrt[3]{2}, \quad \Psi(\delta) = \sqrt{-3};$$

k'_2 in fact represents K in Kronecker's model.

It is not difficult to verify that there is an isomorphism between k_2 and k'_2; it is $\Phi : k_2 \mapsto k'_2$ defined by

$$\Phi(\gamma) = \frac{-\beta - \beta\delta}{2}$$

and it is not difficult to verify that Φ maps the third root of f in k_2 to the third root of f in k'_2:

$$\Phi(-\beta - \gamma) = -\Phi(\beta) - \Phi(\gamma) = -\beta + \frac{\beta}{2} + \frac{\beta\delta}{2} = \frac{-\beta + \beta\delta}{2}.$$

Definition 5.5.2. *We say that $K \supseteq k$ is a* splitting field *of $f(X) \in k[X]$ over k, if f factors into linear polynomials in $K[X]$, while f has no linear factorization in each subfield K', $k \subset K' \subset K$.*

Lemma 5.5.3. *Let f be an irreducible polynomial in $k[X]$. Let $\Phi : k \mapsto k'$ be a field isomorphism and let us denote by $\Phi : k[X] \mapsto k'[X]$ its polynomial extension.*

Let K and K' be two fields, $k \subset K$, $k' \subset K'$.
Let α be a root of f in K, α' a root of $f' := \Phi(f)$ in K'.
Then there is a unique field isomorphism $\Psi : k[\alpha] \mapsto k'[\alpha']$ such that

$$\Psi(\alpha) = \alpha', \Psi(a) = \Phi(a), \forall a \in k.$$

Proof Since $\Phi : k[X] \mapsto k'[X]$ is an isomorphism, the irreducibility of f is equivalent to the irreducibility of f'.
Moreover it is clear that the two fields $k[X]/f(X)$ and $k'[X]/f'(X)$ are isomorphic.
The thesis follows since $k[\alpha]$ is isomorphic to $k[X]/f(X)$ and $k'[\alpha']$ is isomorphic to $k'[X]/f'(X)$.
Uniqueness is obvious. ♄

Proposition 5.5.4. *Let $f(X) \in k[X]$. Let $\Phi : k \mapsto k'$ be a field isomorphism, let us denote by $\Phi : k[X] \mapsto k'[X]$ its polynomial extension and let $f' := \Phi(f)$. Let K (respectively K') be a splitting field of f (respectively f') over k (respectively k'). Then there is a field isomorphism $\Xi : K \mapsto K'$ such that $\Xi(a) = \Phi(a), \forall a \in k$.*
If

$$f(X) = c(X - \alpha_1)\ldots(X - \alpha_n)$$

is the factorization of f in $K[X]$, then the factorization of f' in $K'[X]$ is

$$f'(X) = \Phi(c)(X - \Xi(\alpha_1))\ldots(X - \Xi(\alpha_n)).$$

Proof The argument is by induction on the degree of f.
If f is linear, then the splitting fields are respectively the isomorphic fields k and k' and there is nothing to prove.

Assume $\deg(f) = n > 1$ and let $g \in k[X]$ be an irreducible factor of f; then $g' := \Phi(g) \in k'[X]$ is an irreducible factor of f'. Let α_1 be a root of g in K and α_1' be a root of g' in K'. Because of Lemma 5.5.3, there is a field isomorphism $\Psi : k[\alpha_1] \mapsto k'[\alpha_1']$ which extends Φ and such that $\Psi(\alpha_1) = \alpha_1'$. We also denote by $\Psi : k[\alpha_1][X] \mapsto k'[\alpha_1'][X]$ its polynomial extension.

In $k[\alpha_1][X]$ we have the factorization $f(X) = (X - \alpha_1)h(X)$; by means of Ψ we then have $f' = (X - \alpha_1')\Psi(h)$. Also, K is a splitting field of h over $k[\alpha_1]$ and K' is a splitting field of $\Psi(h)$ over $k'[\alpha_1']$.

Since $\deg(h) = n-1$, by inductive application of the proposition, we conclude that there is a field isomorphism $\Xi : K \mapsto K'$ such that

$$\Xi(a) = \Psi(a) = \Phi(a), \forall a \in k,$$

$$\Xi(\alpha_1) = \Psi(\alpha_1) = \alpha_1'.$$

Moreover if

$$h(X) = c(X - \alpha_2)\ldots(X - \alpha_n)$$

is the factorization of h in $K[X]$, then the factorization of h' in $K'[X]$ is

$$h'(X) = \Phi(c)(X - \Xi(\alpha_2))\ldots(X - \Xi(\alpha_n));$$

as a consequence

$$f(X) = c(X - \alpha_1)(X - \alpha_2)\ldots(X - \alpha_n)$$

is the factorization of f in $K[X]$, and

$$f'(X) = \Phi(c)(X - \Xi(\alpha_1))(X - \Xi(\alpha_2))\ldots(X - \Xi(\alpha_n))$$

is the factorization of f' in $K'[X]$. ♄

Corollary 5.5.5. *Let $f(X) \in k[X]$. Let K and K' be splitting fields of f. Then there is a k-isomorphism $\Xi : K \mapsto K'$.*

If

$$f(X) = c(X - \alpha_1)\ldots(X - \alpha_n)$$

is the factorization of f in $K[X]$, then the factorization of f in $K'[X]$ is

$$f(X) = c(X - \Xi(\alpha_2))\ldots(X - \Xi(\alpha_n)).$$

♄

Theorem 5.5.6. *Let $f(X) \in k[X]$. Then there is a unique (up to k-isomorphisms) splitting field K of f.*

Moreover $[K : k] \leq n!$.

Proof Existence is the content of Theorem 5.2.3; uniqueness is the content of Corollary 5.5.5; so we are left to prove $[K : k] \leq n!$.

However, with the notations of the proof of Theorem 5.2.3, we have that

$$[k_i : k_{i-1}] = \deg(g_{i-1}) \leq \deg(f_{i-1}) = n - i + 1,$$

whence the thesis.

6
Intermezzo: Sylvester

The classical setting for solving univariate polynomial equations is a domain D, in whose polynomial ring $D[X]$ we consider a 'generic' polynomial $f(X) \in D[X]$ of degree n:

$$f(X) = a_0 X^n + a_1 X^{n-1} + \cdots + a_i X^{n-i} + \cdots + a_{n-1} X + a_n. \qquad (6.1)$$

If Q denotes the quotient field Q of D, Theorem 5.5.6 allows us to consider the splitting field $K \supset Q$, of f, in which f contains n – not necessarily different – roots $\alpha_1, \ldots, \alpha_n \in K$ such that

$$f(X) = a_0 \prod_{j=1}^{n} (X - \alpha_j). \qquad (6.2)$$

The setting, in which most of the classical (pre-Abel–Ruffini) research on 'solving' was developed, is

$$\mathbb{Z} = D \subset Q = \mathbb{Q} \subset K = \mathbb{C},$$

based on the Euler Conjecture. It was in this setting that deep work on 'solving' was performed which reached a peak with Lagrange's results from which blossomed Galois Theory and the Abel–Ruffini Theorem. In the same setting I have to quote at least two analyses which are useful today:

- Gauss related the factorization of $f(X) \in D[X]$ over D with that over Q (Section 6.1);
- Newton, starting from the obvious remark that the coefficients a_i of f – assuming wlog $a_0 = 1$ – are symmetric on the roots α_j, introduced the notion of *symmetric functions* on the roots α_j, proving that they can be expressed as polynomials on the coefficients a_i (Section 6.2, Section 6.3).

The nineteenth century English algebra school continued and extended the approach started by Newton: their approach posed and solved questions such as

given a 'generic' polynomial Equation 6.1 is it possible to find a 'universal' formula in terms of the coefficients a_i which allows us to decide whether their roots are simple?

given the 'generic' polynomial Equation 6.1 and a further 'generic' polynomial

$$g(X) = b_0 X^m + b_1 X^{m-1} + \cdots + b_i X^{m-i} + \cdots + b_{m-1} X + b_m$$

is it possible to find a 'universal' formula in terms of the coefficients a_i, b_j which allows us to decide whether f and g have a common root?

In both cases the answer is positive and leads to the notions of the discriminant of a polynomial (Section 6.5) and the resultant of two polynomials (Section 6.6). It is important to understand here the approach and the technique introduced by the English school in their aim to find 'universal' solutions, which will therefore first be discussed in Section 6.4.

Finally, a deeper study (Section 6.7) of the properties of the resultant will introduce essential tools which we will use to deal with real solutions of equations.

6.1 Gauss Lemma

Let D be a unique factorization domain and let Q be its fraction field; so Q is effective if D is an effective domain.

Definition 6.1.1. *Let $f(X) = \sum_{i=0}^{n} a_i X^{n-i} \in D[X]$.*
The content *of f, $\mathrm{Cont}(f)$, is*

$$\mathrm{Cont}(f) := \gcd(a_0, \ldots, a_d);$$

f is called primitive *if $\mathrm{Cont}(f) = 1$.*

Lemma 6.1.2.

(1) *If $f(X) \in Q[X]$ there is a primitive polynomial*

$$g(X) := \mathrm{Prim}(f) \in D[X]$$

which is associate to f.

(2) *Let f and g be primitive polynomials in $D[X]$, then f and g are associate if and only if there is a unit $u \in D$ such that $f = ug$.*

Proof

(1) Let $f = \sum_{i=0}^{n} b_i^{-1} a_i X^{n-i}$, with $b_i, a_i \in D$, $b_i \neq 0$, $\gcd(a_i, b_i) = 1$. Let $b := \text{lcm}_i(b_i)$, $a := \gcd_i(a_i)$. Then $g := a^{-1}bf$ is in $D[X]$ and is associate to f.
Moreover let $c_i, d_i \in D$ be such that $a d_i = a_i$, and $b = b_i c_i$, so that $g = \sum_{i=0}^{n} c_i d_i X^{n-i}$.
Assume $\text{Cont}(g) = \gcd_i(c_i d_i) \neq 1$ and let e be an irreducible factor of $\gcd_i(c_i d_i)$.
Since $\gcd_i(d_i) = 1$, there is j such that e divides c_j, and therefore e divides b; since $\gcd_i(c_i) = 1$, there is k such that e does not divide c_k. Therefore e divides b_k (since it divides b) and a_k (since it divides $c_k d_k$ and so d_k).
Since $\gcd(a_k, b_k) = 1$, we conclude $e = 1$ and so
$$\text{Cont}(g) = \gcd_i(c_i d_i) = 1,$$
proving g is primitive.
(2) If f and g are associate, there are $a, b \in D$, $\gcd(a, b) = 1$, such that $af = bg$. Then a divides $\text{Cont}(g) = 1$ and b divides $\text{Cont}(f) = 1$; therefore both are units in D and so is $u := a^{-1}b$. ♄

Lemma 6.1.3 (Gauss Lemma). *If $f, g \in D[X]$ are primitive, then fg is primitive.*

Proof By contradiction: let

$$f(X) := \sum_{i=0}^{n} a_i X^{n-i}, \quad g(X) := \sum_{i=0}^{m} b_i X^{m-i}, \quad fg := \sum_{i=0}^{n+m} c_i X^{n+m-i}.$$

Let $e \in D$ be an irreducible factor of $\text{Cont}(fg)$; since

$$\text{Cont}(f) = \text{Cont}(g) = 1,$$

e does not divide all coefficients of f, nor of g. So let s be the least index such that a_s is not a multiple of e, and t the least index such that b_t is not a multiple of e.
Then all the summands of $c_{s+t} = \sum_{i=0}^{s+t} a_i b_{s+t-i}$, except $a_s b_t$, are divisible by e, so c_{s+t} is not divisible by e, giving the desired contradiction. ♄

Corollary 6.1.4. *If $f, g \in D[X]$, then $\mathrm{Cont}(fg) = \mathrm{Cont}(f)\mathrm{Cont}(g)$.*

Proof Let $c := \mathrm{Cont}(f), d := \mathrm{Cont}(g)$, so that

$$fg = cd\,\mathrm{Prim}(f)\,\mathrm{Prim}(g).$$

Since $\mathrm{Prim}(f)\,\mathrm{Prim}(g)$ is primitive, then $cd = \mathrm{Cont}(fg)$. ♮

Corollary 6.1.5. *Let $f \in D[X]$ be a primitive polynomial. Let $g, h \in Q[X]$ be such that $f = gh$. Then*

$$f = u\,\mathrm{Prim}(g)\,\mathrm{Prim}(h)$$

for a unit $u \in D$.

Proof There are $a, b \in Q$ such that $g = a\,\mathrm{Prim}(g)$, $h = b\,\mathrm{Prim}(h)$. Then

$$f = (ab)\,\mathrm{Prim}(g)\,\mathrm{Prim}(h).$$

Since f is primitive and $f_0 := \mathrm{Prim}(g)\,\mathrm{Prim}(h)$ is also by the Gauss Lemma, then ab is a unit in D. ♮

Corollary 6.1.6. *Let $g \in D[X]$ be a primitive polynomial and let $f \in D[X]$, $h \in Q[X]$ be such that $f = gh$ in $Q[X]$. Then $h \in D[X]$.*

Proof In $D[X]$ we have $f = \mathrm{Cont}(h)\,\mathrm{Prim}(h)g$; since $\mathrm{Prim}(h)g$ is primitive, we deduce $\mathrm{Prim}(f) = \mathrm{Prim}(h)g$ and $\mathrm{Cont}(h) = \mathrm{Cont}(f)$; since $f \in D[X]$, then $\mathrm{Cont}(h) \in D$ and $h = \mathrm{Cont}(h)\mathrm{Prim}(h) \in D[X]$. ♮

Corollary 6.1.7. *Let f be a primitive polynomial in $D[X]$. Then f is irreducible in $D[X]$ iff it is such in $Q[X]$.*

Proof Assume f is reducible in $Q[X]$, and let $g, h \in Q[X]$ be such that $f = gh$. By Corollary 6.1.5, then $\mathrm{Prim}(g)$ and $\mathrm{Prim}(h)$ are proper factors of f in $D[X]$.

Conversely, assume f is reducible in $D[X]$ and let $g, h \in D[X]$ be such that $f = gh$. Then $\deg(g) > 0$, since otherwise g would be an irreducible factor of $\mathrm{Cont}(f) = 1$. By the same argument, $\deg(h) > 0$ and $f = gh$ is a proper factorization in $Q[X]$. ♮

6.1 Gauss Lemma

Remark 6.1.8. The Gauss Lemma allows us to establish that the polynomial factorizations in $Q[X]$ and $D[X]$ are essentially 'the same', as will be shown in the next theorem.

In particular, this result applies to the cases

$D := \mathbb{Z}, Q := \mathbb{Q}$,
$D := K[X_1, \ldots, X_n], Q := K(X_1, \ldots, X_n)$

and allows us to reduce

factorizations of rational polynomials to those of integer polynomials;
factorizations of univariate polynomials over transcendental extensions, to multivariate polynomial factorizations.

Theorem 6.1.9.

(1) *Let $f \in D[X]$ and let*

$f = \prod_{i=1}^{r} p_i^{e_i}$ *be a factorization in $Q[X]$ into irreducible factors,*
$\mathrm{Cont}(f) := \prod_{i=1}^{s} c_i^{d_i}$ *be a factorization in D into irreducible factors.*

Then, denoting $q_i := \mathrm{Prim}(p_i)$, for all i,

$$f = \prod_{i=1}^{s} c_i^{d_i} \prod_{i=1}^{r} q_i^{e_i}$$

is a factorization into irreducible factors in $D[X]$.

(2) *Conversely let $f \in Q[X]$ and let*

$$\mathrm{Prim}(f) = \prod_{i=1}^{r} p_i^{e_i}$$

be a factorization in $D[X]$ into irreducible factors.
Then, there is $u \in Q$ such that

$$uf = \prod_{i=1}^{r} p_i^{e_i}$$

is a factorization in $Q[X]$ into irreducible factors.

Proof

(1) By Corollary 6.1.5 we have

$$\mathrm{Prim}(f) = \prod_{i=1}^{r} q_i^{e_i}$$

and so
$$f = \prod_{i=1}^{s} c_i^{d_i} \prod_{i=1}^{r} q_i^{e_i}.$$

Since the c_i are irreducible in D, they are irreducible in $D[X]$; since the q_i are irreducible in $Q[X]$, they are so in $D[X]$ too.

(2) Since $\mathrm{Prim}(f)$ is primitive, then $\deg(p_i) > 0, \forall i$; moreover p_i, being irreducible in $D[X]$, is so in $Q[X]$ too; the thesis then holds where u is the element such that $uf = \mathrm{Prim}(f)$. ♄

Corollary 6.1.10. $D[X]$ *is a unique factorization domain.*

Proof By Theorem 6.1.9 each element of $D[X]$ has a factorization into irreducible factors. So we have to prove uniqueness (up to order and to associates).

Let $f \in D[X]$; up to a unit in D, there is a unique factorization $f = cg$, with $c \in D, g \in D[X]$ a primitive polynomial; c has a unique factorization since D is a unique factorization domain; g has a unique factorization in $D[X]$, since any such factorization is a factorization in $Q[X]$. ♄

6.2 Symmetric Functions

Let
$$f(X) = a_0 X^n + a_1 X^{n-1} + \cdots + a_i X^{n-i} + \cdots + a_{n-1} X + a_n$$
and let $\alpha_1, \ldots, \alpha_n$ be its roots so that
$$\sum_{i=0}^{n} a_i X^{n-i} = a_0 \prod_{j=1}^{n} (X - \alpha_j) \tag{6.3}$$
from which we obtain – when $a_0 = 1$:

$$\begin{aligned}
-a_1 &:= \alpha_1 + \alpha_2 + \cdots + \alpha_n, \\
+a_2 &:= \alpha_1\alpha_2 + \alpha_1\alpha_3 + \cdots + \alpha_1\alpha_n + \alpha_2\alpha_3 + \cdots + \alpha_{n-1}\alpha_n, \\
-a_3 &:= \alpha_1\alpha_2\alpha_3 + \cdots + \alpha_{n-2}\alpha_{n-1}\alpha_n, \\
&\ \ldots \\
(-1)^{n-1} a_{n-1} &:= \alpha_1\alpha_2\alpha_3 \ldots \alpha_{n-2}\alpha_{n-1} + \alpha_1\alpha_2\alpha_3 \ldots \alpha_{n-2}\alpha_n \\
&\ + \cdots + \alpha_1\alpha_3 \ldots \alpha_{n-2}\alpha_{n-1}\alpha_n \\
&\ + \alpha_2\alpha_3 \ldots \alpha_{n-2}\alpha_{n-1}\alpha_n, \\
(-1)^n a_n &:= \alpha_1\alpha_2\alpha_3 \ldots \alpha_{n-2}\alpha_{n-1}\alpha_n.
\end{aligned}$$

This remark and the obvious fact that the a_is are stable under any permutation of the roots α_j leads us to introduce

6.2 Symmetric Functions

Definition 6.2.1. *A polynomial* $f \in D[X_1, \ldots, X_n]$, *D a domain, is called a symmetric function iff, for each permutation* π *of* $\{1, \ldots, n\}$,

$$f(X_1, \ldots, X_i, \ldots, X_n) = f(X_{\pi(1)}, \ldots, X_{\pi(i)}, \ldots, X_{\pi(n)}).$$

The *elementary symmetric functions* of $X_1, \ldots, X_n$ are the symmetric functions

$$\sigma_1 := X_1 + X_2 + \cdots + X_n,$$
$$\sigma_2 := X_1 X_2 + X_1 X_3 + \cdots + X_1 X_n + X_2 X_3 + \cdots + X_{n-1} X_n,$$
$$\ldots$$
$$\sigma_{n-1} := X_1 X_2 X_3 \ldots X_{n-2} X_{n-1} + \cdots + X_2 X_3 \ldots X_{n-2} X_{n-1} X_n,$$
$$\sigma_n := X_1 X_2 X_3 \ldots X_{n-2} X_{n-1} X_n.$$

Remark 6.2.2. In order to prove the Fundamental Theorem on Symmetric Functions, which claims that any symmetric function in $D[X_1, \ldots, X_n]$ can be expressed as a polynomial in[1] $D[\sigma_1, \ldots, \sigma_n]$, we need to introduce some notation and remarks.

Each polynomial $\phi \in D[\sigma_1, \ldots, \sigma_n] \subset D[X_1, \ldots, X_n]$ is a symmetric function. In particular a term $\sigma_1^{a_1} \cdots \sigma_n^{a_n}$ is a homogeneous polynomial in $D[X_1, \ldots, X_n]$ of degree $a_1 + 2a_2 + \cdots + na_n$.

We will call $a_1 + 2a_2 + \cdots + na_n$ the *weight* of the term $\sigma_1^{a_1} \cdots \sigma_n^{a_n}$.

The notion of weight is generalized to a polynomial $\phi \in D[\sigma_1, \ldots, \sigma_n]$ to be, as usual, the maximal weight of the terms occurring in ϕ and it is clearly an upper bound of the degree of ϕ as a polynomial in $D[X_1, \ldots, X_n]$.

We denote **T** the semigroup of terms of $D[X_1, \ldots, X_n]$ and we order **T** by the lexicographical ordering $>$, such that $X_1 > X_2 > \cdots > X_n$, given by:

$$X_1^{a_1} \cdots X_r^{a_r} < X_1^{b_1} \cdots X_r^{b_r} \iff \text{there exists } j : a_j < b_j$$
and $a_i = b_i$, for all $i < j$.

---

[1] With the notation $D[\sigma_1, \ldots, \sigma_n]$ where $\sigma_i \in D[X_1, \ldots, X_n]$ we denote the subring $D[\sigma_1, \ldots, \sigma_n] \subset D[X_1, \ldots, X_n]$ obtained by the 'adjunction' to D of the elements $\sigma_1, \ldots, \sigma_n$, generalizing the operation discussed in Section 5.3. That is, $D[\sigma_1, \ldots, \sigma_n]$ denotes the subring which is the image of the morphism

$$\Phi : D[Y_1, \ldots, Y_n] \mapsto D[X_1, \ldots, X_n]$$
$$\Phi(f) = f(\sigma_1, \ldots, \sigma_n);$$

cf. also the discussion in Section 6.4

For a non-zero polynomial $f = \sum_{t \in \mathbf{T}} c_t t$ we denote

$$T(f) := \max_<\{t : c_t \neq 0\}, \quad \mathrm{lc}_<(f) := c_{T(f)}.$$

We need now to generalize this definition on $D[\sigma_1, \ldots, \sigma_n]$ with a twist. Therefore, with $\mathbf{S}$ denoting the semigroup of terms of $D[\sigma_1, \ldots, \sigma_n]$, the twist consists in remarking that any term $\mathbf{t} \in \mathbf{S}$ can be expressed as

$$\mathbf{t} = \sigma_1^{a_1-a_2} \sigma_2^{a_2-a_3} \cdots \sigma_{n-2}^{a_{n-2}-a_{n-1}} \sigma_{n-1}^{a_{n-1}-a_n} \sigma_n^{a_n}$$

for suitable elements a_i and defining an ordering $\prec$ on $\mathbf{S}$ by

$$\sigma_1^{a_1-a_2} \cdots \sigma_{n-1}^{a_{n-1}-a_n} \sigma_n^{a_n} \prec \sigma_1^{b_1-b_2} \cdots \sigma_{n-1}^{b_{n-1}-b_n} \sigma_n^{b_n}$$
$$\iff X_1^{a_1} \cdots X_n^{a_n} < X_1^{b_1} \cdots X_n^{b_n}$$

so that for a non-zero polynomial

$$\phi = \sum_{\mathbf{t} \in \mathbf{S}} c_\mathbf{t} \mathbf{t},$$

we denote

$$S(\phi) := \max_\prec\{\mathbf{t} : c_\mathbf{t} \neq 0\}, \quad \mathrm{lc}_\prec(\phi) := c_{S(\phi)}.$$

With this notation, if we are given a symmetric function

$$\phi \in D[\sigma_1, \ldots, \sigma_n] \subset D[X_1, \ldots, X_n],$$

we can interpret it as an element in $D[\sigma_1, \ldots, \sigma_n]$, where we consider the term $S(\phi)$, or as an element in $D[X_1, \ldots, X_n]$, where we consider the term $T(\phi)$. In this setting we have:

Lemma 6.2.3. *Let*

$$\phi \in D[\sigma_1, \ldots, \sigma_n] \subset D[X_1, \ldots, X_n];$$

then

$$T(\phi) = X_1^{a_1} X_2^{a_2} \cdots X_n^{a_n} \iff S(\phi) = \sigma_1^{a_1-a_2} \cdots \sigma_{n-1}^{a_{n-1}-a_n} \sigma_n^{a_n}.$$

♄

Theorem 6.2.4. (Fundamental Theorem on Symmetric Functions).
A symmetric function $f \in D[X_1, \ldots, X_n]$ can be expressed in a unique way as a polynomial in $\sigma_1, \ldots, \sigma_n$.

Proof

Existence: Let $f \in D[X_1, \ldots, X_n]$ be a symmetric function and let

$$T(f) = X_1^{a_1} X_2^{a_2} \cdots X_{n-1}^{a_{n-1}} X_n^{a_n}.$$

Among the terms $X_1^{a_{\pi(1)}} X_2^{a_{\pi(2)}} \cdots X_{n-1}^{a_{\pi(n-1)}} X_n^{a_{\pi(n)}}$ where π runs among the permutations of $\{1, \ldots, n\}$, $X_1^{a_1} X_2^{a_2} \cdots X_{n-1}^{a_{n-1}} X_n^{a_n}$ is the maximal term with respect to $<$ iff

$$a_1 \geq a_2 \geq \cdots \geq a_{n-1} \geq a_n;$$

therefore,

$$\psi := \mathrm{lc}_<(f) \sigma_1^{a_1-a_2} \sigma_2^{a_2-a_3} \cdots \sigma_{n-2}^{a_{n-2}-a_{n-1}} \sigma_{n-1}^{a_{n-1}-a_n} \sigma_n^{a_n}$$

satisfies $T(\psi) = T(f)$. As a consequence,

$$g := f - \psi \in D[X_1, \ldots, X_n]$$

is such that $T(g) < T(f)$.

We can therefore conclude that a finite number of rewritings allow us to compute a function $\phi \in D[\sigma_1, \ldots, \sigma_n]$ such that

$$\phi(\sigma_1, \ldots, \sigma_n) = f(X_1, \ldots, X_n) \text{ in } D[X_1, \ldots, X_n].$$

Uniqueness: Let us assume that there are two polynomials

$$\phi_1, \phi_2 \in D[\sigma_1, \ldots, \sigma_n]$$

such that

$$\phi_1(\sigma_1, \ldots, \sigma_n) = f(X_1, \ldots, X_n) = \phi_2(\sigma_1, \ldots, \sigma_n) \text{ in } D[X_1, \ldots, X_n].$$

We need to show that

$$\phi_1(\sigma_1, \ldots, \sigma_n) = \phi_2(\sigma_1, \ldots, \sigma_n) \text{ in } D[\sigma_1, \ldots, \sigma_n].$$

It is, of course, sufficient to show, for each polynomial

$$\phi(\sigma_1, \ldots, \sigma_n) \in D[\sigma_1, \ldots, \sigma_n],$$

that

$$\phi(\sigma_1, \ldots, \sigma_n) \neq 0 \text{ in } D[\sigma_1, \ldots, \sigma_n] \implies \phi(\sigma_1, \ldots, \sigma_n) \neq 0$$
$$\text{in } D[X_1, \ldots, X_n].$$

Let then $S(\phi) := \sigma_1^{a_1-a_2} \sigma_2^{a_2-a_3} \cdots \sigma_{n-2}^{a_{n-2}-a_{n-1}} \sigma_{n-1}^{a_{n-1}-a_n} \sigma_n^{a_n}$; as a consequence $T(\phi) = X_1^{a_1} X_2^{a_2} \cdots X_{n-1}^{a_{n-1}} X_n^{a_n}$ which proves the claim. ♄

Example 6.2.5. As an example we will verify Newton's formula

$$s_3 = \sigma_1^3 - 3\sigma_1\sigma_2 + 3\sigma_3,$$

where $s_3 := X_1^3 + X_2^3 + X_3^3$.

Since $T(s_3) = X_1^3$, we choose $\psi = \sigma_1^3$, getting

$$g := s_3 - \sigma_1^3 = -3(X_1^2 X_2 + X_1 X_2^2 + X_1^2 X_3 + X_2^2 X_3 + X_1 X_3^2 + X_2 X_3^2) - 6X_1 X_2 X_3.$$

Therefore $T(g) = X_1^2 X_2$ and $\psi = -3\sigma_1\sigma_2$, yielding

$$g := s_3 - \sigma_1^3 + 3\sigma_1\sigma_2 = -3X_1 X_2 X_3$$

so that

$$s_3 = \sigma_1^3 - 3\sigma_1\sigma_2 + 3\sigma_3$$

Corollary 6.2.6. *A symmetric function* $h \in D(X_1, \ldots, X_n)$ *can be expressed in a unique way as a rational function in* $\sigma_1, \ldots, \sigma_n$.

Proof Let $(f(X_1, \ldots, X_n))/(g(X_1, \ldots, X_n))$, $g \neq 0$, be a symmetric function, and let $G(X_1, \ldots, X_n)$ be the product of all the polynomials $g(X_{\pi(1)}, \ldots, X_{\pi(n)})$, where π runs over all the permutations except the identity one.

Therefore gG is a symmetric polynomial and, since $f/g = fG/gG$ is a symmetric function, so is fG; therefore there are polynomials $\phi, \gamma \in D(\sigma_1, \ldots, \sigma_n)$, $\gamma \neq 0$ such that

$$\begin{aligned} \phi(\sigma_1, \ldots, \sigma_n) &= f(X_1, \ldots, X_n)G(X_1, \ldots, X_n), \\ \gamma(\sigma_1, \ldots, \sigma_n) &= g(X_1, \ldots, X_n)G(X_1, \ldots, X_n), \\ \frac{f(X_1, \ldots, X_n)}{g(X_1, \ldots, X_n)} &= \frac{\phi(\sigma_1, \ldots, \sigma_n)}{\gamma(\sigma_1, \ldots, \sigma_n)}. \end{aligned}$$

♃

6.3 Newton's Theorem

Example 6.2.5 is just the most elementary instance of Newton's Theorem, which relates different important symmetric functions.

Before discussing that, let us introduce some notation which will be used throughout this section. If $h_d(X_1, \ldots, X_n) \in D[X_1, \ldots, X_n]$ is a symmetric function of degree d in $X_1, \ldots, X_n$, we will denote either $h(d, v)$ or $h_d(\mathbf{X}_v)$ to be the symmetric function

$$h(d, v) = h_d(\mathbf{X}_v) := h_d(X_1, \ldots, X_v, 0, \ldots, 0) \in D[X_1, \ldots, X_v]$$

of degree d in $X_1, \ldots, X_\nu$, extending the notation to include also $h(0, \nu) = h_0(\mathbf{X}_\nu) = 1$.

In particular $\sigma(d, \nu) = \sigma_d(\mathbf{X}_\nu)$ are the dth elementary symmetric functions of the ν variables $X_1, \ldots, X_\nu$, so that $\sigma(d, n) = \sigma_d$ – which we will extend to $\sigma(d, n) = 0$ for $d > n$ – and

Corollary 6.3.1. *We have:*

(1) *for all d, $1 \leq d \leq \nu$, $\sigma_d(\mathbf{X}_\nu) = \sigma_d(\mathbf{X}_{\nu-1}) + \sigma_{d-1}(\mathbf{X}_{\nu-1})X_\nu$;*

(2) $\sigma_d(\mathbf{X}_{\nu-1}) = \sigma_d(\mathbf{X}_\nu) + \sum_{j=1}^{d-1}(-1)^j X_\nu^j \sigma_{d-j}(\mathbf{X}_\nu) + (-1)^d X_\nu^d$, $1 \leq d < \nu$;

(3) $\sigma_\nu(\mathbf{X}_\nu) + \sum_{j=1}^{\nu-1}(-1)^j X_\nu^j \sigma_{\nu-j}(\mathbf{X}_\nu) + (-1)^\nu X_\nu^\nu = 0$;

(4) $D[\sigma_1(\mathbf{X}_{\nu-1}), \ldots, \sigma_{\nu-1}(\mathbf{X}_{\nu-1})] \subset D[\sigma_1(\mathbf{X}_\nu), \ldots, \sigma_{\nu-1}(\mathbf{X}_\nu), X_\nu]$;

(5) $D[\sigma_1(\mathbf{X}_\nu), \ldots, \sigma_\nu(\mathbf{X}_\nu)] \subset D[\sigma_1(\mathbf{X}_{\nu-1}), \ldots, \sigma_{\nu-1}(\mathbf{X}_{\nu-1}), X_\nu]$.

♄

Definition 6.3.2.

The Waring functions *are*

$$s_d(\mathbf{X}_\nu) := \sum_{i=1}^{\nu} X_i^d;$$

when n is fixed we will freely write $s_d := s_d(\mathbf{X}_n)$.
The locator polynomial[2] $L(\mathbf{X}_\nu, Z) \in D[X_1, \ldots, X_\nu][Z]$ *is*

$$L(\mathbf{X}_\nu, Z) = \prod_{i=1}^{\nu}(1 - X_i Z) = 1 + \sum_{j=1}^{\nu}(-1)^j \sigma_j(\mathbf{X}_\nu) Z^j.$$

The complete sums *in $D[X_1, \ldots, X_\nu]$, $h_d(\mathbf{X}_\nu)$ are the polynomials consisting of the sum of all terms of degree d in $D[X_1, \ldots, X_\nu]$.*

The Gröbnerian symmetric functions *are the polynomials*

$$g_d(\mathbf{X}_{\nu-d+1}) := h_d(\mathbf{X}_{\nu-d+1}) \in D[X_1, \ldots, X_{\nu-d+1}] \subset D[X_1, \ldots, X_\nu].$$

Example 6.3.3. For $\nu = 3$ we have

$$\begin{aligned} s_d(\mathbf{X}_3) &= X_1^3 + X_2^3 + X_3^3, \\ h_2(\mathbf{X}_3) &= X_1^2 + X_1 X_2 + X_1 X_3 + X_2^2 + X_2 X_3 + X_3^2, \\ g_1(\mathbf{X}_3) &= X_1 + X_2 + X_3, \\ g_2(\mathbf{X}_2) &= X_1^2 + X_1 X_2 + X_2^2, \\ g_3(\mathbf{X}_1) &= X_1^3. \end{aligned}$$

---

[2] I borrow the terminology from Coding Theory.

Lemma 6.3.4 (Newton). *We have*

$$-L(\mathbf{X}_\nu, Z) \sum_{d=1}^{\infty} \mathsf{s}_d(\mathbf{X}_\nu) Z^d = -Z \sum_{i=1}^{\nu} X_i \prod_{\substack{j=1 \\ j \neq i}}^{\nu} (1 - X_j Z) = ZL'(\mathbf{X}_\nu, Z).$$

Proof Note that

$$\sum_{d=1}^{\infty} \mathsf{s}_d(\mathbf{X}_\nu) Z^d = \sum_{d=1}^{\infty} \sum_{i=1}^{\nu} X_i^d Z^d = \sum_{i=1}^{\nu} \sum_{d=1}^{\infty} (X_i Z)^d = \sum_{i=1}^{\nu} \frac{X_i Z}{1 - X_i Z}$$

so that

$$\begin{aligned}
-L(\mathbf{X}_\nu, Z) \sum_{d=1}^{\infty} \mathsf{s}_d(\mathbf{X}_\nu) Z^d &= -\prod_{j=1}^{\nu}(1 - X_j Z) \sum_{i=1}^{\nu} \frac{X_i Z}{1 - X_i Z} \\
&= -Z \sum_{i=1}^{\nu} X_i \prod_{\substack{j=1 \\ j \neq i}}^{\nu} (1 - X_j Z) \\
&= ZL'(\mathbf{X}_\nu, Z).
\end{aligned}$$

♄

Corollary 6.3.5. *We have*

$$ZL'(\mathbf{X}_\nu, Z) + L(\mathbf{X}_\nu, Z) \sum_{d=1}^{\infty} \mathsf{s}_d(\mathbf{X}_\nu) Z^d = 0. \tag{6.4}$$

♄

Corollary 6.3.6 (Newton's formula). *We have:*

(1) $\mathsf{s}_j(\mathbf{X}_\nu) + \sum_{\lambda=1}^{j-1}(-1)^\lambda \mathsf{s}_{j-\lambda}(\mathbf{X}_\nu)\sigma_\lambda(\mathbf{X}_\nu) + (-1)^j j\sigma_j(\mathbf{X}_\nu) = 0$, $j \leq \nu$;
(2) $\mathsf{s}_j(\mathbf{X}_\nu) + \sum_{\lambda=1}^{j-1}(-1)^\lambda \mathsf{s}_{j-\lambda}(\mathbf{X}_\nu)\sigma_\lambda(\mathbf{X}_\nu) = 0$, $j > \nu$.

Proof We only have to equate to 0 the coefficients of each power of Z in Equation 6.4. ♄

Remark 6.3.7. In particular

$$\begin{aligned}
\mathsf{s}_1(\mathbf{X}_\nu) &= \sigma_1(\mathbf{X}_\nu), \\
\mathsf{s}_2(\mathbf{X}_\nu) &= \sigma_1^2(\mathbf{X}_\nu) - 2\sigma_2(\mathbf{X}_\nu), \\
\mathsf{s}_3(\mathbf{X}_\nu) &= \sigma_1^3(\mathbf{X}_\nu) - 3\sigma_1(\mathbf{X}_\nu)\sigma_2(\mathbf{X}_\nu) + 3\sigma_3(\mathbf{X}_\nu).
\end{aligned}$$

6.3 Newton's Theorem

Corollary 6.3.8. *If* char$(Q) = 0$ *or* char$(Q) > v$, *then* $D[\sigma_1(\mathbf{X}_v), \ldots, \sigma_v(\mathbf{X}_v)] = D[\mathsf{s}_1(\mathbf{X}_v), \ldots, \mathsf{s}_v(\mathbf{X}_v)]$. ♮

Lemma 6.3.9. *Let* $f_d(\mathbf{X}_v) \in D[X_1, \ldots, X_v]$ *be symmetric functions satisfying, for all* $j \leq v$, *the relations*

$$F_j := a_j \sigma_j(\mathbf{X}_v) + \sum_{\lambda=1}^{j-1} f_{j-\lambda}(\mathbf{X}_v) h_{j\lambda}(\mathbf{X}_v) + f_j(\mathbf{X}_v) = 0,$$

for suitable $h_{j\lambda}(\mathbf{X}_v) \in D[X_1, \ldots, X_v]$, $a_j \in D \setminus \{0\}$.
Then

$$(f_1(\mathbf{X}_v), \ldots, f_v(\mathbf{X}_v)) = (\sigma_1(\mathbf{X}_v), \ldots, \sigma_v(\mathbf{X}_v)).$$

Proof The equation $F_j = 0$ allows us to deduce that

$$\sigma_j(\mathbf{X}_v) \in (f_1(\mathbf{X}_v), \ldots, f_v(\mathbf{X}_v))$$

so that

$$(f_1(\mathbf{X}_v), \ldots, f_v(\mathbf{X}_v)) \supset (\sigma_1(\mathbf{X}_v), \ldots, \sigma_v(\mathbf{X}_v)).$$

Since $F_j = 0$ also proves

$$f_j(\mathbf{X}_v) \in \big(f_1(\mathbf{X}_v), \ldots, f_{j-1}(\mathbf{X}_v), \sigma_j(\mathbf{X}_v)\big),$$

we can deduce inductively, for all $j \leq v$, that

$$f_j(\mathbf{X}_v) \in \big(\sigma_1(\mathbf{X}_v), \ldots, \sigma_j(\mathbf{X}_v)\big)$$

and the converse inclusion

$$(f_1(\mathbf{X}_v), \ldots, f_v(\mathbf{X}_v)) \subset (\sigma_1(\mathbf{X}_v), \ldots, \sigma_v(\mathbf{X}_v)).$$
♮

Corollary 6.3.10. *If* char$(Q) = 0$ *or* char$(Q) > v$, *then* $(\sigma_1(\mathbf{X}_v), \ldots, \sigma_v(\mathbf{X}_v)) = (\mathsf{s}_1(\mathbf{X}_v), \ldots, \mathsf{s}_v(\mathbf{X}_v))$. ♮

Proposition 6.3.11. *We have*

(1) $\sum_{d=0}^{\infty} (-1)^d \mathsf{h}_d(\mathbf{X}_v) Z^d = \prod_{j=1}^{v} (1 - X_j Z)^{-1}$;
(2) $(-1)^j \sigma_j(\mathbf{X}_v) + \sum_{\lambda=1}^{j-1} (-1)^\lambda \mathsf{h}_{j-\lambda}(\mathbf{X}_v) \sigma_\lambda(\mathbf{X}_v) + \mathsf{h}_j(\mathbf{X}_v) = 0$,
 for all j;
(3) $D[\sigma_1(\mathbf{X}_v), \ldots, \sigma_v(\mathbf{X}_v)] = D[\mathsf{h}_1(\mathbf{X}_v), \ldots, \mathsf{h}_v(\mathbf{X}_v)]$;
(4) $(\sigma_1(\mathbf{X}_v), \ldots, \sigma_v(\mathbf{X}_v)) = (\mathsf{h}_1(\mathbf{X}_v), \ldots, \mathsf{h}_v(\mathbf{X}_v))$.

Proof

(1) Obvious.
(2) Since $\sum_{d=0}^{\infty}(-1)^d h_d(\mathbf{X}_\nu) = L(\mathbf{X}_\nu, Z)^{-1}$.
(3) Obvious.
(4) As a corollary of Lemma 6.3.9.

♌

Lemma 6.3.12. *We have*

(1) $h_{d+1}(\mathbf{X}_\mu) = h_{d+1}(\mathbf{X}_{\mu-1}) + X_\mu h_d(\mathbf{X}_\mu)$, *for all* d, μ;
(2) $(h_1(\mathbf{X}_\nu), \ldots, h_\nu(\mathbf{X}_\nu)) = (g_1(\mathbf{X}_\nu), \ldots, g_{d+1}(\mathbf{X}_{\nu-d}), \ldots, g_\nu(\mathbf{X}_1))$.

Proof

(1) Obvious.
(2) As a consequence we have

$$(h_d(\mathbf{X}_{\nu-d+1}), h_{d+1}(\mathbf{X}_{\nu-d+1}), \ldots, h_\nu(\mathbf{X}_{\nu-d+1}))$$
$$= (h_d(\mathbf{X}_{\nu-d+1}), h_{d+1}(\mathbf{X}_{\nu-d}), \ldots, h_\nu(\mathbf{X}_{\nu-d}))$$
$$= (g_d(\mathbf{X}_{\nu-d+1}), h_{d+1}(\mathbf{X}_{\nu-d}), \ldots, h_\nu(\mathbf{X}_{\nu-d})).$$

So that we can deduce

$$(h_1(\mathbf{X}_\nu), h_2(\mathbf{X}_\nu), \ldots, h_\nu(\mathbf{X}_\nu))$$
$$= (g_1(\mathbf{X}_\nu), h_2(\mathbf{X}_{\nu-1}), \ldots, h_\nu(\mathbf{X}_{\nu-1}))$$
$$= \cdots$$
$$= (g_1(\mathbf{X}_\nu), \ldots, g_d(\mathbf{X}_{\nu-d+1}), h_{d+1}(\mathbf{X}_{\nu-d}), \ldots, h_\nu(\mathbf{X}_{\nu-d}))$$
$$= (g_1(\mathbf{X}_\nu), \ldots, g_{d+1}(\mathbf{X}_{\nu-d}), h_{d+2}(\mathbf{X}_{\nu-d-1}), \ldots, h_\nu(\mathbf{X}_{\nu-d-1}))$$
$$= \cdots$$
$$= (g_1(\mathbf{X}_\nu), \ldots, g_{d+1}(\mathbf{X}_{\nu-d}), \ldots, g_\nu(\mathbf{X}_1)).$$

♌

This allows us to deduce the following result

Corollary 6.3.13. *Under the same notation,*

$(\sigma_1(\mathbf{X}_\nu), \ldots, \sigma_\nu(\mathbf{X}_\nu)) = (g_1(\mathbf{X}_\nu), \ldots, g_{d+1}(\mathbf{X}_{\nu-d}), \ldots, g_\nu(\mathbf{X}_1))$;
$T(g_{d+1}(\mathbf{X}_{\nu-d})) = X_{\nu-d}^d$, *for all d, under lexicographical ordering such that*

$$X_1 < X_2 < \cdots < X_\nu.$$

♌

6.3 Newton's Theorem

from this, those familiar with the Gröbner bases will easily deduce – when D is a field – that [3]

Fact 6.3.14. *The set of the Gröbnerian symmetric functions*

$$\mathcal{G} := \{g_1(\mathbf{X}_\nu), \ldots, g_{d+1}(\mathbf{X}_{\nu-d}), \ldots, g_\nu(\mathbf{X}_1)\}$$

is the Gröbner basis of $(\sigma_1, \ldots, \sigma_\nu)$ *under lexicographical ordering such that* $X_1 < X_2 < \cdots < X_\nu$. ♄

This of course poses the question, how are the symmetric functions σ_d Gröbner-reduced by $\mathcal{G}$? The solution is given by the formula

Proposition 6.3.15.

$$\text{for all } d, \nu, d \leq \nu, \sigma_d(\mathbf{X}_\nu) + \sum_{i=1}^{d-1}(-1)^i g_i(\mathbf{X}_{\nu-i+1})\sigma_{d-i}(\mathbf{X}_{\nu-i})$$
$$+ (-1)^d g_d(\mathbf{X}_{\nu-d+1}). \tag{6.5}$$

Proof Let us note the obvious relations

$$\sigma_d(\mathbf{X}_\nu) = \sigma_d(\mathbf{X}_{\nu-1}) + X_\nu \sigma_{d-1}(\mathbf{X}_{\nu-1})$$
$$g_d(\mathbf{X}_{\nu-d+1}) = g_d(\mathbf{X}_{\nu-d}) + X_{\nu-d+1} g_{d-1}(\mathbf{X}_{\nu-d+1}),$$

which allow the following inductive reduction

$$\sigma_d(\mathbf{X}_\nu) + \sum_{i=1}^{d-1}(-1)^i g_i(\mathbf{X}_{\nu-i+1})\sigma_{d-i}(\mathbf{X}_{\nu-i}) + (-1)^d g_d(\mathbf{X}_{\nu-d+1})$$

$$= \sigma_d(\mathbf{X}_{\nu-1}) + X_\nu \sigma_{d-1}(\mathbf{X}_{\nu-1})$$

$$+ \sum_{i=1}^{d-1}(-1)^i g_i(\mathbf{X}_{\nu-i})\sigma_{d-i}(\mathbf{X}_{\nu-i})$$

$$+ \sum_{i=1}^{d-1}(-1)^i g_{i-1}(\mathbf{X}_{\nu-i+1}) X_{\nu-i+1} \sigma_{d-i}(\mathbf{X}_{\nu-i})$$

$$+ (-1)^d g_d(\mathbf{X}_{\nu-d}) + (-1)^d g_{d-1}(\mathbf{X}_{\nu-d+1}) X_{\nu-d+1}$$

---

[3] In fact, these polynomials were frequently found by M. Sala in the Gröbner basis of 0-dimensional ideals (related to a coding theory problem) whose set of roots was symmetric; he therefore proposed adding these polynomials to the basis in order to obtain a faster 'heuristic' result. Due to Fact 6.3.14, this approach is definitely not heuristic and should be recommended.

When I saw his polynomials, I immediately got the impression that they were somehow related to the Gröbner basis of symmetric functions and it required just a few hand computations to deduce (6.5) from which Fact 6.3.14 comes immediately.

Proving them was a more difficult problem and I am indebted to M. Sala, E. Briand and L. Gonzalez-Vega for their help.

$$= \sigma_d(\mathbf{X}_{\nu-1}) + X_\nu \left(\sigma_{d-1}(\mathbf{X}_{\nu-1}) - g_0(\mathbf{X}_\nu)\sigma_{d-1}(\mathbf{X}_{\nu-1}) \right)$$

$$+ \sum_{i=1}^{d-1} g_i(\mathbf{X}_{\nu-i})(-1)^i \left(\sigma_{d-i}(\mathbf{X}_{\nu-i}) - X_{\nu-i}\sigma_{d-i-1}(\mathbf{X}_{\nu-i-1}) \right)$$

$$+ (-1)^d g_d(\mathbf{X}_{\nu-d})$$

$$= \sigma_d(\mathbf{X}_{\nu-1}) + \sum_{i=1}^{d-1} (-1)^i g_i(\mathbf{X}_{\nu-i})\sigma_{d-i}(\mathbf{X}_{\nu-i-1}) + (-1)^d g_d(\mathbf{X}_{\nu-d}).$$

We are therefore reduced to the cases $d = \nu - 1$ and to the formulas

$$\sigma_d(\mathbf{X}_d) + \sum_{i=1}^{d-1} (-1)^i g_i(\mathbf{X}_{d-i+1})\sigma_{d-i}(\mathbf{X}_{d-i}) + (-1)^d g_d(\mathbf{X}_1),$$

which we derive in the same way using

$$\begin{aligned} \sigma_d(\mathbf{X}_d) &= X_d \sigma_{d-1}(\mathbf{X}_{d-1}), \\ g_i(\mathbf{X}_{d-i+1}) &= g_i(\mathbf{X}_{d-i}) + X_{d-i+1} g_{i-1}(\mathbf{X}_{d-i+1}), \\ g_d(\mathbf{X}_1) &= X_1 g_{d-1}(\mathbf{X}_1), \end{aligned}$$

and which allow the following inductive reduction

$$\sigma_d(\mathbf{X}_d) + \sum_{i=1}^{d-1} (-1)^i g_i(\mathbf{X}_{d-i+1})\sigma_{d-i}(\mathbf{X}_{d-i}) + (-1)^d g_d(\mathbf{X}_1)$$

$$= X_d \sigma_{d-1}(\mathbf{X}_{d-1}) + \sum_{i=1}^{d-1} (-1)^i g_i(\mathbf{X}_{d-i})\sigma_{d-i}(\mathbf{X}_{d-i})$$

$$+ \sum_{i=1}^{d-1} (-1)^i g_{i-1}(\mathbf{X}_{d-i+1}) X_{d-i+1} \sigma_{d-i}(\mathbf{X}_{d-i})$$

$$+ (-1)^d g_{d-1}(\mathbf{X}_1) X_1$$

$$= X_d \left(\sigma_{d-1}(\mathbf{X}_{d-1}) - g_0(\mathbf{X}_d)\sigma_{d-1}(\mathbf{X}_{d-1}) \right)$$

$$+ \sum_{i=1}^{d-2} (-1)^i g_i(\mathbf{X}_{d-i}) \left(\sigma_{d-i}(\mathbf{X}_{d-i}) - X_{d-i}\sigma_{d-i-1}(\mathbf{X}_{d-i-1}) \right)$$

$$+ (-1)^d g_{d-1}(\mathbf{X}_1) (X_1 - \sigma_1(\mathbf{X}_1))$$
$$= 0$$

♄

6.4 The Method of Indeterminate Coefficients

The technique applied by the English algebra school, in order to have a means of discussing 'generic' polynomials and roots, which extended Newton's idea

6.4 The Method of Indeterminate Coefficients

of interpreting the coefficients of the polynomial as a symmetric function of its roots, consisted of introducing indeterminate coefficients a_i (respectively a_0, α_j), considering the domain $\mathbb{D} := \mathbb{Z}[a_0, a_1, \ldots a_n]$ (respectively $\mathbb{D} := \mathbb{Z}[a_0, \alpha_1, \ldots \alpha_n]$) and the polynomial $f(X) \in \mathbb{D}[X]$ defined by Equation 6.1 (respectivly Equation 6.2) and analysing it in order to obtain a 'universal' solution $\Delta \in \mathbb{D}$.

Then, when we have a given, specific domain D and a polynomial $f(X) \in D[X]$, we just make an *ansatz*, substituting within the obtained expression $\Delta \in \mathbb{D}$ the corresponding given elements a_i (respectively α_j) and interpreting the coefficients in $\mathbb{Z}$ by their image in the prime field of Q.

In other words we choose f by fixing a domain morphism $\Xi : \mathbb{D}[X] \mapsto K[X]$.

Example 6.4.1. To explain this, we only have to recall what is usual for quadratic polynomials. If we are given a quadratic polynomial

$$f(X) = aX^2 + bX + c \in D[X]$$

we know that it has distinct roots when the discriminant

$$\Delta := b^2 - 4ac$$

is non-zero in which case the roots – if $\mathrm{char}(Q) \neq 2$ – are

$$-\frac{b \pm \sqrt{\Delta}}{2a}.$$

Therefore if we are given

the polynomial $f(X) := X^2 + 6X + 9 \in \mathbb{Q}[X]$, substituting in the expression Δ the *ansatz* $a = 1, b = 6, c = 9$, we get, in $\mathbb{Q}$, $\Delta = 6^2 - 4 \cdot 1 \cdot 9 = 0$ so that the roots are not distinct;

the polynomial $f(X) := X^2 - 3X + 2 \in \mathbb{Q}[X]$, substituting in the expression Δ the *ansatz* $a = 1, b = -3, c = 2$, we get, in $\mathbb{Q}$, $\Delta = 3^2 - 4 \cdot 1 \cdot 2 = 1$ so that the roots are distinct and are given by

$$-\frac{b \pm \sqrt{\Delta}}{2} = -\frac{-3 \pm \sqrt{1}}{2} = \begin{cases} 1 \\ 2 \end{cases};$$

the polynomial $f(X) := X^2 + X - 1 \in \mathbb{Z}_5[X]$, substituting in the expression $\Delta = b^2 + ac$ the *ansatz* $a = 1, b = 1, c = -1$, we get, in $\mathbb{Z}_5$, $\Delta = 0$ so that the roots are not distinct; in fact $f(X) = (X + 3)^2$;

the polynomial $f(X) := X^2 - 2X + 2 \in \mathbb{Z}_5[X]$, substituting in the expression $\Delta = b^2 + ac$ the *ansatz* $a = 1, b = -2, c = 2$, we get, in $\mathbb{Z}_5$, $\Delta = 1$

so that the roots are distinct and are given by

$$-\frac{b \pm \sqrt{\Delta}}{2} = -3(-2 \pm \sqrt{1}) = \begin{cases} -1 \\ 3 \end{cases};$$

in fact $(X - 3)(X + 1) = f(X)$.

Remark 6.4.2. To show another example of the method of indeterminate coefficients, let us verify that the fact that f has distinct roots when the discriminant $\Delta := b^2 - 4ac$ is non-zero, also holds (as we will prove in Proposition 6.5.4) in the case $\mathrm{char}(k) = 2$ where[4] $\Delta = b^2$ in $\mathbb{Z}_2[a, b, c]$. In fact $f = aX^2 + bX + c \in \mathbb{Z}_2[a, b, c][X]$ has roots α, β if and only if

$$f = a(X - \alpha)(X - \beta) \in \mathbb{Z}_2[\alpha, \beta, a, b, c][X]$$

i.e.

$$b = a(\alpha - \beta), \quad c = a\alpha\beta$$

in $\mathbb{Z}_2[\alpha, \beta, a, b, c]$; it is then clear that

$$\alpha = \beta \iff b = a(\alpha - \beta) = 0 \iff b^2 = \Delta = 0.$$

Note that we have derived in this case the formula of Definition 6.5.3:

$$\Delta = b^2 = a^2(\alpha - \beta)^2.$$

Remark 6.4.3. In this setting we can (and will) consider also polynomials $f(X) \in D[X]$ defined by Equation 6.1 of degree $m < n$ just by making the ansatz $a_i = 0$, for all $i, m < i \leq n$.

6.5 Discriminant

Let

$$f(X) = a_0 X^n + a_1 X^{n-1} + \cdots + a_i X^{n-i} + \cdots + a_{n-1} X + a_n$$

and $\alpha_1, \ldots, \alpha_n$ be its roots so that Equation 6.3 holds.

There is an obvious function which vanishes iff the generic polynomial f has a multiple root, i.e.

$$\prod_{i>j}(\alpha_i - \alpha_j).$$

To transform it as a symmetric function we consider

$$\prod_{i>j}(X_i - X_j)^2$$

---

[4] But where the formula which gives the roots as $-(b \pm \sqrt{\Delta})/2a$ cannot, of course, be applied.

6.5 Discriminant

which is therefore an element in $\mathbb{Z}[\sigma_1, \ldots, \sigma_n] = \mathbb{Z}[s_1, \ldots, s_n]$. Its expression can be obtained thanks to

Proposition 6.5.1.

$$\prod_{i>j}(X_i - X_j)^2 = \begin{vmatrix} n & s_1 & \cdots & s_{n-1} \\ s_1 & s_2 & \cdots & s_n \\ \vdots & \vdots & \ddots & \vdots \\ s_{n-1} & s_n & \cdots & s_{2n-2} \end{vmatrix}.$$

Proof It is known that $\prod_{i>j}(X_i - X_j)$ is the Vandermonde determinant[5]:

$$\prod_{i>j}(X_i - X_j) = \begin{vmatrix} 1 & 1 & \cdots & 1 \\ X_1 & X_2 & \cdots & X_n \\ X_1^2 & X_2^2 & \cdots & X_n^2 \\ \vdots & \vdots & \ddots & \vdots \\ X_1^{n-1} & X_2^{n-1} & \cdots & X_n^{n-1} \end{vmatrix},$$

---

[5] I record here a proof of the formula, which is obtained by induction, transforming the matrix by subtracting from each row the preceding row multiplied by X_1:

$$\begin{vmatrix} 1 & 1 & \cdots & 1 \\ X_1 & X_2 & \cdots & X_n \\ X_1^2 & X_2^2 & \cdots & X_n^2 \\ \vdots & \vdots & \ddots & \vdots \\ X_1^{n-1} & X_2^{n-1} & \cdots & X_n^{n-1} \end{vmatrix}$$

$$= \begin{vmatrix} 1 & 1 & \cdots & 1 \\ 0 & (X_2 - X_1) & \cdots & (X_n - X_1) \\ 0 & X_2(X_2 - X_1) & \cdots & X_n(X_n - X_1) \\ \vdots & \vdots & \ddots & \vdots \\ 0 & X_2^{n-2}(X_2 - X_1) & \cdots & X_n^{n-2}(X_n - X_1) \end{vmatrix}$$

$$= \begin{vmatrix} (X_2 - X_1) & (X_3 - X_1) & \cdots & (X_n - X_1) \\ X_2(X_2 - X_1) & X_3(X_3 - X_1) & \cdots & X_n(X_n - X_1) \\ \vdots & \vdots & \ddots & \vdots \\ X_2^{n-2}(X_2 - X_1) & X_3^{n-2}(X_3 - X_1) & \cdots & X_n^{n-2}(X_n - X_1) \end{vmatrix}$$

$$= \prod_{i>1}(X_i - X_1) \begin{vmatrix} 1 & 1 & \cdots & 1 \\ X_2 & X_3 & \cdots & X_n \\ \vdots & \vdots & \ddots & \vdots \\ X_2^{n-2} & X_3^{n-2} & \cdots & X_n^{n-2} \end{vmatrix}.$$

so that we have

$$\prod_{i>j}(X_i - X_j)^2 = \begin{vmatrix} 1 & 1 & \cdots & 1 \\ X_1 & X_2 & \cdots & X_n \\ X_1^2 & X_2^2 & \cdots & X_n^2 \\ \vdots & \vdots & \ddots & \vdots \\ X_1^{n-1} & X_2^{n-1} & \cdots & X_n^{n-1} \end{vmatrix}^2$$

$$= \begin{vmatrix} 1 & \cdots & 1 \\ X_1 & \cdots & X_n \\ X_1^2 & \cdots & X_n^2 \\ \vdots & \ddots & \vdots \\ X_1^{n-1} & \cdots & X_n^{n-1} \end{vmatrix} \cdot \begin{vmatrix} 1 & X_1 & \cdots & X_1^{n-1} \\ 1 & X_2 & \cdots & X_2^{n-1} \\ 1 & X_3 & \cdots & X_3^{n-1} \\ \vdots & \vdots & \ddots & \vdots \\ 1 & X_n & \cdots & X_n^{n-1} \end{vmatrix}$$

$$= \begin{vmatrix} n & s_1 & \cdots & s_{n-1} \\ s_1 & s_2 & \cdots & s_n \\ \vdots & \vdots & \ddots & \vdots \\ s_{n-1} & s_n & \cdots & s_{2n-2} \end{vmatrix}.$$

♄

Remark 6.5.2. As a consequence we can represent $\prod_{i>j}(X_i - X_j)^2$ as a polynomial in $\mathbb{Z}[\sigma_1, \ldots, \sigma_n]$. Since the leading term of $\prod_{i>j}(X_i - X_j)^2$ is $X_2^2 X_3^4 \cdots X_n^{2n-2}$, the leading term of its representation in $D[\sigma_1, \ldots, \sigma_n]$ is $\sigma_2^2 \cdots \sigma_{n-1}^2 \sigma_n^2$.

Therefore, if we substitute each occurrence of σ_i by $(-1)^i \frac{a_i}{a_0}$, in the representation of $\prod_{i>j}(X_i - X_j)^2$ in $\mathbb{Z}[\sigma_1, \ldots, \sigma_n]$, the denominator is a_0^{2n-2}. Multiplying by it, we therefore obtain a representation of $\prod_{i>j}(\alpha_i - \alpha_j)^2$ as a polynomial in $\mathbb{Z}[a_0, a_1, \ldots, a_n]$.

This justifies the introduction of

Definition 6.5.3. *Let*

$$f(X) = a_0 X^n + a_1 X^{n-1} + \cdots + a_i X^{n-i} + \cdots + a_{n-1} X + a_n$$

and $\alpha_1, \ldots, \alpha_n$ be its roots.

The discriminant *of f is the polynomial* $\mathrm{Disc}(f) \in \mathbb{Z}[a_0, \ldots, a_n]$ *which represents the symmetric function*

$$a_0^{2n-2} \prod_{i>j}(\alpha_i - \alpha_j)^2 \in \mathbb{Z}[a_0, \alpha_1, \ldots, \alpha_n].$$

Proposition 6.5.4. *Let $f \in D[X]$. Then f is squarefree iff* $\mathrm{Disc}(f) \neq 0$. ♄

Example 6.5.5. When $n = 2$ we have

$$\begin{aligned}\operatorname{Disc}(f) &= a_0^2(\alpha_1 - \alpha_2)^2 \\ &= a_0^2(\alpha_1^2 - 2\alpha_1\alpha_2 + \alpha_2^2) \\ &= a_0^2(\mathsf{s}_2 - 2\sigma_2) \\ &= a_0^2(\sigma_1^2 - 4\sigma_2) \\ &= a_1^2 - 4a_0 a_2.\end{aligned}$$

A similar computation when $n = 3$ yields

$$\operatorname{Disc}(f) = a_1^2 a_2^2 - 4a_0 a_2^3 - 4a_1^3 a_3 - 27 a_0^2 a_3^2 + 18 a_0 a_1 a_2 a_3.$$

In the same context as the previous sections, let us now consider the symmetric polynomial

$$\prod_{i > j}(X - \alpha_i - \alpha_j - t\alpha_i\alpha_j) \in \mathbb{Z}[a_0, \alpha_1, \ldots, \alpha_n][X, t]$$

so that it is an element in $\mathbb{Z}[a_0, a_1, \ldots, a_n][t, X]$.

Definition 6.5.6. *We will call the* Laplace–Gauss Resolvent *of the polynomial $f(X) \in k[X]$ that polynomial*[6]

$$\mathfrak{LG}(f) := \prod_{i > j}(X - \alpha_i - \alpha_j - t\alpha_i\alpha_j) \in \mathbb{Z}[a_0, \alpha_1, \ldots, \alpha_n][X, t] \subset K[X, t];$$

for any $\lambda \in K$, we will also use the notation: $\mathfrak{LG}_\lambda(f)(X) := \mathfrak{LG}(f)(\lambda, X)$ and we will omit the dependence on f if there is no ambiguity.

Proposition 6.5.7. *With the notation above, when k is infinite and $0 \neq \operatorname{Disc}(f) \in D$ we have:*

$0 \neq \operatorname{Disc}(\mathfrak{LG}) \in D[t];$
there are infinitely many $\lambda \in K$ such that $0 \neq \operatorname{Disc}(\mathfrak{LG}_\lambda) \in K$.

Proof We only have to note that there are infinitely many $\lambda \in K$ such that

$$\alpha_i + \alpha_j + \lambda\alpha_i\alpha_j \neq \alpha_k + \alpha_l + \lambda\alpha_k\alpha_l, \text{ for all } i, j, k, l : i \neq l \text{ or } j \neq k,$$

since we only have to remove for each $i, j, k, l : i \neq l$ or $j \neq k$ the solution of the linear equation

$$\alpha_i + \alpha_j + t\alpha_i\alpha_j = \alpha_k + \alpha_l + t\alpha_k\alpha_l.$$

♄

---

[6] Recall that $K \supset D$ is a field which contains all the roots of f.

6.6 Resultants

Let us consider the domain

$$\mathbb{D} := \mathbb{Z}[a_0, \ldots, a_n, b_0, \ldots, b_m]$$

and the 'generic' polynomials

$$f(X) = a_0 X^n + a_1 X^{n-1} + \cdots + a_i X^{n-i} + \cdots + a_{n-1} X + a_n,$$

$$g(X) = b_0 X^m + a_1 X^{mn-1} + \cdots + b_i X^{m-i} + \cdots + b_{m-1} X + b_m,$$

in $\mathbb{D}[X]$.

Definition 6.6.1. *The* Sylvester matrix *of f and g is the $n+m$ square matrix*

$$\begin{pmatrix}
a_0 & a_1 & a_2 & \cdots & a_n & 0 & 0 & 0 & \cdots & 0 & 0 & 0 \\
0 & a_0 & a_1 & \cdots & a_{n-1} & a_n & 0 & 0 & \cdots & 0 & 0 & 0 \\
0 & 0 & a_0 & \cdots & a_{n-2} & a_{n-1} & a_n & 0 & \cdots & 0 & 0 & 0 \\
\vdots & \vdots & \vdots & \ddots & \vdots & \vdots & \vdots & \vdots & \ddots & \vdots & \vdots & \vdots \\
0 & 0 & 0 & \cdots & a_0 & a_1 & a_2 & a_3 & \cdots & a_{n-1} & a_n & 0 \\
0 & 0 & 0 & \cdots & 0 & a_0 & a_1 & a_2 & \cdots & a_{n-2} & a_{n-1} & a_n \\
b_0 & b_1 & b_2 & \cdots & b_{m-2} & b_{m-1} & b_m & 0 & \cdots & 0 & 0 & 0 \\
0 & b_0 & b_1 & \cdots & b_{m-3} & b_{m-2} & b_{m-1} & b_m & \cdots & 0 & 0 & 0 \\
\vdots & \vdots & \vdots & \ddots & \vdots & \vdots & \vdots & \vdots & \ddots & \vdots & \vdots & \vdots \\
0 & 0 & 0 & \cdots & b_3 & b_4 & b_5 & b_6 & \cdots & b_m & 0 & 0 \\
0 & 0 & 0 & \cdots & b_2 & b_3 & b_4 & b_5 & \cdots & b_{m-1} & b_m & 0 \\
0 & 0 & 0 & \cdots & b_1 & b_2 & b_3 & b_4 & \cdots & b_{m-2} & b_{m-1} & b_m
\end{pmatrix}.$$

Definition 6.6.2. *The* (Sylvester) resultant *of f and g, $\mathrm{Res}(f, g)$, is the determinant of the Sylvester matrix of f and g.*

Let us make an *ansatz* by specifying a morphism $\Xi : \mathbb{D}[X] \mapsto D[X]$. By a standard abuse of notation, we identify the elements in $\mathbb{D}[X]$ with their images by Ξ in $D[X]$; when there is ambiguity we just specify the domain in which we are considering them.

Theorem 6.6.3. *Let $f, g \in D[X]$ and let us assume that at least one among a_0 and b_0 does not vanish in D, i.e. either $\Xi(a_0) \neq 0$ or $\Xi(b_0) \neq 0$.*

Then the following conditions are equivalent:

(1) $\gcd(f, g) := h(X) \in D[X] \setminus D$;
(2) *there are $p, q \in D[X]$ (not both being zero) such that*

$$\deg(p) < m, \deg(q) < n,$$
$$pf = qg;$$

(3) $\mathrm{Res}(f, g)$ *vanishes in D.*

6.6 Resultants

Proof

(1) $\Longrightarrow$ (2) We only have to take $p = g/h$ and $q = f/h$.

(2) $\Longrightarrow$ (1) By our assumption we can assume that $\deg(f) = n$ since $a_0 \neq 0$ – otherwise we interchange f and g.

Each irreducible factor of f must divide qg; it is impossible, however, that all of them divide q since this implies that f divides q, contradicting the relation $\deg(q) < n = \deg(f)$.

Therefore, at least one irreducible factor of f divides g.

(2) $\Longleftrightarrow$ (3) The existence of polynomials p, q (not both being null) satisfying the conditions required in (2) is equivalent to the existence of elements (not all zero)

$$c_0, \ldots, c_{m-1}, d_0, \ldots d_{n-1} \in D :$$

$$\sum_{i=0}^{m-1} c_{m-i-1} X^i \cdot \sum_{i=0}^{n} a_{n-i} X^i = \sum_{i=0}^{n-1} -d_{n-i-1} X^i \cdot \sum_{i=0}^{m} b_{m-i} X^i, \quad (6.6)$$

i.e. a non-null solution of the $n + m$ homogeneous linear equations which can be obtained by equating the coefficients of the powers X^i, $0 \leq i \leq m+n-1$ in the left and right sides of Equation 6.6. Such linear equations yield the transpose of the Sylvester matrix. ♮

Proposition 6.6.4. *Let $f, g \in D[X]$. The following conditions are equivalent:*

(1) Either $\gcd(f, g) = h(X) \in D[X] \setminus D$ or a_0 and b_0 vanish in D;

(2) $\mathrm{Res}(f, g)$ vanishes in D.

Proof If at least one of the leading coefficients does not vanish in D, the equivalence follows by the preceding theorem. If $a_0 = b_0 = 0$, then the resultant obviously vanishes: the first column is null. ♮

Corollary 6.6.5. *Let $f, g \in D[X]$. The following conditions are equivalent:*

(1) $\gcd(f, g) = 1$, $a_0 \neq 0$, and $b_0 \neq 0$;

(2) $\mathrm{Res}(f, g) \neq 0$ in D. ♮

Corollary 6.6.6. *Let $p \in D$ be a prime and let $\pi : D[X] \mapsto D/(p)[X]$ denote the canonical projection.*

Let $f, g \in D[X]$ be such that

$$\mathrm{Res}(f, g) \neq 0, \quad a_0 \neq 0, \quad b_0 \neq 0$$

and
$$\mathrm{lc}(f) = a_0 \not\equiv 0 \not\equiv b_0 = \mathrm{lc}(g) \,(\mathrm{mod}\, p);$$
then the following conditions are equivalent:

(1) $\mathrm{Res}(f, g) \not\equiv 0 \,(\mathrm{mod}\, p)$*;*
(2) $\gcd(\pi(f), \pi(g)) = 1$*.*

Proof In fact, denoting $\mathfrak{r} := \mathrm{Res}(f, g)$ we have $\mathrm{Res}(\pi(f), \pi(g)) = \pi(\mathfrak{r})$.
♄

The argument of the proof of Theorem 6.6.1(2) $\iff$ (3) can be adapted to prove

Proposition 6.6.7. *There are* $p, q \in D[X]$ *such that*

$\deg(p) < m$, $\deg(q) < n$,
$pf + qg = \mathrm{Res}(f, g)$.

Proof The existence of polynomials p, q satisfying the required conditions is equivalent to the existence of elements
$$c_0, \cdots, c_{m-1}, d_0, \cdots d_{n-1} \in D:$$

$$\sum_{i=0}^{m-1} c_{m-i-1} X^i \cdot \sum_{i=0}^{n} a_{n-i} X^i + \sum_{i=0}^{n-1} d_{n-i-1} X^i \cdot \sum_{i=0}^{m} b_{m-i} X^i = \mathrm{Res}(f, g)$$
(6.7)

i.e. a non-null solution of the $n + m$ homogeneous linear equations which can be obtined by equating the coefficients of the powers $X^i, 0 \le i \le m + n - 1$ in the left and right sides of Equation 6.6. Since such linear equations yield the transpose of the Sylvester matrix, the solutions c_i and d_i can therefore be obtained by Cramer's formula as a proper subdeterminant of the Sylvester matrix.
♄

Due to its interest, let us specify some of the main properties of the resultant of the two polynomials $f, g \in \mathbb{D}[X]$.

Proposition 6.6.8. *Let* $f, g \in \mathbb{D}[X]$. *Then* $\mathrm{Res}(f, g)$ *is homogeneous of degree* m *in the variables* a_i *and homogeneous of degree* n *in the variables* b_i.
It contains $a_0^m b_m^n$ *among the terms.*

Proof The statements follow easily from the expansion of the determinant of the Sylvester matrix; for the last statement note that $a_0^m b_m^n$ is the principal diagonal. ♄

6.7 Resultants and Roots

In this section let us assume we are given two 'generic' polynomials

$$f(X) = a_0 \prod_{i=1}^{n}(X - \alpha_i) = a_0 X^n + a_1 X^{n-1} + \cdots + a_n,$$

$$g(X) = b_0 \prod_{j=1}^{m}(X - \beta_j) = b_0 X^n + b_1 X^{n-1} + \cdots + b_m,$$

in $\mathbb{Z}[a_0, \alpha_1, \ldots, \alpha_n, b_0, \beta_1, \ldots, \beta_m]$.

Proposition 6.7.1. *With the notation above,*

$$\operatorname{Res}(f, g) = a_0^m b_0^n \prod_{i=1}^{n} \prod_{j=1}^{m} (\alpha_i - \beta_j) = a_0^m \prod_{i=1}^{n} g(\alpha_i) = (-1)^{mn} b_0^n \prod_{j=1}^{m} f(\beta_j).$$

Proof Since a_i (respectively b_i) is the product of a_0 (respectively b_0) and a symmetric function of the α_js (respectively β_js) and $\operatorname{Res}(f, g)$ is homogeneous of degree n in the a_js and of degree m in the b_js, we have that $\operatorname{Res}(f, g)$ is the product of $a_0^m b_0^n$ and a function in $D[\alpha_1, \ldots, \alpha_n, \beta_1, \ldots, \beta_m]$ which is a symmetric function both in the α_is and in the β_js.

Moreover $\operatorname{Res}(f, g)$ vanishes if the two polynomials have a common root, i.e. if there exists $i, j : \alpha_i = \beta_j$. As a consequence $\operatorname{Res}(f, g)$ is divisible by $\alpha_i - \beta_j$, for all i, j, and so by

$$\mathsf{R} := a_0^m b_0^n \prod_{i=1}^{n} \prod_{j=1}^{m} (\alpha_i - \beta_j).$$

Therefore

$$a_0^m \prod_{i=1}^{n} g(\alpha_i) = a_0^m \prod_{i=1}^{n} b_0 \prod_{j=1}^{m} (\alpha_i - \beta_j) = a_0^m b_0^n \prod_{i=1}^{n} \prod_{j=1}^{m} (\alpha_i - \beta_j) = \mathsf{R}. \quad (6.8)$$

In the same way we conclude that

$$\mathsf{R} := (-1)^{mn} b_0^n \prod_{j=1}^{m} f(\beta_j). \quad (6.9)$$

The two results allow us to conclude that

$$\mathsf{R} \in D[a_0, a_1, \ldots, a_n, b_0, b_1, \ldots, b_m]$$

and that it is homogeneous of degree n in the b_is, homogeneous of degree m in the a_is and contains $a_0^m b_0^n$ among its terms.

Since $\mathrm{Res}(f, g)$ has the same properties, by Corollary 6.6.4, and since R divides it, the result follows. ♄

Corollary 6.7.2. $\mathrm{Res}(f, g)$ *is irreducible in* $D[a_0, a_1, \ldots, a_n, b_0, b_1, \ldots, b_m]$.

Proof Assume

$$\mathrm{Res}(f, g) = \phi \psi$$

is a non-trivial factorization in $D[a_0, a_1, \ldots, a_n, b_0, b_1, \ldots, b_m]$. Therefore ϕ and ψ can be expressed as symmetric functions in the α_is and in the β_js.

Since in

$$D[a_0, \alpha_1, \ldots, \alpha_n, b_0, \beta_1, \ldots, \beta_m] \supset D[a_0, a_1, \ldots, a_n, b_0, b_1, \ldots, b_m]$$

we have the factorization

$$\mathrm{Res}(f, g) = a_0^m b_0^n \prod_{i=1}^{n} \prod_{j=1}^{m} (\alpha_i - \beta_j),$$

one of the two factors, say ϕ, is divisible by $(\alpha_1 - \beta_1)$ and, by symmetry, by each factor $(\alpha_i - \beta_j)$. Therefore ψ can be divided only by a_0 or b_0.

However, as an element in $D[a_0, a_1, \ldots, a_n, b_0, b_1, \ldots, b_m]$, $\mathrm{Res}(f, g)$ is divisible by neither a_0 – in its expansion there are terms that do not depend on a_0 – nor b_0. So we get the required contradiction. ♄

Corollary 6.7.3. *Let*

$$f(X) = a_0 \prod_{i=1}^{n} (X - \alpha_i) = a_0 X^n + a_1 X^{n-1} + \cdots + a_n.$$

Then $\mathrm{Res}(f, f') = a_0 \mathrm{Disc}(f)$.

Proof In fact we have

$$f'(X) = \sum_{i=1}^{n} a_0 (X - \alpha_1) \cdots (X - \alpha_{i-1})(X - \alpha_{i+1}) \cdots (X - \alpha_n),$$

and

$$f'(\alpha_i) = a_0 (\alpha_i - \alpha_1) \cdots (\alpha_i - \alpha_{i-1})(\alpha_i - \alpha_{i+1}) \cdots (\alpha_i - \alpha_n),$$

6.7 Resultants and Roots

from which

$$\begin{aligned}
a_0 \text{Disc}(f) &= a_0^{2n-1} \prod_{i>j}(\alpha_i - \alpha_j)^2 = a_0^{2n-1} \prod_{i \neq j}(\alpha_i - \alpha_j) \\
&= a_0^{n-1} \prod_i a_0 \prod_{j \neq i}(\alpha_i - \alpha_j) = a_0^{n-1} \prod_i f'(\alpha_i) \\
&= \text{Res}(f, f').
\end{aligned}$$

♄

Another corollary of Proposition 6.7.1 allows us to solve the following:

Problem 6.7.4. Given two polynomials whose roots are respectively α and $\beta \neq 0$, can we compute a polynomial whose root is $\alpha + \beta$ (respectively $\alpha - \beta, \alpha\beta, \alpha/\beta$)?

Corollary 6.7.5 (Loos). *Let D be a domain, let $f(X), g(X) \in D[X]$ be such that $g(0) \neq 0$, let Q the fraction field of D, let $K \supset Q$ the splitting field of fg, and let $\alpha_1, \ldots, \alpha_n \in K$ (respectively $\beta_1, \ldots, \beta_m \in K$) be such that*

$$f(X) = a_0 \prod_{i=1}^{n}(X - \alpha_i) = a_0 X^n + a_1 X^{n-1} + \cdots + a_n,$$

$$g(X) = b_0 \prod_{j=1}^{m}(X - \beta_j) = b_0 X^n + b_1 X^{n-1} + \cdots + b_m.$$

Then

(1) $\text{Res}(f(Y-X), g(X)) = (-1)^{mn} a_0^m b_0^n \prod_{i=1}^{n} \prod_{j=1}^{m}(Y - (\alpha_i + \beta_j))$,
(2) $\text{Res}(f(Y+X), g(X)) = (-1)^{mn} a_0^m b_0^n \prod_{i=1}^{n} \prod_{j=1}^{m}(Y - (\alpha_i - \beta_j))$,
(3) $\text{Res}(X^n f(Y/X), g(X)) = (-1)^{mn} a_0^m b_0^n \prod_{i=1}^{n} \prod_{j=1}^{m}(Y - (\alpha_i \beta_j))$,
(4) $\text{Res}(f(YX), g(X)) = a_0^m b_m^n \prod_{i=1}^{n} \prod_{j=1}^{m}(Y - (\alpha_i/\beta_j))$,

where the resultants are computed in the domain $D[Y]$.

Proof By Equation 6.9 we have

(1)
$$\begin{aligned}
\text{Res}(f(Y-X), g(X)) &= (-1)^{mn} b_0^n \prod_{j=1}^{m} f(Y - \beta_j) \\
&= (-1)^{mn} a_0^m b_0^n \prod_{i=1}^{n} \prod_{j=1}^{m}(Y - \alpha_i - \beta_j).
\end{aligned}$$

(2)
$$\operatorname{Res}(f(Y+X), g(X)) = (-1)^{mn} b_0^n \prod_{j=1}^{m} f(Y+\beta_j)$$
$$= (-1)^{mn} a_0^m b_0^n \prod_{i=1}^{n} \prod_{j=1}^{m} (Y - \alpha_i + \beta_j).$$

(3)
$$\operatorname{Res}(X^n f(Y/X), g(X)) = (-1)^{mn} b_0^n \prod_{j=1}^{m} \beta_j^n f(Y/\beta_j)$$
$$= (-1)^{mn} a_0^m b_0^n \prod_{i=1}^{n} \prod_{j=1}^{m} (Y - \alpha_i \beta_j).$$

(4)
$$\operatorname{Res}(f(YX), g(X)) = (-1)^{mn} b_0^n \prod_{j=1}^{m} f(Y\beta_j)$$
$$= (-1)^{mn} a_0^m b_0^n \prod_{i=1}^{n} \prod_{j=1}^{m} (Y\beta_j - \alpha_i)$$
$$= a_0^m \prod_{i=1}^{n} \left((-1)^m b_0 \prod_{j=1}^{m} \beta_j \right) \prod_{j=1}^{m} (Y - \alpha_i/\beta_j)$$
$$= a_0^m \prod_{i=1}^{n} b_m \prod_{j=1}^{m} (Y - \alpha_i/\beta_j).$$

♄

7
Galois I: Finite Fields

> Je vais répondre à quelques erreurs de l'accusateur public. Il m'a d'abord objecté mes réponses dans l'instruction et l'omission du correctif 's'il trahit'. Je dois dire que j'ai mieux aimé céder au voeu du juge d'instruction, que de m'exposer à rester trois au quatre mois en prison. J'avoue d'ailleurs qu'il y a eu peut-être un peu de malice dans mon fait; vous ne vous figurez pas la joie du commissaire de police, quand'il a cru avoir découvert en moi un conspirateur. Peu s'en est fallu qu'il n'ait cru sa fortune faite; il doit être un peu détrompé.
>
> E. Galois [1]

In this chapter I would like to discuss the consequences on finite fields of Kronecker's theory.

The most important result is the possibility of reaching a complete taxonomy of finite fields: for each power of a prime $q = p^n$ there is a unique field $GF(q)$, called the *Galois field*, with cardinality q, and there are no other finite fields except these (Section 7.1).

This characterization of the Galois fields allows us to easily describe all their splitting fields; it becomes elementary to prove that $GF(q^e)$ is the splitting field of each irreducible polynomial $f(X) \in GF(q)[X]$ such that $\deg(f) = e$ (Section 7.2).

---

[1] This declaration was made by Galois during the trial in which he was charged for having toasted Louis Philippe showing a dagger.
It is published in the *Gazette des Tribunaux*, 1822, 16 June 1831 and I took it from L. Toti Rigatelli, *Evariste Galois*, Birkhäuser.
This biography advances the theory that Galois himself set up his duel and his death in the hope that this apparent political killing of an outstanding republican could start a revolution.

From this we can deduce the factorization of $X^{q^\mu} - X$ in $GF(q)[X]$: it is the product of all the irreducible monic polynomials in $GF(q)[X]$ whose degree divides μ. This remark allows us to deduce a partial factorization for squarefree polynomials over a finite field, the *distinct degree factorization* (Section 7.3); distinct degree factorization together with distinct power decomposition permits us to deduce easily a partial factorization of a polynomial in $GF(q)[X]$, which is used as preprocessing in the factorization algorithm in $GF(q)[X]$.

In the eighteenth century, there was strong interest and research in the *roots of unity*, i.e. the solutions of the equations $g_n(X) = X^n - 1$, since they were not just an instance of but also a crucial step in root extraction. The existence of *primitive nth roots* of unity, i.e. a root of $g_n(X)$ which generates all the others, was proved by Gauss; this result leads to an alternative representation of the finite fields $GF(q)$, which we introduced as the splitting field of $X^q - X$: it is also the set of all the powers of a primitive nth root (Section 7.4).

There are therefore two possible representations in order to compute with finite fields: Kronecker's model and Gauss' suggestion of using the primitive roots as a basis for a logarithmic system; I will briefly discuss them in Section 7.5.

In the next section (Section 7.6) I will discuss the factorization of $g_n(X) = X^n - 1$ in each field, introducing the *nth cyclotomic polynomial*, which is the polynomial whose roots are the primitive nth roots of unity: clearly g_n is the product of all the dth cyclotomic polynomials, where d runs through the factors of n. Since each cyclotomic polynomial is a polynomial over a prime field, to complete the analysis of the factorization of $g_n(X) = X^n - 1$, we should restrict ourself to that case: we will prove that the cyclotomic polynomials are irreducible over $\mathbb{Q}[X]$ and we will limit ourselves to giving some structural results for prime finite fields. However, due to its important applications in computer science (mainly in error correcting codes) I will briefly discuss, through an example, how to factorize $g_n(X) = X^n - 1$ in $\mathbb{Z}_2[X]$ using idempotents (Section 7.7).

7.1 Galois Fields

First we will recall the properties we have already proved in Section 4.2.

Proposition 7.1.1. *Let F be a finite field, $\mathrm{card}(F) = q$, $p := \mathrm{char}(F)$. Then:*

$p \neq 0$ and $q = p^n$ for some n.
For all $a, b \in F$, $(a+b)^p = a^p + b^p$ and $(a+b)^q = a^q + b^q$.
For all $f, g \in F[X]$, $(f+g)^p = f^p + g^p$ and $(f+g)^q = f^q + g^q$.

7.1 Galois Fields

$a^q = a$, for all $a \in F$.

Each element a of F has a pth root, namely $b := a^{p^{n-1}}$.

$F = \{b^p : b \in F\}$ – i.e. F is perfect.

Denoting, for each $m \in \mathbb{N}$, $\Phi_m : F \mapsto F$ the map defined by $\Phi_m(a) = a^{p^m}$, then

- each Φ_m is an automorphism;
- Φ_n is the identity;
- for all $m_1, m_2 \in \mathbb{N}$, $\Phi_{m_1} = \Phi_{m_2} \iff m_1 \equiv m_2 \pmod{n}$.

$\boxed{\text{ħ}}$

We already know that a finite field has cardinality $q = p^n$ for a prime p; Kronecker's theory allows us to prove more: it guarantees that, for each prime p and each integer n, there exists a *unique* finite field of such cardinality q.

To obtain this result, let us first discuss the splitting field F of the polynomial $X^q - X \in \mathbb{Z}_p[X]$, where p is a prime and $q = p^n$.

Proposition 7.1.2. *Let p be a prime and let $q = p^n$.*
Let F be the splitting field of $X^q - X \in \mathbb{Z}_p[X]$ and let
$$R := \{\alpha \in F : \alpha^q - \alpha = 0\}.$$
Then R is a field of q elements and coincides with F.

Proof $0, 1 \in R$.
If $a, b \in R$, then
$$(a - b)^q = a^q - b^q = a - b,$$
$$(ab)^q = a^q b^q = ab,$$
so that both $a - b$ and ab are in R.
If $a \in R \setminus \{0\}$ then
$$\left(a^{-1}\right)^q = \left(a^q\right)^{-1} = a^{-1},$$
so that $a^{-1} \in R$.
Therefore R is a field; since R contains, by its definition, all roots of $X^q - X$, it must coincide with F.
Since $D(X^q - X) = -1$, then all the roots of $X^q - X$ are simple and so there are q of them. $\boxed{\text{ħ}}$

Theorem 7.1.3. *For each prime number p, and for each $n \in \mathbb{N}$, there is a unique (up to isomorphism) finite field $GF(q)$ of cardinality $q = p^n$, which is called the* **Galois field** *with q elements.*

Such a field is the splitting field of $X^q - X \in \mathbb{Z}_p[X]$ and is the set of its roots.

Proof In Proposition 7.1.2 we have proved that F, the splitting field of
$$X^q - X \in \mathbb{Z}_p[X],$$
has q elements.

We now have to show that any field K with q elements is isomorphic with F. If K is a field of cardinality q, then, by the Little Fermat Theorem, for all $\alpha \in K$, $\alpha^q = \alpha$, so the q elements of K are all roots (and so all the roots) of $X^q - X$, and therefore K is a splitting field of $X^q - X$, whence the result by Theorem 5.5.6. ♄

Remark 7.1.4. This result gives a complete characterization of all finite fields: for each integer q which is a power of a prime integer there is a unique field, $GF(q)$, whose cardinality is q; for any other integer n, there is no field with such cardinality.

Before discussing the relation between the Galois fields and the splitting fields of the polynomials $f(X) \in \mathbb{Z}_p[X]$, we need first to answer a natural question: what is the relation between all the fields $GF(p^n)$?

The answer can easily be deduced by either of the following elementary results:

Lemma 7.1.5. *Let $s = p^d$, $r = s^e$. Then*

(1) $X^s - X$ divides $X^r - X$;
(2) let K be a field and let $\alpha \in K$ be such that $\alpha^s = \alpha$; then $\alpha^r = \alpha$.

Proof

(1) Since
$$r - 1 = s^e - 1 = (s-1) \sum_{i=0}^{e-1} s^i,$$
setting $m := \sum_{i=0}^{e-1} s^i$ we have
$$X^{r-1} - 1 = \left(X^{s-1}\right)^m - 1 = (X^{s-1} - 1) \sum_{i=0}^{m-1} X^{(s-1)i}.$$

(2) From $\alpha^s = \alpha$ in K, we deduce
$$\alpha^{s^2} = (\alpha^s)^s = \alpha^s = \alpha.$$
So we can easily derive inductively that $\alpha^{s^e} = \alpha$, for all $e \geq 1$:
$$\alpha^{s^e} = \left(\alpha^{s^{e-1}}\right)^s = \alpha^s = \alpha;$$
therefore $\alpha^r = \alpha^{s^e} = \alpha$. ♮

Corollary 7.1.6. *Let $r = p^m$, $s = p^d$. Then*
$$GF(s) \subset GF(r) \iff d \text{ divides } m.$$

Proof If d divides m, then $m = de$ and $r = p^{de} = s^e$. By the above Lemma, we deduce that each element of $GF(s)$, being a root of $X^s - X$, is a root of $X^r - X$ and so an element in $GF(r)$; therefore $GF(s) \subset GF(r)$.
Conversely, if $GF(s) \subset GF(r)$, then $GF(r)$ is a $GF(s)$-vector space so r is a power of s, $p^m = r = s^e = p^{de}$, $m = de$. ♮

7.2 Roots of Polynomials over Finite Fields

As a consequence of Theorem 7.1.3, denoting
$$s := p^d, \quad m = ed, \quad r = p^m = s^e,$$
we have

Corollary 7.2.1. *If $f(X) \in GF(s)[X]$ is irreducible, $\deg(f) = e$, then the field $GF(s)[X]/f(X)$ is isomorphic to $GF(r)$, which, therefore, contains one root of f.* ♮

But we will soon even show that $GF(r)$ contains *all* the roots of *all* irreducible polynomials in $GF(s)[X]$ of degree e, and in particular ($s = p$, $r = q = p^n$) $GF(q)$ contains all the roots of all polynomials in $\mathbb{Z}_p[X]$ of degree n.

Theorem 7.2.2. *Let $q := p^n$.*
Then $X^{q^\mu} - X \in GF(q)[X]$ is the product of all monic irreducible polynomials in $GF(q)[X]$ whose degree divides μ; each such polynomial is a simple factor of $X^{q^\mu} - X$ in $GF(q)[X]$.

Proof Let $f(X) \in GF(q)[X]$ be a monic irreducible polynomial, such that $\delta := \deg(f)$ divides μ. Let $K := GF(q)[X]/f(X)$, $\pi : GF(q)[X] \mapsto K$ be the canonical projection, and let $\alpha = \pi(X)$, so that f is the minimal polynomial of α.

Let e be such that $\mu = \delta e$, and let $m := n\mu$, $d := n\delta$,
$$s := q^\delta = p^d, \quad r := q^\mu = p^m$$
so that
$$r = p^m = p^{de} = s^e.$$
Since K has cardinality s, we have $\alpha^s = \alpha$ and therefore, by Lemma 7.1.5.(2), $\alpha^r = \alpha$. So α is a root of $X^{q^\mu} - X$, which therefore is a multiple of f.

Conversely let f be a monic irreducible factor of $X^{q^\mu} - X$ in $GF(q)[X]$ and let $\delta := \deg(f)$. Let $K := GF(q)[X]/f(X)$ which has q^δ elements, $\pi : GF(q)[X] \mapsto K$ be the canonical projection, and let $\alpha = \pi(X)$. Then α is a root of $X^{q^\mu} - X$. So $K = GF(q)[\alpha] \subset GF(q^\mu)$. Then
$$\mu = [GF(q^\mu) : GF(q)] = [GF(q^\mu) : K][K : GF(q)] = [GF(q^\mu) : K]\delta,$$
i.e. δ divides μ. ♄

Corollary 7.2.3. *Let $f(X) \in GF(q)[X]$ be a monic irreducible polynomial in $GF(q)[X]$, $\deg(f) = \mu$.*

Then $GF(q)[X]/f(X)$ is the splitting field of f and is isomorphic to $GF(q^\mu)$.

Proof $K := GF(q)[X]/f(X)$ is a $GF(q)$-vector space of dimension μ, so it has q^μ elements. It contains a root of f, so is contained in its splitting field, which in turn is contained in the splitting field of $X^{q^\mu} - X$, i.e. $GF(q^\mu)$. Since both have q^μ elements they coincide. ♄

Corollary 7.2.4. *$GF(q^\mu)$ contains all the roots of all the polynomials of $GF(q)[X]$ whose degree divides μ.* ♄

Finally we can describe the structure of an irreducible polynomial over a finite field: let $f(X) \in GF(q)[X]$ be a monic irreducible polynomial in $GF(q)[X]$, $\deg(f) = \mu$, so that all its roots live in $GF(q^\mu)$. In this setting we have

Lemma 7.2.5. *With the notation above, let $\alpha \in GF(q^\mu)$ be a root of f. Then α^q is a root of f.*

7.3 Distinct Degree Factorization

Proof
Let $f(X) := \sum_{i=0}^{\mu} a_i X^i$; then

$$f(\alpha^q) = \sum_{i=0}^{\mu} a_i (\alpha^q)^i = \sum_{i=0}^{\mu} (a_i)^q (\alpha^i)^q = \left(\sum_{i=0}^{\mu} a_i \alpha^i \right)^q = f(\alpha)^q = 0.$$

♄

Corollary 7.2.6. *Let $f(X) \in GF(q)[X]$ be a monic irreducible polynomial in $GF(q)[X]$, $\deg(f) = \mu$, and let $\alpha \in GF(q^\mu)$ be a root of f.*
Then the set of the roots of f is exactly

$$\{\alpha, \alpha^q, \alpha^{q^2}, \ldots, \alpha^{q^i}, \ldots, \alpha^{q^{\mu-1}}\}$$

and all of them are simple.

Proof We only have to prove that for $i, j \in \{1, \ldots, \mu\}$

$$\alpha^{q^i} = \alpha^{q^j} \implies i = j.$$

In fact if $\alpha^{q^i} = \alpha^{q^j}$ we have

$$\alpha^{q^{\mu+i-j}} = \left(\alpha^{q^i}\right)^{q^{\mu-j}} = \left(\alpha^{q^j}\right)^{q^{\mu-j}} = \alpha^{q^\mu} = \alpha;$$

therefore f divides $X^{q^{\mu+i-j}} - X$ in $GF(q)[X]$, and μ divides $\mu + i - j$ by Theorem 7.2.2, hence $i = j$. ♄

7.3 Distinct Degree Factorization

Theorem 7.2.2 suggests the introduction of a partial polynomial factorization over finite fields, which is a tool in the polynomial factorization algorithm:

Proposition 7.3.1. *Let k be a finite field and let $f(X) \in k[X]$ be a squarefree polynomial. Then there are unique (up to associates) polynomials $g_1, \ldots, g_s \in k[X]$ such that*

(1) $f = g_1 g_2 \cdots g_s$;
(2) g_i is the product of irreducible polynomials of degree i. ♄

Definition 7.3.2. *We will call the factorization*

$$f = g_1 g_2 \cdots g_s$$

Fig. 7.1. Distinct Degree Factorization Algorithm

$[g_1, \ldots, g_s] := \text{\bf DistinctDegreeFactorization}(f)$
where
  k is a finite field of characteristic p
  $f \in k[X]$ is squarefree
  $f = g_1 \ldots g_s$ is the distinct degree factorization of f
$d := 1$
Repeat
  $q := p^d$
  $g_d := \gcd(f, X^q - X)$
  $f := \text{\bf Quot}(f, g_d)$
  $d := d + 1$
until $f = 1$

---

of a squarefree polynomial $f(X) \in k[X]$, where k is a finite field, the distinct degree factorization *of f.*

Algorithm 7.3.3. Because of Theorem 7.2.2, the algorithm in Figure 7.1 will clearly compute the distinct degree factorization of a squarefree polynomial $f(X) \in k[X]$, where k is a finite field.

The requirement that f is squarefree is not an essential restriction: given a polynomial f, it is sufficient to apply first a distinct power factorization and then we can apply a distinct degree factorization to each factor.

Correctness of the algorithm is based on the fact that at the beginning of each **Repeat** loop iteration, all factors h of f such that $\deg(h) < d$ have been removed by division with g_i, $i < d$; so the factors of f have degree at least d, while the ones of $X^q - X$ have degree at most d; the common factors, collected in g_d, are therefore the factors of f with degree exactly d.

Algorithm 7.3.4. The algorithm can be improved as follows: computing either $\gcd(f, X^q - X)$ or $\gcd(f, \text{\bf Rem}(X^q - X, f))$ is obviously the same. The latter computation requires one polynomial division less, provided we have a faster way of computing $\text{\bf Rem}(X^q - X, f)$ than actually performing the division.

Such an algorithm can be obtained, since, denoting

$$R_d(X) := \text{\bf Rem}(X^{p^d}, f), r_j(X) := \text{\bf Rem}(X^{pj}, f),$$

the following holds:

$R_d(X) := \text{\bf Rem}(R_{d-1}(X^p), f),$
$R(X) = \sum_j a_j X^j \implies \text{\bf Rem}(R(X^p), f) = \sum_j a_j r_j(X),$
$\text{\bf Rem}(X^{p^d} - X, f) = \text{\bf Rem}(f, R^d - X),$
$r_{j+1} = \text{\bf Rem}(r_j r_1, f).$

Fig. 7.2. Distinct Degree Factorization Algorithm

$[g_1, \ldots, g_s] := \textbf{DistinctDegreeFactorization}(f)$
where
    k is a finite field of characteristic p
    $f \in k[X]$ is squarefree
    $f = g_1 \ldots g_s$ is the distinct degree factorization of f
$r_1 := \textbf{Rem}(X^p, f)$
$r := r_1$
$d := 1$
While $d < \deg(f)$ **do**
    $r := r_d r_1$
    $d := d + 1$
    $r_d := \textbf{Rem}(r, f)$
$R := r_1$
$d := 1$
Repeat
    $g_d := \gcd(f, R - X)$
    $f := \textbf{Quot}(f, g_d)$
    $d := d + 1$
    $R(X) := \textbf{Rem}(R(X^p), f)$
until $f = 1$

Therefore, it is sufficient to precompute r_j, for all $i < \deg(f)$, and then, for each d, substitute the computation of $\gcd(f, X^{p^d} - X)$ with the ones of $R_d(X) := \textbf{Rem}(R_{d-1}(X^p), f)$, and $\textbf{Rem}(f, R^d - X)$.

The difference is that computing directly $\textbf{Rem}(X^{p^d} - X, f)$ can cost up to $\mathcal{O}\left(p^d \deg(f)\right)$ arithmetical operations, while computing both R_d and $\textbf{Rem}(f, R_d - X)$ costs $\mathcal{O}\left(\deg(f)^2\right)$ arithmetical operations, with a definite computational advantage.

The improved version is presented in Figure 7.2.

7.4 Roots of Unity and Primitive Roots

Let K be a field.

Definition 7.4.1. *For each $n \in \mathbb{N}$, the splitting field $K^{(n)}$ over K of the polynomial $g_n(X) := X^n - 1 \in K[X]$ is called the nth* cyclotomic field *over K. Each root of $X^n - 1$ in $K^{(n)}$ is called an nth* root of unity.

Let $\mathcal{R}^{(n)} \subset K^{(n)}$ be the set of all the nth roots of unity.

Remark 7.4.2. Let $p := \text{char}(K) \neq 0$ be a factor of n and let r, e be such that

$$n = rp^e, \quad \gcd(r, p) = 1.$$

Then
$$X^n - 1 = (X^r - 1)^{p^e}.$$

As a consequence, throughout this section we will implicitly restrict ourselves to the case in which $\operatorname{char}(K)$ does not divide n.

Definition 7.4.3. *Let $\alpha \in \mathcal{R}^{(n)}$. The order of α is*
$$\operatorname{ord}(\alpha) := \min\{\nu \in \mathbb{N} : \alpha^\nu = 1\}.$$

Lemma 7.4.4. *We have*

(1) $\mathcal{R}^{(n)}$ is a multiplicative group;
(2) $\operatorname{card}(\mathcal{R}^{(n)}) = n$;
(3) for all $\alpha \in \mathcal{R}^{(n)}$, $\alpha^a = 1 \iff \operatorname{ord}(\alpha)$ divides a;
(4) for all $\alpha \in \mathcal{R}^{(n)}$, $\operatorname{ord}(\alpha)$ divides n;
(5) for all $\alpha \in \mathcal{R}^{(n)}$, $\alpha^a = \alpha^b = 1, c = \gcd(a, b) \implies \alpha^c = 1$;
(6) for all $\alpha, \beta \in \mathcal{R}^{(n)}$, $\operatorname{ord}(\alpha\beta) = \operatorname{lcm}(\operatorname{ord}(\alpha), \operatorname{ord}(\beta))$;
(7) for all $\alpha \in \mathcal{R}^{(n)}$, $\operatorname{ord}(\alpha^k) = \frac{\operatorname{ord}(\alpha)}{\gcd(k, \operatorname{ord}(\alpha))}$.

Proof

(1) It is sufficient to show that $\alpha\beta^{-1} \in \mathcal{R}^{(n)}$, for each $\alpha, \beta \in \mathcal{R}^{(n)}$:
$$\left(\alpha\beta^{-1}\right)^n = \alpha^n(\beta^n)^{-1} = 1.$$

(2) $D(X^n - 1) = nX^{n-1} \neq 0$ since $\operatorname{char}(K)$ does not divide n. Therefore $X^n - 1$ has only simple roots and $\operatorname{card}(\mathcal{R}^{(n)}) = n$.

(3) Let q, r be such that $a = \operatorname{ord}(\alpha)q + r, r < \operatorname{ord}(\alpha)$. Then
$$1 = \alpha^a = \left(\alpha^{\operatorname{ord}(\alpha)}\right)^q \alpha^r = \alpha^r,$$
contradicting the minimality of $a := \operatorname{ord}(\alpha)$.

(4) This follows from the above result since $\alpha^n = 1$.

(5) There are $d, e \in \mathbb{N}$ such that $c = ad + be$. Therefore
$$\alpha^c = \left(\alpha^a\right)^d \left(\alpha^b\right)^e = 1.$$

(6) Let $c, d, e \in \mathbb{N}$ be such that
$$c = \operatorname{lcm}(\operatorname{ord}(\alpha), \operatorname{ord}(\beta)) = d\operatorname{ord}(\alpha) = e\operatorname{ord}(\beta).$$
Then
$$(\alpha\beta)^c = \left(\alpha^{\operatorname{ord}(\alpha)}\right)^d \left(\beta^{\operatorname{ord}(\beta)}\right)^e = 1.$$

7.4 Roots of Unity and Primitive Roots

Conversely, if
$$c := \operatorname{ord}(\alpha\beta) \neq \operatorname{lcm}(\operatorname{ord}(\alpha), \operatorname{ord}(\beta)),$$
there is d such that $C = dc$ is a multiple of $\operatorname{ord}(\alpha)$ and not of $\operatorname{ord}(\beta)$ (or conversely); then
$$1 = (\alpha\beta)^C = (\alpha)^C(\beta)^C = (\beta)^C$$
giving a contradiction.

(7) Let $a = \operatorname{ord}(\alpha)$ and $c = \gcd(k, a)$. Then
$$(\alpha^k)^{a/c} = (\alpha^a)^{k/c} = 1.$$
On the other hand, if m is such that $(\alpha^k)^m = 1$, then km is a multiple of $a = c\frac{a}{c}$ and, since $\gcd(\frac{a}{c}, k) = 1$, $\frac{a}{c}$ divides m.
So $a/c = \operatorname{ord}(\alpha^k)$.

$\square$

Theorem 7.4.5 (Primitive Root Theorem).
There is $\alpha \in \mathcal{R}^{(n)}$, such that

$\operatorname{ord}(\alpha) = n$,
for all $\beta \in \mathcal{R}^{(n)}$, β is a power of α.

Such an α is called a primitive nth root of unity.

Proof If we show the existence[2] of an element $\alpha \in \mathcal{R}^{(n)}$ such that $\operatorname{ord}(\alpha) = n$, the general result then follows easily, since the powers α^i, $0 \leq i \leq n-1$, are distinct and so coincide with the n elements of $\mathcal{R}^{(n)}$.
To prove the existence of such α, let
$$n = p_1^{e_1} \cdots p_h^{e_h}$$
be a factorization in prime factors and let
$$q_i := \frac{n}{p_i}, \quad r_i := \frac{n}{p_i^{e_i}}.$$
Then, for all i, there exists $\beta_i \in \mathcal{R}^{(n)}$ such that $\beta_i^{q_i} \neq 1$, since the q_ith roots of unity are at most $q_i < n$.
Let $\gamma_i := \beta_i^{r_i}$; then $\operatorname{ord}(\gamma_i) = p_i^{e_i}$; in fact
$$\gamma_i^{p_i^{e_i}} = \beta_i^{p_i^{e_i} r_i} = \beta_i^n = 1,$$

---

[2] As an example of the construction implied by this result, the reader is directed to Example 7.4.10.

while
$$1 = \gamma_i^{p_i^{e_i-1}} = \beta_i^{n/p_i} = \beta_i^{q_i}$$
would contradict the assumption $\beta_i^{q_i} \neq 1$.
Then $\alpha := \prod_i \gamma_i$ is such that $\operatorname{ord}(\alpha) = \prod_i p_i^{e_i}$. ♮

Corollary 7.4.6. *m divides n $\iff \mathcal{R}^{(m)} \subset \mathcal{R}^{(n)}$;*

Proof If $n = dm$ and $\gamma \in \mathcal{R}^{(m)}$, then $\gamma^n = (\gamma^m)^d = 1$.
Conversely, let $\alpha \in \mathcal{R}^{(m)} \subset \mathcal{R}^{(n)}$ be a primitive mth root of unity; then $m = \operatorname{ord}(\alpha)$ divides n. ♮

Corollary 7.4.7. *For each prime p there is a* primitive root *or* generator[3] $g \in \mathbb{Z}_p \setminus \{0\}$ *such that*

for all $h \in \mathbb{Z}_p \setminus \{0\}$, there exists $a \in \mathbb{Z}_{p-1} : h = g^a$ in $\mathbb{Z}_p$.

Proof If $K = \mathbb{Z}_p$ we have $K \setminus \{0\} = K^{(p-1)} \setminus \{0\} = \mathcal{R}^{(p-1)}$. ♮

Historical Remark 7.4.8. While this result was probably well known in the eighteenth century, Gauss – who claimed to have given the first rigorous proof, the same proof we used in Theorem 7.4.5 – surely has the merit of having pointed out the main significance of this result: it allows us to compute with logarithms in $\mathbb{Z}_p$: multiplication in $\mathbb{Z}_p \setminus \{0\} = \mathcal{R}^{(p-1)}$ is transformed to addition in $\mathbb{Z}_{p-1}$. As Gauss put it in *Disquisitiones Arithmeticae*, Section 57:

Insignis haec proprietas permagnae est utilitatis, operationesque arithmeticas, ad congruentias pertinentes, haud parum sublevare potest, simili fere modo, ut logarithmorum introductio operationes arithemeticae vulgaris.

The interpretation of Theorem 7.4.5 in terms of prime fields given by Corollary 7.4.7 applies, of course, to any finite field, giving a different characterization of them than the one implied by Theorem 7.1.3:

Corollary 7.4.9. *Let $GF(q)$ be a finite field. Then there is $\alpha \in GF(q)$ such that, for all $\beta \in GF(q) \setminus \{0\}$, β is a power of α.*
Such an α is called a primitive element.

---

[3] The term originally introduced by Euler is 'primitive root'; to avoid confusion with other notions which stem from Gauss and have the term 'primitive elements', it is usual to use 'generator' for Euler's concept.

7.4 Roots of Unity and Primitive Roots

Proof In fact $GF(q) \setminus \{0\} = \mathcal{R}^{(q-1)}$ over its prime field. ♄

Example 7.4.10. To give an example of the construction contained in the proof of Theorem 7.4.5, let us consider a representation of $GF(64)$ whose prime field is $k_0 := \mathbb{Z}_2$. To build it, let us note that

$$f(X) = X^3 + X + 1 \in k_0[X]$$

is irreducible – since $f(x) \neq 0$, for all $x \in \mathbb{Z}_2$ – so we obtain the field

$$GF(8) = k_1 = k_0[X]/f(X) = k_0[\delta] \text{ where } \delta^3 + \delta + 1 = 0.$$

Let us now consider

$$g(X) = X^2 + X + 1 \in k_1[X]$$

which is irreducible, since $g(x) \neq 0$, for all $x \in k_1$, so that we obtain

$$GF(64) = k_1[X]/g(X) = k_1[\epsilon] = k_0[\delta, \epsilon] \text{ where } \epsilon^2 + \epsilon + 1 = 0.$$

We have $q - 1 = 63 = 3^2 7$ and our task is to find β_1 and β_2 such that $\beta_1^{21} \neq 1$ and $\beta_2^9 \neq 1$.

Note that, since $\delta^7 = 1$, we have $\delta^9 = \delta^2 \neq 1$; also, setting $\gamma := \delta + \epsilon$ we have

$$\begin{array}{rclclcl}
\gamma^2 & & & = & \delta^2 + \epsilon^2 & = & \delta^2 + \epsilon + 1, \\
\gamma^4 & = & (\gamma^2)^2 & = & \delta^4 + \epsilon^2 + 1 & = & \delta^2 + \delta + \epsilon, \\
\gamma^8 & = & (\gamma^4)^2 & = & \delta^4 + \delta^2 + \epsilon^2 & = & \delta + \epsilon + 1, \\
\gamma^{16} & = & (\gamma^8)^2 & = & \delta^2 + \epsilon^2 + 1 & = & \delta^2 + \epsilon, \\
\gamma^{20} & = & \gamma^{16}\gamma^4 & = & \delta^4 + \delta^3 + \delta\epsilon + \epsilon^2 & = & \delta^2 + \delta\epsilon + \epsilon, \\
\gamma^{21} & = & \gamma^{20}\gamma & = & \delta^3 + \delta\epsilon^2 + \delta\epsilon + \epsilon^2 & = & \epsilon,
\end{array}$$

we can take $\beta_1 := \epsilon + \delta$, $\beta_2 := \delta$ and so

$$\alpha = \beta_1^7 \beta_2^9 = \delta^2 \epsilon + \delta^2 + \delta\epsilon + 1.$$

Corollary 7.4.11. *Let $r = p^m$, d be a factor of m, thus $m = de$, and $s := p^d$. Then $GF(r)$ is a simple algebraic extension of $GF(s)$.*

In particular ($r := q, s := p$) $GF(q)$ is a simple algebraic extension of $\mathbb{Z}_p$.

Proof Let α be a primitive element of $GF(r)$ and let

$$K := GF(s)[\alpha] \subseteq GF(r).$$

Since for all $\beta \in GF(r)$, there exists $a : \beta = \alpha^a$, then $GF(r) = GF(s)[\alpha]$. ♄

132   Galois I: Finite Fields

Theorem 7.4.5 also has an important consequence, which is related to many applications of finite fields:

Corollary 7.4.12. *For each $s := p^d$ and $e \in \mathbb{N}$ there is an irreducible polynomial in $GF(s)[X]$ of degree e.*

In particular ($d := 1, e := m$) there is an irreducible polynomial in $\mathbb{Z}_p[X]$ of degree m, for all $m \in \mathbb{N}$.

Proof Let $m := de$ and $r = p^m$. Let α be a primitive element of $GF(r)$. Since $GF(r) = GF(s)[\alpha]$, let f be the minimal polynomial of α over $GF(s)$; then $GF(r) = GF(s)[\alpha] = GF(s)[X]/f(X)$.
Since $[GF(r) : GF(s)] = e$, then $\deg(f) = e$. ♄

Example 7.4.13. In order to compute the minimal polynomial of α over $GF(s)$, we only have to find the linear dependence between $1, \alpha, \alpha^2, \ldots, \alpha^e$ (cf. Section 8.3.8).

Let us apply this technique to Example 7.4.10 in order to find the minimal polynomial of $\alpha \in GF(64)$ and therefore an irreducible polynomial of degree 6 in $\mathbb{Z}_2$. Here is the computation:

$$
\begin{aligned}
\alpha^0 &= +1 \\
\alpha^1 &= +1 &&&& +\epsilon\delta && +\delta^2 && +\epsilon\delta^2 \\
\alpha^2 &= +1 &&&& +\epsilon\delta && +\delta^2 \\
\alpha^3 &= && +\epsilon &&&&&& +\epsilon\delta^2 \\
\alpha^4 &= +1 && && +\delta &&&&& +\epsilon\delta^2 \\
\alpha^5 &= +1 &&&& +\epsilon\delta &&&&& +\epsilon\delta^2 \\
\alpha^6 &= +1 && +\epsilon && +\delta && +\epsilon\delta && +\delta^2 && +\epsilon\delta^2
\end{aligned}
$$

from which we deduce that the polynomial

$$h(X) := X^6 + X^4 + X^3 + X + 1 \in \mathbb{Z}_2[X]$$

is irreducible and is the minimal polynomial of α so that $GF(64)$ has the representation

$$GF(64) = \mathbb{Z}_2[X]/h(X) = \mathbb{Z}_2[\alpha] \text{ where } \alpha^6 + \alpha^4 + \alpha^3 + \alpha + 1 = 0.$$

Note also that

$$\delta = \alpha^4 + \alpha^2 + \alpha + 1, \quad \epsilon = \alpha^3 + \alpha^2 + \alpha.$$

It is then easy to verify that f and g are the minimal polynomials of δ and ϵ respectively, by computing in $\mathbb{Z}_2[\alpha]$: for instance, while of course $\{1, \epsilon\}$ are linearly independent, since $\epsilon^2 = \alpha^6 + \alpha^4 + \alpha^2 = \alpha^3 + \alpha^2 + \alpha + 1$, $\{1, \epsilon, \epsilon^2\}$ are linearly dependent satisfying $1 + \epsilon + \epsilon^2$.

7.5 Representation and Arithmetics of Finite Fields

By the results given above, we have at least two ways of representing a finite field $GF(q)$, $q = p^n$:

♭ The first way is via **Kronecker's model**, producing an irreducible polynomial $f(X) \in \mathbb{Z}_p[X]$ of degree n, whose existence is guaranteed by Corollary 7.4.12, and using the representation

$$GF(q) \cong \mathbb{Z}_p[X]/f(X) = \mathbb{Z}_p[\alpha].$$

♯ By the **Primitive Root Theorem** we know that there exists a primitive element ξ of $GF(q)$, so that

$$GF(q) = \left\{ \xi^i : i \in \{1, \ldots, p^n - 1\} \right\} \cup \{0\}.$$

Since the Theorem does not give any hint of how to 'find' such a ξ, we should discuss how to 'find' it, but, of course, we should first understand what 'finding' means in this context.

Remark 7.5.1. As we pointed out in 7.4.8, the main advantage of this theorem is that it allows us to compute with logarithms in $GF(q)$.

In fact let us assume we are given two elements $\beta, \gamma \in GF(q) \setminus \{0\}$

♭ in Kronecker's model, where they are represented by two polynomials $g_\beta(X), g_\gamma(X) \in \mathbb{Z}_p[X]$ such that

$$g_\beta(\alpha) = \beta, \quad g_\gamma(\alpha) = \gamma;$$

♯ via a primitive root, through two indices $\text{ind}(\beta)$, $\text{ind}(\gamma)$ such that

$$\beta = \xi^{\text{ind}(\beta)}, \quad \gamma = \xi^{\text{ind}(\gamma)},$$

and that we want to compute their product $\delta := \beta\gamma$:

♭ in Kronecker's model, we have to compute, by the Division Algorithm, the polynomial

$$g_\delta := \mathbf{Rem}(g_\beta g_\gamma, f),$$

knowing that it satisfies the relation $g_\delta(\alpha) = \delta$.

♯ via a primitive root, we only have to compute

$$\text{ind}(\delta) = \text{ind}(\beta) + \text{ind}(\gamma),$$

since

$$\xi^{\text{ind}(\delta)} = \xi^{\text{ind}(\beta)}\xi^{\text{ind}(\gamma)} = \beta\gamma = \delta.$$

It is clear that the computation is much easier via a primitive root representation. However, if we need to compute $\epsilon := \beta + \gamma$,

♭ in Kronecker's model, we only have to compute the polynomial

$$g_\epsilon := g_\beta + g_\gamma$$

which obviously satisfies the relation $g_\epsilon(\alpha) = \epsilon$.

♯ via a primitive root, we have essentially no other way than computing $\xi^{\text{ind}(\beta)} + \xi^{\text{ind}(\gamma)}$ and apparently this is possible only if we know the minimal polynomial of ξ.

The discussion above clarifies what we meant by 'finding' a primitive root ξ: to represent it via its minimal polynomial[4] $f(X) \in \mathbb{Z}_p$.

What is more relevant is that, according to Theorem 7.6.14,

$$GF(q) \cong \mathbb{Z}_p[X]/f(X) = \mathbb{Z}_p[\xi],$$

so that the two representations discussed here coincide.

Therefore, what we need to do is just (according to Gauss' suggestion) build a logarithmic table, which returns the representation $r(i) := r_i(\xi)$ of ξ^i, for all $i \in \mathbb{Z}_{q-1}$, where

$$r_i(X) = \textbf{Rem}(X^i, f) \in \mathbb{Z}_p[X]/f(X).$$

This is obtained by a computation similar to the one discussed in Algorithm 7.3.4, i.e. via the formula

$$r_i(X) = \textbf{Rem}(Xr_{i-1}(X), f);$$

in this way we obtain a biunivocal correspondence

$$r : \mathbb{Z}_{q-1} \mapsto GF(q) \setminus \{0\}.$$

With a classical abuse of notation, which we will need in Remark 7.5.3, let us associate a 'logarithmic value' to $0 \in GF(q)$ by introducing a symbol $\star$ and generalizing the additive operation of $\mathbb{Z}_{q-1}$ to $\mathbb{Z}_{q-1} \cup \{\star\}$ by the rules

$$\star = \star + \star = a + \star = \star + a, \text{ for all } a \in \mathbb{Z}_{q-1};$$

---

[4] We will show in Section 7.6 that f can be obtained by factorizing the cyclotomic polynomial $\Phi_n(X) \in \mathbb{Z}_p$. Any such factor is a minimal polynomial of a primitive root.

Table 7.1. *Logarithmic table for $GF(16)$*

i	r(i)	i	r(i)
1	ξ	$\star$	0
2	ξ^2	0	1
3	ξ^3	14	$\xi^3 + 1$
4	$\xi + 1$	13	$\xi^3 + \xi^2 + 1$
5	$\xi^2 + \xi$	12	$\xi^3 + \xi^2 + \xi + 1$
6	$\xi^3 + \xi^2$	11	$\xi^3 + \xi^2 + \xi$
7	$\xi^3 + \xi + 1$	10	$\xi^2 + \xi + 1$
8	$\xi^2 + 1$	9	$\xi^3 + \xi$

then we can generalize $r(\cdot)$ to a biunivocal correspondence

$$r : \mathbb{Z}_{q-1} \cup \{\star\} \mapsto GF(q)$$

by putting $r(\star) = 0$ and, continuing the abuse of notation, $\xi^\star = 0$, $\xi^0 := \xi^{q-1} = 1$.

Example 7.5.2. Let us build a logarithmic table for $GF(16)$: choosing as irreducible polynomial $f(X) := X^4 + X + 1$ we get the logarithmic table of Table. 7.1.

Remark 7.5.3. This logarithmic table allows us to perform direct sums via Kronecker's model and indirect products via the table.

By symmetry we expect that we could perform direct products via primitive root representation and indirect sums by a suitable table.

In fact the *Zech logarithm function* $\mathcal{Z} : \mathbb{Z}_{q-1} \cup \{\star\} \mapsto \mathbb{Z}_{q-1} \cup \{\star\}$ satisfies the relation

$$\xi^{\mathcal{Z}(i)} = \xi^i + 1$$

and therefore reduces sum computation to table consultation via

$$\xi^i + \xi^j = \xi^i(\xi^{j-i} + 1) = \xi^i \xi^{\mathcal{Z}(j-i)} = \xi^{i+\mathcal{Z}(j-i)}.$$

The Zech table (7.2) is simply built by consulting the logarithmic table 7.1.

7.6 Cyclotomic Polynomials

In this section we will use the same setting and notation of Section 7.4; therefore K is a field; for each $n \in \mathbb{N}$ we denote $g_n(X) := X^n - 1 \in K[X]$; $K^{(n)}$ is the nth cyclotomic field, i.e. the splitting field over K of $g_n(X)$; and $\mathcal{R}^{(n)} \subset K^{(n)}$ is the set of all the nth roots of unity. Moreover $\mathcal{S}^{(n)} \subset \mathcal{R}^{(n)}$

Table 7.2. *Zech table for $GF(16)$*

i	Z(i)	i	Z(i)
1	4	⋆	0
2	8	0	⋆
3	14	14	3
4	1	13	6
5	10	12	11
6	13	11	12
7	9	10	5
8	2	9	7

will denote the set of the primitive nth roots of unity and we introduce the following

Definition 7.6.1. *The polynomial*

$$\Phi_n := \prod_{\alpha \in \mathcal{S}^{(n)}} X - \alpha \in K^{(n)}[X]$$

is called the nth cyclotomic polynomial over K.

Let us first note the following

Definition 7.6.2. *For each $n \in \mathbb{N}$ the* Euler totient function $\phi(n)$ *is the cardinality of the set*

$$\{j \in \mathbb{N} : 1 \leq j \leq n, \gcd(n, j) = 1\},$$

which allows us to compute the cardinality of $\mathcal{S}^{(n)}$ and a partial factorization of $g_n(X)$ over K, via

Lemma 7.6.3. *Let α be a primitive nth root of unity.*
Let $i \in \{1, \ldots, n\}$ and $n = ed$ be a factorization. Then

$$\operatorname{ord}(\alpha^i) = d \iff \gcd(i, n) = e.$$

Proof We have

$$\begin{aligned}
\operatorname{ord}(\alpha^i) = d &\iff \alpha^{id} = 1, \alpha^{i\delta} \neq 1, \text{ for all } \delta, \epsilon > 1 : d = \epsilon\delta \\
&\iff n \mid id, n \nmid i\delta, \text{ for all } \delta, \epsilon > 1 : d = \epsilon\delta \\
&\iff ed \mid id, ed \nmid i\delta, \text{ for all } \delta, \epsilon > 1 : d = \epsilon\delta \\
&\iff e \mid i, e\epsilon \nmid i, \text{ for all } \delta, \epsilon > 1 : d = \epsilon\delta \\
&\iff \gcd(i, n) = e.
\end{aligned}$$

♄

7.6 Cyclotomic Polynomials

Corollary 7.6.4. *We have*

(1) $\mathcal{R}^{(n)} = \bigcup_{d\mid n} \mathcal{S}^{(d)}$.
(2) *for all* $\alpha \in \mathcal{R}^{(n)} : \alpha \in \mathcal{S}^{(n)} \iff \operatorname{ord}(\alpha) = n$.
(3) *for all* $\alpha \in \mathcal{S}^{(n)} : \alpha^j \in \mathcal{S}^{(n)} \iff \gcd(n, j) = 1$.
(4) $\operatorname{card}(\mathcal{S}^{(n)}) = \phi(n)$.

♄

Corollary 7.6.5. *For each n we have:*

$$X^n - 1 = \prod_{d\mid n} \Phi_d(X).$$

♄

Corollary 7.6.5 allows us to obtain a partial factorization of g_n in terms of the cyclotomic polynomials. Therefore, in order to factorize g_n we need to compute the cyclotomic polynomials and their factorization.

The computation of them is obtained by

Lemma 7.6.6. *If p is prime and* $\gcd(p, m) = 1$

(1) $\Phi_{mp}(X) = \Phi_m(X^p)/\Phi_m(X)$;
(2) $\Phi_{mp^e}(X) = \Phi_{mp^{e-1}}(X^p)$ *if* $e > 1$;
(3) $\Phi_{mp^e}(X) = \Phi_{mp}(X^{p^{e-1}})$.

Proof

(1) The proof being by induction, we can assume

$$\Phi_d(X)\Phi_{pd}(X) = \Phi_d(X^p), \quad \text{for all } d < m, \gcd(d, p) = 1,$$

from which, and from the obvious relation $g_{mp}(X) = g_m(X^p)$, we deduce:

$$\begin{aligned}
\Phi_{mp}(X) &= \frac{g_{mp}(X)}{\Phi_m(X) \prod_{\substack{d\mid m \\ d\neq m}} \Phi_d(X)\Phi_{pd}(X)} \\
&= \frac{\prod_{d\mid m} \Phi_d(X^p)}{\Phi_m(X) \prod_{\substack{d\mid m \\ d\neq m}} \Phi_d(X^p)} \\
&= \frac{\Phi_m(X^p)}{\Phi_m(X)}.
\end{aligned}$$

(2) The proof being by induction on m, we can assume

$$\Phi_{dp^e}(X) = \Phi_{dp^{e-1}}(X^p), \text{ for all } d < m, \gcd(d, p) = 1,$$

from which we deduce:

$$\begin{aligned}
\prod_{d \mid mp^e} \Phi_d(X) &= g_{mp^e}(X) = g_{mp^{e-1}}(X^p) = \prod_{d \mid mp^{e-1}} \Phi_d(X^p) \\
&= \Phi_{mp^{e-1}}(X^p) \prod_{d \mid mp^{e-2}} \Phi_d(X^p) \prod_{\substack{d \mid mp^{e-2} \\ d \neq mp^{e-2}}} \Phi_{dp}(X^p) \\
&= \Phi_{mp^{e-1}}(X^p) g_{mp^{e-2}}(X^p) \prod_{\substack{d \mid mp^{e-2} \\ d \neq mp^{e-2}}} \Phi_{dp^2}(X) \\
&= \Phi_{mp^{e-1}}(X^p) g_{mp^{e-1}}(X) \prod_{\substack{d \mid mp^{e-2} \\ d \neq mp^{e-2}}} \Phi_{dp^2}(X) \\
&= \Phi_{mp^{e-1}}(X^p) \prod_{\substack{d \mid mp^e \\ d \neq mp^e}} \Phi_d(X)
\end{aligned}$$

from which we deduce

$$\Phi_{mp^e}(X) = \Phi_{mp^{e-1}}(X^p).$$

(3) By iterative application of (2).

□

Remark 7.6.7. From Corollary 7.6.5 and Lemma 7.6.6, we can deduce an obvious procedure for computing Φ_n, for each n.

We illustrate it, computing Φ_{36} over $\mathbb{Z}$:

$$\begin{aligned}
\Phi_{36}(X) &= \Phi_{12}(X^3); \\
\Phi_{12}(X) &= \Phi_4(X^3)/(\Phi_4(X)); \\
\Phi_4(X) &= \Phi_2(X^2); \\
\Phi_2(X) &= g_2(X)/(X-1);
\end{aligned}$$

so that

$$\begin{aligned}
\Phi_2(X) &= X + 1; \\
\Phi_4(X) &= X^2 + 1; \\
\Phi_4(X^3)/(\Phi_4(X)) &= X^4 - X^2 + 1; \\
\Phi_{12}(X) &= X^4 - X^2 + 1; \\
\Phi_{36}(X) &= X^{12} - X^6 + 1.
\end{aligned}$$

7.6 Cyclotomic Polynomials

An elementary consequence of this procedure is that:

Corollary 7.6.8. *The cyclotomic polynomials over K are elements of*

$\mathbb{Z}_p$ *if* $\text{char}(K) = p \neq 0$;
$\mathbb{Z}$ *if* $\text{char}(K) = 0$.

♄

Having a procedure for computing the cyclotomic polynomials, we now need to study their factorization. An important result is given by

Lemma 7.6.9. *Let $\xi \in \mathbb{Q}^{(n)}$ be a primitive nth root of unity and let $f(X) \in \mathbb{Z}[X]$ be the primitive polynomial associated to the minimal polynomial of ξ. Let p be a prime such that $\gcd(p, n) = 1$, then ξ^p is a root of f.*

Proof Let $g(X) \in \mathbb{Z}[X]$ be the primitive polynomial associated to the minimal polynomial of ξ^p; our aim is to show that $g(X)$ and $f(X)$ are associate, so that ξ^p is a root of f.
Otherwise, since $X^n - 1$ is divisible by the irreducible and distinct factors f and g there is a polynomial $h(X)$ such that

$$X^n - 1 = f(X)g(X)h(X);$$

also the polynomial $G(X) := g(X^p)$ has ξ as a root so that there is $k(X)$ such that $G(X) = f(X)k(X)$.
Since f and g are primitive, we know that both are in $\mathbb{Z}[X]$; since both $X^n - 1$ and G are in $\mathbb{Z}[X]$, by the Gauss Lemma (Corollary 6.1.6) we can conclude that both h and k are in $\mathbb{Z}[X]$.
Therefore we can interpret all equalities in $\mathbb{Z}_p[X]$; to do so let $-_p : \mathbb{Z}[X] \mapsto \mathbb{Z}_p[X]$ be the projection. We have

$$f_p(X)k_p(X) = G_p(X) = g_p(X^p) = g_p^p(X)$$

so that each irreducible factor $\phi(X) \in \mathbb{Z}_p[X]$ of f_p is also a factor of g_p and, since $X^n - 1 = f_p(X)g_p(X)h_p(X)$, we know ϕ^2 is a factor of $X^n - 1$ in $\mathbb{Z}_p[X]$ and $\phi(X)$ is a factor of nX^{n-1} in $\mathbb{Z}_p[X]$.
Since $n \not\equiv 0 \pmod{p}$ we have reached an absurd result: in fact, on the one hand $\phi = X$ and on the other hand it divides $X^n - 1$. The contradiction is due to the assumption that $g(X)$ and $f(X)$ are not associate.

♄

Proposition 7.6.10. $\Phi_n \in \mathbb{Z}[X]$ *is irreducible over $\mathbb{Q}$, so that $[\mathbb{Q}^{(n)} : \mathbb{Q}] = \phi(n)$.*

Proof Let $\xi \in \mathbb{Q}^{(n)}$ be a primitive nth root of unity and let $f(X)$ be the primitive polynomial associated to the minimal polynomial of ξ; since f is irreducible, we only need to prove that each primitive root ξ^ν is a root of f, so that $\Phi_n(X) = f(X)$ is irreducible.

Each such ν is a product of (not necessarily distinct) primes, $\nu = \prod_{i=1}^m p_i$; since ξ^ν is primitive, $\gcd(n,\nu) = 1$ so that $\gcd(n, p_i) = 1$.

Setting $\nu_j := \prod_{i=1}^j p_i$, the lemma above allows us to prove that ξ^ν is a root of f by induction, since

ξ is a root of f, and

if $\xi^{\nu_{j-1}}$ is a root of f so also is $\xi^{\nu_j} = (\xi^{\nu_{j-1}})^{p_j}$.

□

Remark 7.6.11. The structure of the factorization of the cyclotomic polynomial $g_n(X)$ in the finite field $GF(q)$ is a consequence of the following discussion, where the restriction to the case $\gcd(n, q) = 1$ is justified, since, Remark 7.4.2 allows a reduction to that case.

Definition 7.6.12. *Let n, q be such that $\gcd(q, n) = 1$. The multiplicative order of q mod n, is*

$$d := \min\{k \in \mathbb{N} : q^k \equiv 1 (\bmod\ n)\}.$$

Proposition 7.6.13. *Let $\alpha \in GF(q)$ and $n := \mathrm{ord}(\alpha)$ so that $\gcd(q, n) = 1$; let d be the multiplicative order of q mod n.*

Then $\alpha^{q^d} = \alpha$ and the d elements

$$\alpha, \alpha^q, \ldots, \alpha^{q^i}, \ldots, \alpha^{q^{d-1}}$$

are distinct.

Proof In fact

$$\alpha^{q^i} = \alpha^{q^j} \iff \alpha^{q^i - q^j} = 1$$
$$\iff (q^i - q^j) \mid n$$
$$\iff q^i \equiv q^j (\bmod\ n)$$
$$\iff q^{i-j} = 1(\bmod\ n)$$
$$\iff d \mid i - j.$$

□

Theorem 7.6.14. *If $K = GF(q)$, with $\gcd(q, n) = 1$, let d be the multiplicative order of q mod n.*

If $\xi \in K^{(n)}$ is any primitive nth root of unity and $f(X)$ is its minimal polynomial, then

$$K^{(n)} = GF(q^d) \cong GF(q)[X]/f(X) = GF(q)[\xi].$$

Proof Let ξ be any primitive nth root of unity and f its minimal polynomial; by Proposition 7.6.13, the smallest field extension of $GF(q)$ containing ξ is

$$GF(q^d) \cong GF(q)[X]/f(X) = GF(q)[\xi];$$

thus $\deg(f) = d$.
Since this holds for each primitive root we deduce that $K^{(n)} = GF(q^d)$ and the result follows. ♮

Corollary 7.6.15. *If $K = GF(q)$ with $\gcd(q, n) = 1$, and d is the multiplicative order of q mod n, then Φ_n factors into $\frac{\phi(n)}{d}$ distinct monic irreducible polynomials in $K[X]$ of the same degree d, $K^{(n)}$ is the splitting field of each such factor and $[K^{(n)} : K] = d$.* ♮

Proposition 7.6.16. *The $(q-1)$th cyclotomic field over any of its subfields is $GF(q)$.*

Proof The set of all roots of $X^{q-1} - 1$ is $GF(q)$; therefore $X^{q-1} - 1$ splits in $GF(q)$ but not in any of its subfields. ♮

7.7 Cycles, Roots and Idempotents

Remark 7.7.1. Theorem 7.6.14 informs us that each factor of the vth cyclotomic polynomial over $GF(q)$ – $\gcd(v, q) = 1$ – has d roots, where d is the multiplicative order of q mod v.

Moreover Corollary 7.2.6 (cf. also Proposition 7.6.13) informs us that the roots are exactly

$$\alpha, \alpha^q, \alpha^{q^2}, \ldots, \alpha^{q^i}, \ldots, \alpha^{q^{d-1}}.$$

This leads us to consider, for each $v \in \mathbb{N}$, the permutation

$$\pi_q : \mathbb{Z}_v \mapsto \mathbb{Z}_v$$

defined by

$$\pi_q(i) := qi \,(\text{mod } v)$$

and the cycles $C_1, \ldots, C_s$ of this permutation.

In fact each of these cycles corresponds to the set of the roots of each factor of $X^\nu - 1$.

Example 7.7.2. Let us consider the case $\nu = 15, q = 2$; we know that

$GF(16)$ is the splitting field of $X^{15} - 1 \in \mathbb{Z}_2[X]$;

and is generated over $\mathbb{Z}_2[X]$ by a primitive 15th root of unity, which we will denote α;

moreover $X^{15} - 1 \in \mathbb{Z}_2[X]$ is the product of the irreducible polynomials in $\mathbb{Z}_2[X]$ of degrees 1, 2 and 4; and

it factorizes in terms of cyclotomic polynomials as:

$$X^{15} - 1 = \Phi_1(X)\Phi_3(X)\Phi_5(X)\Phi_{15}(X);$$

since the multiplicative order of 2 mod 15 is 4 we also know that $\Phi_{15}(X)$ factorizes in two factors of degree 4.

We have:

$$\begin{aligned}
\Phi_1(X) &= X + 1, \\
\Phi_3(X) &= X^2 + X + 1, \\
\Phi_5(X) &= X^4 + X^3 + X^2 + X + 1, \\
\Phi_{15}(X) &= X^8 + X^7 + X^5 + X^4 + X^3 + X + 1.
\end{aligned}$$

Before factorizing $\Phi_{15}(X)$, let us consider the cycles of π_2 in $\mathbb{Z}_{15}$ which are:

$$\begin{aligned}
\mathbf{C}_1 &= \{1, 2, 4, 8\}, \\
\mathbf{C}_2 &= \{3, 6, 9, 12\}, \\
\mathbf{C}_3 &= \{5, 10\}, \\
\mathbf{C}_4 &= \{7, 11, 13, 14\}, \\
\mathbf{C}_5 &= \{0\}.
\end{aligned}$$

Clearly

$\mathbf{C}_2$ corresponds to the 5th root of unity – for each $i \in \mathbf{C}_2$, $(\alpha^i)^5 = 1$ – and so corresponds to $\Phi_5(X)$,

$\mathbf{C}_3$ corresponds to the 3rd root of unity and to $\Phi_3(X)$,

while $\mathbf{C}_5$ corresponds to $\Phi_1(X)$,

and $\mathbf{C}_1$ and $\mathbf{C}_4$ correspond to the two factors of Φ_{15};

7.7 Cycles, Roots and Idempotents

but we can also note that

$$i \in C_1 \iff -i \in C_4.$$

This means that in the factorization $\Phi_{15}(X) = g(X)h(X)$ we have

$$\beta \text{ is a root of } g \iff \beta^{-1} \text{ is a root of } h,$$

i.e.

$$h(X) = X^4 g\left(\frac{1}{X}\right).$$

This remark allows us to factorize $\Phi_{15}(X)$ by setting

$$g(X) = X^4 + aX^3 + bX^2 + cX + 1, \quad h(X) = X^4 + cX^3 + bX^2 + aX + 1$$

from which we get

$$\begin{aligned}\Phi_{15}(X) = {} & X^8 + (a+c)X^7 + acX^6 + (ab+bc+a+c)X^5 \\ & + (a+b+c)X^4 + (ab+bc+a+c)X^3 + acX^2 + (a+c)X + 1\end{aligned}$$

which requires us to solve in $\mathbb{Z}_2$ the system

$$\begin{cases} a+c &= 1 \\ ac &= 0 \\ ab+bc+a+c &= 1 \\ a+b+c &= 1. \end{cases}$$

Noting that by $ac = 0$ we can wlog deduce, by symmetry, $a = 0$, we get

$$\begin{cases} a &= 0 \\ b &= 0 \\ c &= 1 \end{cases}$$

and

$$\begin{aligned} g(X) &= X^4 + X + 1, \\ h(X) &= X^4 + X^3 + 1. \end{aligned}$$

By symmetry we can freely associate either of the two factors to the cycle C_1; let us choose g.

So we have the following correspondence between the cycles $\mathbf{C}_i$ and factors m_i of $m := X^{15} - 1$:

i	$\mathbf{C}_i$	m_i
1	$\{1, 2, 4, 8\}$	$X^4 + X + 1$
2	$\{3, 6, 9, 12\}$	$X^4 + X^3 + X^2 + X + 1$
3	$\{5, 10\}$	$X^2 + X + 1$
4	$\{7, 11, 13, 14\}$	$X^4 + X^3 + 1$
5	$\{0\}$	$X + 1$

Remark 7.7.3. Let $D := \mathbb{Z}_2[X]$,

$$m(X) := X^\nu - 1 = \prod_{i=1}^{n} m_i(X)$$

be a factorization in D into irreducible factors, and $R := D/m$.

In this context, Remark 2.7.8 gives us bijections between

the element $i \in \{1, \ldots, n\}$,

the factor m_i,

the cofactor $h_i = \prod_{j \neq i} m_j$,

the subring $R_i = D/(m_i)$,

the idempotent e_i,

the ideal $\mathcal{J}_i := (h_i) = (e_i) \cong R_i = D/(m_i)$,

where $\{e_1, \ldots, e_n\}$ is the primitive set of idempotents of R.

In the setting of finite fields we can add to these bijections another item:

the cycle $\mathbf{C}_i$

which is highly useful in the setting we are studying, i.e. $D := \mathbb{Z}_2[X]$. In fact, while in the general case of a finite prime field, we cannot easily factorize the cyclotomic polynomials, the cycles $\mathbf{C}_i$ allow us to factorize them in $\mathbb{Z}_2[X]$ by means of the following result:

Proposition 7.7.4. *Using the notation of Rem. 7.7.3, let*

$$\{c_1, \ldots, c_r\} \subset \{0, 1, \ldots, \nu - 1\}$$

and $c(X) := \sum_{i=0}^{r} X^{c_i} \in R$. Then the following are equivalent

(1) $c(X)$ is an idempotent;

(2) $(c_1, \ldots, c_r)$ is invariant under the permutation π_2.

7.7 Cycles, Roots and Idempotents

Proof It is sufficient to remark that in $\mathbb{Z}_2[X]/(X^v - 1)$ we have

$$c(X)^2 = \sum_{i=0}^{r} X^{2c_i} = \sum_{i=0}^{r} X^{\pi_2(c_i)}.$$

♄

For each set **C**, invariant under the permutation π_2, $\mathcal{P}(\mathbf{C})$ denotes the idempotent polynomial $\mathcal{P}(\mathbf{C})(X) := \sum_{i=0}^{r} X^{c_i} \in R$.

Algorithm 7.7.5. Since the set of all the idempotents of R is a $\mathbb{Z}_2$-vector space and $\{\mathcal{P}(\mathbf{C}_1), \ldots, \mathcal{P}(\mathbf{C}_n)\}$ is a linear basis of it, the above result allows us to compute a factorization of the cyclotomic polynomial $m(X) = X^v - 1 \in \mathbb{Z}_2[X]$ as follows:

by the algorithm discussed below (cf. Algorithm 7.7.6), it is possible to compute the primitive set of idempotents $\{c_1, \ldots, c_n\}$ of R from the set

$$\{\mathcal{P}(\mathbf{C}_1), \ldots, \mathcal{P}(\mathbf{C}_n)\};$$

since, for each i, $(c_i) = (h_i)$ in R, $h_i = \gcd(c_i, m)$, the factorization of $m(X)$ is obtained by the gcd computation $m_i := \frac{m}{\gcd(c_i, m)}$.

Algorithm 7.7.6. To complete the algorithm outlined above, we need an algorithm which produces the primitive set of idempotents for R, from a given linear basis

$$\{\eta_i, 1 \leq i \leq n\}$$

of the set of all the idempotents of R.

Such an algorithm works iteratively as follows: let us assume we have obtained a set $\{\varepsilon_1, \ldots, \varepsilon_m\}$ of m idempotents, such that

$1 = \sum_{j=1}^{m} \varepsilon_j,$
$\varepsilon_i \varepsilon_j = 0,$ if $i \neq j,$
ε_i is an idempotent;

then, for some i, k, let us compute

$$\epsilon' := \varepsilon_i \eta_k, \quad \epsilon'' := \varepsilon_i (\eta_k + 1),$$

which satisfy:

$$\epsilon' + \epsilon'' + \sum_{\substack{j=1\\j\neq i}}^{m} \varepsilon_j = \varepsilon_i(\eta_k + \eta_k + 1) + \sum_{\substack{j=1\\j\neq i}}^{m} \varepsilon_j = \sum_{j=1}^{m} \varepsilon_j = 1;$$

for all $j \neq i$, $\varepsilon_j \epsilon' = \varepsilon_j \varepsilon_i \eta_k = 0$;
for all $j \neq i$, $\varepsilon_j \epsilon'' = \varepsilon_j \varepsilon_i (\eta_k + 1) = 0$;
$\epsilon' \epsilon'' = \varepsilon_i^2 \eta_k(\eta_k + 1) = \varepsilon_i(\eta_k^2 + \eta_k) = 0$,

so that the set

$$\{\varepsilon_1, \ldots, \varepsilon_{i-1}, \epsilon', \epsilon'', \varepsilon_{i+1}, \ldots, \varepsilon_m\}$$

consists of $m+1$ idempotents satisfying the conditions above, provided that neither $\epsilon' = 0$ nor $\epsilon'' = 0$ holds.

But, if this happens for all choices i, k, this would imply that $m = n$ and $\{\varepsilon_1, \ldots, \varepsilon_m\}$ is a primitive set of idempotents[5], so that the algorithm has successfully terminated.

Example 7.7.7. Continuing Example 7.7.2, let us compute the primitive set of idempotents for R, starting with $\eta_i := \mathcal{P}(\mathbf{C}_i)$.

We will freely denote $\mathbf{C}'\mathbf{C}''$ to identify the single set[6] $\mathbf{C}$ such that

$$\mathcal{P}(\mathbf{C}) = \mathcal{P}(\mathbf{C}')\mathcal{P}(\mathbf{C}'').$$

Initially we can use the set $\{\mathcal{P}(\mathbf{C}_{11}), \mathcal{P}(\mathbf{C}_{12})\}$ where

$$\begin{aligned}\mathbf{C}_{11} &:= \{1, 2, 4, 8\} &= \eta_1\\ \mathbf{C}_{12} &:= \{1, 2, 4, 8, 0\} &= \eta_1 + 1.\end{aligned}$$

While

$$\mathbf{C}_{11}\eta_2 = \{1, 2, 4, 8\}\{3, 6, 9, 12\} = \{1, 2, 4, 8\}$$

and so $\mathbf{C}_{11}(\eta_2 - 1) = 0$, we find

$$\mathbf{C}_{12}\eta_2 = \{1, 2, 4, 8, 0\}\{3, 6, 9, 12\} = \{1, 2, 3, 4, 6, 8, 9, 12\},$$

---

[5] In fact, because $\sum_{j=1}^{m} \varepsilon_j = 1$, if this happens for all choices i, k we will deduce that all the n elements η_k are linear combinations in $\mathbb{Z}_2$ of the m elements c_j.
Since the n elements η_k are a linear basis of the set of all the idempotents of R, this implies the claim.

[6] To compute such $\mathbf{C}$ we only have to compute the set

$$\{i + j \pmod{\nu - 1} : i \in \mathbf{C}', j \in \mathbf{C}''\}$$

and eliminate any pairs of identical elements, e.g.

$$\begin{aligned}\{1, 2, 4, 8\}\{3, 6, 9, 12\} &= \{4, 5, 7, 11, 7, 8, 10, 14, 10, 11, 13, 2, 13, 14, 1, 5\}\\ &= \{1, 2, 4, 8\}.\end{aligned}$$

7.7 Cycles, Roots and Idempotents

obtaining the set $\{\mathcal{P}(C_{21}), \mathcal{P}(C_{22}), \mathcal{P}(C_{23})\}$, where

$$
\begin{aligned}
C_{21} &:= \{1, 2, 4, 8\} & &= \eta_1 \\
C_{22} &:= \{1, 2, 3, 4, 6, 8, 9, 12\} & &= \eta_1 + \eta_2 \\
C_{23} &:= \{3, 6, 9, 12, 0\} & &= \eta_2 + 1.
\end{aligned}
$$

The computation

$$C_{21}\eta_3 = \{1, 2, 4, 8\}\{5, 10\} = \{3, 6, 7, 9, 11, 12, 13, 14\}$$

yields $\{\mathcal{P}(C_{31}), \mathcal{P}(C_{32}), \mathcal{P}(C_{33}), \mathcal{P}(C_{34})\}$, where

$$
\begin{aligned}
C_{31} &:= \{3, 6, 7, 9, 11, 12, 13, 14\} & &= \eta_2 + \eta_4 \\
C_{32} &:= \{1, 2, 3, 4, 6, 7, 8, 9, 11, 12, 13, 14\} & &= \eta_1 + \eta_2 + \eta_4 \\
C_{33} &:= \{1, 2, 3, 4, 6, 8, 9, 12\} & &= \eta_1 + \eta_2 \\
C_{34} &:= \{3, 6, 9, 12, 0\} & &= \eta_2 + 1.
\end{aligned}
$$

The computation

$$C_{34}\eta_3 = \{3, 6, 9, 12, 0\}\{5, 10\} = \{1, 2, 4, 5, 7, 8, 10, 11, 13, 14\}$$

yields $\{\mathcal{P}(C_{41}), \mathcal{P}(C_{42}), \mathcal{P}(C_{43}), \mathcal{P}(C_{44}), \mathcal{P}(C_{45})\}$, where

$$
\begin{aligned}
C_{41} &:= \{3, 6, 7, 9, 11, 12, 13, 14\} & &= \eta_2 + \eta_4 \\
C_{42} &:= \{1, 2, 3, 4, 6, 7, 8, 9, 11, 12, 13, 14\} & &= \eta_1 + \eta_2 + \eta_4 \\
C_{43} &:= \{1, 2, 3, 4, 6, 8, 9, 12\} & &= \eta_1 + \eta_2 \\
C_{44} &:= \{1, 2, 4, 5, 7, 8, 10, 11, 13, 14\} & &= \eta_1 + \eta_3 + \eta_4 \\
C_{45} &:= \{1, 2, 3, 4, 5, 6, 7, 8, 9, 10, 11, 12, 13, 14, 0\} & &= \sum_j \eta_j.
\end{aligned}
$$

Thus having the primitive set of idempotents, greatest common divisor computation allows us to deduce the factorization of $X^{15} - 1$. In fact we have:

$$
\begin{aligned}
(X^{15} - 1)/\gcd(X^{15} - 1, \mathcal{P}(C_{41})) &:= X^4 + X^3 + 1 \\
(X^{15} - 1)/\gcd(X^{15} - 1, \mathcal{P}(C_{42})) &:= X^4 + X^3 + X^2 + X + 1 \\
(X^{15} - 1)/\gcd(X^{15} - 1, \mathcal{P}(C_{43})) &:= X^4 + X + 1 \\
(X^{15} - 1)/\gcd(X^{15} - 1, \mathcal{P}(C_{44})) &:= X^2 + X + 1 \\
(X^{15} - 1)/\gcd(X^{15} - 1, \mathcal{P}(C_{45})) &:= X + 1.
\end{aligned}
$$

The bijections given us by Remark 2.7.8 can therefore be extended and computed to those between

the element $i \in \{1, \ldots, n\}$,
the factor m_i,

the cofactor $h_i = \prod_{j \neq i} m_j$,
the subring $R_i = D/(m_i)$,
the idempotent e_i,
the ideal $\mathcal{J}_i := (h_i) = (e_i) \cong R_i = D/(m_i)$,
the cycle C_i
the permutation invariant set S_i such that $\mathcal{P}(\mathsf{S}_i) = e_i$.

In this setting we have

i	C_i	m_i	S_i
1	$\{1, 2, 4, 8\}$	$X^4 + X + 1$	C_{43}
2	$\{3, 6, 9, 12\}$	$X^4 + X^3 + X^2 + X + 1$	C_{42}
3	$\{5, 10\}$	$X^2 + X + 1$	C_{44}
4	$\{7, 11, 13, 14\}$	$X^4 + X^3 + 1$	C_{41}
5	$\{0\}$	$X + 1$	C_{45}

7.8 Deterministic Polynomial-time Primality Test

On August 6, 2002, while I was checking the proofs of this book, Agrawal, Kayal and Saxena announced their discovery of a deterministic polynomial-time algorithm which determines whether an input number $n \in \mathbb{N}$ is prime or composite and whose complexity is[7]

$$O(\log^{12}(n)(\log\log(n))^d) \text{ for some } d \in \mathbb{N};$$

moreover, if a classical conjecture on the density of Sophie Germain primes holds, the complexity reduces to

$$O(\log^6(n)(\log\log(n))^d) \text{ for some } d \in \mathbb{N};$$

finally the authors conjecture an algorithm having the complexity

$$O(\log^3(n)(\log\log(n))^d) \text{ for some } d \in \mathbb{N}.$$

Before their result the state of the art consisted of

a deterministic polynomial-time algorithm whose correctness was depending on the one of the Extended Riemann Hypothesis,
several probabilistic polynomial-time algorithms, and
a deterministic algorithm which runs in $\log(n)^{O(\log\log\log(n))}$.

---

[7] Logarithms are taken to base 2.

7.8 Deterministic Polynomial-time Primality Test

Apart from a lemma from Number Theory and the complexity analysis, this algorithm just applies the tools presented in this chapter. Therefore I give here a sketch of this result refering the reader to the original paper[8].

The primality test is based on the following identity

$$(X - a)^p \equiv X^p - a \pmod{p}. \tag{7.1}$$

Proposition 7.8.1. *Let a be such that $\gcd(a, p) = 1$. Then p is prime iff Equation 7.1 holds.*

Proof In fact

$$(X - a)^p - X^p + a = \sum_{i=1}^{p-1}(-1)^i \binom{p}{i} a^{p-i} X^i$$

and the formula is trivial if p is prime. If, on the other side, p is composite, $p = \prod_j p_j^{e_j}$, with p_j primes, then each $p_j^{e_j}$ does not divide $\binom{p}{p_j}$ nor p_j divides a; therefore

$$(-1)^{p_j}\binom{p}{p_j} a^{p-p_j} \not\equiv 0 \pmod{p}.$$

♄

Equation 7.1 does not give the required polynomial-time algorithm since the evaluation of $(X - a)^n$ requires us to compute n coefficients and is therefore exponential in $\log(n)$. The idea of the algorithm consists in choosing a 'suitable' prime r and to test in the ring $\mathbb{Z}_n[X]/(X^r - 1)$ the equalities

$$(X - a)^n \equiv X^n - a \pmod{n, X^r - 1}, \text{ for each } a \in \mathbb{N}, \quad a \leq 2\sqrt{r}\log(n),$$

which are true if n is prime and which can be performed in polynomial time in $\log(n)$ and r.

For $r, n \in \mathbb{N}$ such that $\gcd(r, n) = 1$, I will denote by $o_r(n)$ the multiplicative order of $n \bmod r$ (cf. Definition 7.6.12) and I recall that (cf. Corollary 7.6.16) if r and n are prime then $\frac{X^r-1}{X-1}$ factorizes in $\mathbb{Z}_n[X]$ into irreducible factors of degree $o_r(n)$. Also

Lemma 7.8.2. *Let $r, n \in \mathbb{N}$, $m_1, m_2 \in \mathbb{N}$. Then in $\mathbb{Z}_n[X]$*

(1) $m_1 \equiv m_2 \pmod{r} \implies X^{m_1} \equiv X^{m_2} \pmod{X^r - 1}.$

---

[8] M. Agrawal, N. Kayal, N. Saxena, PRIMES *is in* P.

(2) *each $g(X) \in \mathbb{Z}_n[X]$ such that*

$$g^{m_1}(X) \equiv g(X^{m_1}), g^{m_2}(X) \equiv g(X^{m_2}) \pmod{X^r - 1}$$

also satisfies

$$g^{m_1 m_2}(X) \equiv g(X^{m_1 m_2}) \pmod{X^r - 1}.$$

Proof

(1) By assumption there is k such that $m_1 = kr + m_2$; therefore, since

$$X^{kr} = (X^r)^k \equiv 1^k = 1 \pmod{X^r - 1}$$

we have

$$X^{m_1} = X^{kr+m_2} = X^{kr} X^{m_2} \equiv X^{m_2} \pmod{X^r - 1}$$

whence the claim.

(2) The substitution of Y^{m_1} into X in

$$g^{m_2}(X) \equiv g(X^{m_2}) \pmod{X^r - 1}$$

yields

$$g^{m_2}(Y^{m_1}) \equiv g(Y^{m_1 m_2}) \pmod{Y^{m_1 r} - 1}$$

whence

$$g^{m_2}(X^{m_1}) \equiv g(X^{m_1 m_2}) \pmod{X^r - 1}$$

and

$$g^{m_1 m_2}(X) \equiv g^{m_2}(X^{m_1}) \equiv g(X^{m_1 m_2}) \pmod{X^r - 1}.$$

♄

I now specify the meaning of 'suitable' prime:

Fact 7.8.3. *Given $n \in \mathbb{N}$, there are a constant $c > 0$ and primes r and q which satisfy:*

$r \leq c \log^6(n)$;
$q \geq 4\sqrt{r} \log(n)$;
q *is the greatest prime divisor of* $r - 1$;
q *divides* $o_r(n)$.

♄

Note that, since q is the greatest prime divisor of $r - 1$, q divides $o_r(n)$ iff $n^{\frac{r-1}{q}} \not\equiv 1 \pmod{r}$.

7.8 Deterministic Polynomial-time Primality Test

Fig. 7.3. Primality test

---
bool := **Prime**(n)
where
   $n \in \mathbb{N}$
   bool := $\begin{cases} \textbf{true} & \text{if } n \text{ is prime} \\ \textbf{false} & \text{if } n \text{ is composite} \end{cases}$
If $n = a^b, b > 1$ **or** n is even **then**
   bool := **false**
else
   $s := 2$
   Repeat
     Let r be the minimal prime $r > s$
     Let q be the largest prime factor of $r - 1$,
     $s := r$
   until $\gcd(n, r) \neq 1$ **or** $\left(q \geq 4\sqrt{r} \log(n) \text{ and } n^{\frac{r-1}{q}} \not\equiv 1 \pmod{r} \right)$.
   If $\gcd(n, r) \neq 1$ **then**
     bool := **false**
   else
     %% $q \geq 4\sqrt{r}\log(n)$ and $n^{\frac{r-1}{q}} \not\equiv 1 \pmod{r}$
     bool := **true**
     $\ell := \lfloor 2\sqrt{r}\log(n) \rfloor$
     For $a = 1, \ldots, \ell$ **do**
        If $(X - a)^n \not\equiv X^n - a \pmod{n, X^r - 1}$ **then bool** := **false**

Algorithm 7.8.4. It is now possible to present in Figure 7.3 the primality test algorithm.

Clearly if n is prime the algorithm returns **true**. Moreover

each **for** computation is polynomial in r and $\log(n)$,
each **while**-loop performs computations which are polynomials in r and in $\log\log(n)$,
and the **while**-loop performs $\log^6(n)$ iterations,

so that the algorithm has complexity

$$O(\log^e(n)(\log\log(n))^d) \text{ for some } d, e \in \mathbb{N}.$$

We have therefore to deduce a contradiction from the assumption that the algorithm returns **true** for a composite n.

Let us therefore assume that n is composite, $n = \prod_j p_j^{e_j}$, with p_j primes, and let r and q be primes as in Fact 7.8.3.

Since (cf. Lemma 7.4.1) $o_r(n)$ divides $\text{lcm}_i \, o_r(p_i)$ there is necessarily a factor p of n for which q divides $o_r(p)$.

If $h(X)$ denotes any of the irreducible factors of $\frac{X^r-1}{X-1}$ in $\mathbb{Z}_p[X]$, α any root of h and $d := o_r(p) = \deg(h)$, then in the Galois field

$$GF(p^d) \cong \mathbb{Z}_p[X]/h(X) = \mathbb{Z}_p[\alpha]$$

we have

$$(\alpha - a)^n = \alpha^n - a, \forall a, 1 \le a \le \ell = \lfloor 2\sqrt{r}\log(n) \rfloor.$$

Lemma 7.8.5. *With the notations and the assumptions above:*

(1) $2\ell \le q \le d$, $\ell < r$;
(2) *For each* a, b, $1 \le a$, $b \le \ell$, *we have* $a \equiv b \pmod{p} \implies a = b$;
(3) *For any two elements in the set*

$$S := \left\{ \prod_{a=1}^{\ell}(X-a)^{\delta_a} : \delta_a \ge 0, \forall a, \sum_a \delta_a < d \right\} \subset \mathbb{Z}_p[X]$$

we have

$$\prod_{a=1}^{\ell}(X-a)^{\delta_a} = \prod_{a=1}^{\ell}(X-a)^{\gamma_a} \implies \delta_a = \gamma_a, \text{ for all } a;$$

(4) $\#S > \left(\frac{d}{\ell}\right)^{\ell} > n^{2\sqrt{r}}$.

Proof

(1) We have $2\ell \le 4\sqrt{r}\log(n) \le q$ and q is a factor of both d and r.
(2) If $a \ne b$ and $a \equiv b \pmod{p}$ then $p < \ell < r$. The algorithm therefore would have performed the computation $\gcd(n, p) \ne 1$ returning **true** contradicting the assumption.
(3) The result above implies that S consists of distinct elements modulo p each having degree less than h, whence the claim.
(4) As a consequence S consists of

$$\binom{\ell+d-1}{\ell} = \frac{(\ell+d-1)(\ell+d-2)\cdots(d)}{\ell!} > \left(\frac{d}{\ell}\right)^{\ell}$$

elements. Moreover $\frac{d}{\ell} \ge 2$ implies

$$\left(\frac{d}{\ell}\right)^{\ell} \ge 2^{\ell} = 2^{2\sqrt{r}\log(n)} = n^{2\sqrt{r}}.$$

♄

7.8 Deterministic Polynomial-time Primality Test

Corollary 7.8.6. *With the notations and the assumptions above, we have:*

(1) *In $\mathbb{Z}_p[\alpha]$ the group $G := \{\prod_{a=1}^{\ell}(\alpha - a)^{\delta_a} : \delta_a \geq 0, \forall a\}$ is cyclic;*
(2) *$\#G \geq \#S > n^{2\sqrt{r}}$.*

Proof

(1) Clearly G is a group and a subgroup of the group $GF(p^d)/\{0\}$ which (Corollary 7.4.9) is cyclic.
(2) Clearly $G \supset \{f(\alpha) : f(X) \in S\}$; the claim then follows by noting that for each $f(X) \in S$, $\deg(f) < \deg(h)$, so that, for each $f_1, f_2 \in S$ $f_1(\alpha) = f_2(\alpha) \implies f_1(X) = f_2(X)$. ∎

Since $G \subset GF(p^d)$ is cyclic, it has a generator ζ, not necessarily a primitive element of $GF(p^d)$. Let $o := \mathrm{ord}(\zeta)$ denote its order in $GF(p^d)$.

Let

$$g(X) := \prod_{a=1}^{\ell}(X - a)^{\delta_a} \in \mathbb{Z}_p[X],$$

be the unique element for which $\deg(g) < d$ and $\zeta = g(\alpha)$ and let

$$\mathcal{I} := \{m \in \mathbb{N} : g^m(X) \equiv g(X^m) \pmod{X^r - 1}\}.$$

Lemma 7.8.7. *With the notations and the assumptions above:*

(1) *$p \in \mathcal{I}$ and $1 \in \mathcal{I}$;*
(2) *$n \in \mathcal{I}$;*
(3) *$\mathcal{I}$ is multiplicative.*

Proof

(1) Trivial since $g^p(X) = g(X^p)$ in $\mathbb{Z}_p[X]$.
(2) Because

$$(X - a)^n \equiv X^n - a \pmod{n, X^r - 1}, \text{ for all } a, \quad 1 \leq a \leq \ell$$

and g is a product of powers of such elements.
(3) It is a direct consequence of Lemma 7.8.2(2). ∎

Proposition 7.8.8. *For $m_1, m_2 \in \mathcal{I}$,*

$$m_1 \equiv m_2 \pmod{r} \implies m_1 \equiv m_2 \pmod{\mathrm{ord}(\zeta)}.$$

Proof Since $m_1 \equiv m_2 \pmod{r}$ there is k such that $m_1 = kr + m_2$ and (Lemma 7.8.1(1)) $g(\alpha^{m_1}) = g(\alpha^{m_2})$.

Therefore in $GF(p^d)$ we have

$$\zeta^{m_2} = g^{m_2}(\alpha) = g(\alpha^{m_2}) = g(\alpha^{m_1}) = g^{m_1}(\alpha) = \zeta^{m_1} = \zeta^{m_2+kr},$$

whence $\zeta^{kr} = 1$, $kr \equiv 0 \pmod{\mathrm{ord}(\zeta)}$ and $m_1 \equiv m_2 \pmod{\mathrm{ord}(\zeta)}$. ♄

This proposition is the core of the argument, proving, as the authors put it, that *there are 'very few'* $(\leq r)$ *numbers in $\mathcal{I}$ that are less than* $\mathrm{ord}(\zeta)$.

Theorem 7.8.9. *For a composite $n \in \mathbb{N}$ the algorithm returns* **false**.

Proof Let

$$E := \{n^i p^j : 0 \leq i, j \leq \lfloor \sqrt{r} \rfloor\}.$$

By Lemma 7.8.3 we know $E \subseteq \mathcal{I}$. Since $\#E = (1 + \lfloor \sqrt{r} \rfloor)^2 > r$, there are two different pairs (i_1, j_1) and (i_2, j_2) such that $n^{i_1} p^{j_1} \equiv n^{i_2} p^{j_2} \pmod{r}$ so that, by Proposition 7.8.8 one has $n^{i_1} p^{j_1} \equiv n^{i_2} p^{j_2} \pmod{\mathrm{ord}(\zeta)}$ and

$$n^{i_1 - i_2} \equiv p^{j_2 - j_1} \pmod{\mathrm{ord}(\zeta)}.$$

Since $n^{|i_1 - i_2|} < n^{\sqrt{r}}$ and $p^{|j_2 - j_1|} < p^{\sqrt{r}} < n^{\sqrt{r}}$ and since Corollary 7.8.1(2) implies that $\mathrm{ord}(\zeta) > n^{2\sqrt{r}}$, then

$$n^{i_1 - i_2} = p^{j_2 - j_1}.$$

Since p is prime this means that $n = p^b$ for some $b > 1$. As a consequence the first instruction of the algorithm returns **false**. ♄

Fig. 7.4. (Conjectured) Primality test

---
bool := **Prime**(n)
where
  $n \in \mathbb{N}$
  $\mathbf{bool} := \begin{cases} \textbf{true} & \text{if } n \text{ is prime} \\ \textbf{false} & \text{if } n \text{ is composite} \end{cases}$
$s := 2$
Repeat
  Let r be the minimal prime $r > s$
  $s := r$
until $n^2 \not\equiv 1 \pmod{r}$.
bool := **false**
If $(X - 1)^n \equiv X^n - 1 \pmod{n, X^r - 1}$ **then bool** := **true**

7.8 Deterministic Polynomial-time Primality Test

The authors also suggest the following

Conjecture 7.8.10. *If r does not divide n and*

$$(X - 1)^n \equiv X^n - 1 \pmod{n, X^r - 1}$$

then either n is prime or $n^2 \equiv 1 \pmod{r}$.

whose corresponding primality test algorithm (Figure 7.4) has complexity

$$O(\log^3(n)(\log\log(n))^d) \text{ for some } d \in \mathbb{N}.$$

8
Kronecker II: Kronecker's Model

> ...said the Duchess, digging her sharp little chin into Alice's shoulder as she added 'and the moral of that is — "Take care of the sense, and the sounds will take care of themselves."
>
> C.L. Dodgson, *Alice's Adventures in Wonderland*

This chapter is, in one sense, the direct continuation of Chapter 5: Kronecker's constructions and theory, presented there, aimed to give a model for computing the roots of polynomial equations over a field K and, mainly, for the case $\mathbb{Q} \subset K \subset \mathbb{C}$.

In this chapter I analyse this model, which allows us to 'solve' polynomial equations by representing any finite extension K over its prime field k: First (Section 8.1) I discuss the 'philosophy' behind it, and point to its weaknesses, mainly its inability to deal with real problems.

Then I introduce Kronecker's model (Section 8.2) of such fields, and the techniques used to represent their elements and operations (Section 8.3): we will see that, by iteration of the construction discussed in Section 5.1, each finite algebraic extension $K \supset K_0$ can be represented as a quotient of a multivariate polynomial ring $K_0[Z_1, \ldots, Z_r]$ by an ideal generated by a basis $(f_1, \ldots, f_r)$ satisfying specific properties.

I will then show that any such finite algebraic extension, provided it involves at most a single inseparable element, is in fact a simple extension, and so can be represented as a univariate polynomial ring modulo an irreducible polynomial (Section 8.4).

8.1 Kronecker's Philosophy

As we said, Kronecker's solution, in order to get out of the impasse consequent on the Abel–Ruffini Theorem, was to change the sense of 'solving': before

8.1 Kronecker's Philosophy

Kronecker 'solving' meant producing programs whose input was a polynomial equation and the output its roots, i.e. *producing programs which compute the roots of a polynomial equation*; Kronecker proposed to interpret 'solving' as *producing programs which compute* **with** *the roots of a polynomial equation*.

In fact, Kronecker's theory discussed in Chapter 5 supplies a computational model for dealing with any finite extension field K over its prime field k, by giving a representation of such field K that allows us to explicitly:

(1) perform arithmetical operations in K;
(2) 'solve' polynomial equations over K, by which I mean

   (a) to give a representation of a finite algebraic extension $K_1 \supseteq K$ which contains either a root or all the roots of $f(X) \in K[X]$, and on which (1), (2) and (3) can again be effectively performed;
   (b) to explicitly give a root (all roots) of f in K_1;

(3) factorize polynomials in $K[X]$ (which is technically needed for (2) and interesting in itself).

Note that this enunciation implicitly contains the possibility of iteratively expressing a new polynomial $f_1(X) \in K_1[X]$ whose coefficients are functions of a root (all the roots) of $f(X) \in K[X]$ and thereby 'solve' it.

In some sense, therefore, Kronecker's model allows us to 'solve' polynomial equations, by allowing us to perform arithmetical operations over arithmetical expressions of their roots.

Informally speaking, an elementary 'arithmetical expression' over a set of roots $\alpha_1, \ldots, \alpha_n$ is one of:

the assignment of α_i for some i – which consists of giving an *irreducible* polynomial $f_i(X) \in k(\alpha_1, \ldots, \alpha_{i-1})[X]$ such that $f_i(\alpha_i) = 0$,
the sum, difference or the product of two arithmetical expressions,
the inverse of a non-zero arithmetical expression,

and an 'arithmetical operation' is any one of the four elementary operations, extractions of pth roots over fields of finite characteristic p, testing whether an arithmetical expression is 0.

Note that, as a consequence of the assumption that any α_i is assigned by giving it an *irreducible* polynomial $f_i(X)$, an 'arithmetical expression' is nothing more than an algebraic number in $k[\alpha_1, \ldots, \alpha_n]$.

The Splitting Field Theorem (Theorem 5.5.6) guarantees that Kronecker's model gives a faithful representation of any subfield of $\mathbb{C}$ which is a finite extension field of $\mathbb{Q}$ by roots of iteratively expressed polynomials.

However, it points to a negative aspect of Kronecker's model: any finite extension field of $\mathbb{Q}$ by a root of an irreducible polynomial $f(X) \in \mathbb{Q}[X]$ is faithfully represented by the field $\mathbb{Q}[X]/f(X) := \mathbb{Q}[\alpha]$. While we restrict the notion of arithmetical operations as we did before, there is no problem, but if we are willing, or need, to work with real numbers, then problems arise.

Example 8.1.1. Consider $f := X^2 - 2$; then the field $\mathbb{Q}[X]/f(X) := \mathbb{Q}[\alpha]$ has two roots of $f(X)$, which are α and $-\alpha$. Which of them is $\sqrt{2}$?

By the Splitting Field Theorem, $\mathbb{Q}[X]/f(X) := \mathbb{Q}[\alpha]$ represents both $\mathbb{Q}[\sqrt{2}]$ and $\mathbb{Q}[-\sqrt{2}] = \mathbb{Q}[\sqrt{2}]$; in the first representation $\sqrt{2}$ is represented by α, while in the second one it is represented by $-\alpha$!

This is not at all strange, no more than the fact that there is no way that the two imaginary numbers can be distinguished from each other: in fact Lemma 5.5.3 informs us that there is an automorphism $\Psi : \mathbb{Q}[\sqrt{2}] \mapsto \mathbb{Q}[\sqrt{2}]$ which is defined by $\Phi(a + b\sqrt{2}) = a - b\sqrt{2}$, for all $a, b \in \mathbb{Q}$, exactly as the conjugation $z \mapsto \bar{z}$ exits in $\mathbb{C}$.

However, the existence of these two automorphic models of $\mathbb{Q}[\sqrt{2}]$, and therefore of these two ways of interpreting Kronecker's representation $\mathbb{Q}[\alpha]$ of it, has the following consequence: how can we produce a program which allows us to decide whether a given arithmetical expression is positive? If such an algorithm exists and is applied to $\alpha \in \mathbb{Q}[\alpha] = \mathbb{Q}[X]/f(X)$, what will be the solution? Of course α is positive if it represents $\sqrt{2}$ and negative if it represents $-\sqrt{2}$, but, as we said before, it represents both!

Example 8.1.2. Let us briefly point to similar weaknesses in Kronecker's model.

Consider $f := X^3 - 2$; then the field $\mathbb{Q}[X]/f(X) := \mathbb{Q}[\alpha]$ has three roots of $f(X)$, one of which is real while the other two are complex and conjugate. Of course, no algorithm exists which can be applied to $\mathbb{Q}[\alpha]$ and answer the question of whether α is real or complex. Even moreso: in the field k_2 of Example 5.2.5, there is no way of distinguishing the real root among β, γ and $-\beta - \gamma$.

Again, in the same setting as before where $f = X^2 - 2$, the number $\alpha + 1$ is such that $|\alpha + 1| > 1$ if α represents $\sqrt{2}$, while $|\alpha + 1| < 1$ if α represents $-\sqrt{2}$.

In conclusion, before discussing in detail the computational model proposed by Kronecker for dealing effectively with finite extension fields K over its prime field, we have to point out that Kronecker's proposal has at least three problems which, for a while, we need to brush under the carpet:

- We need to be able to factorize any polynomial in $K[X]$: its crucial importance is obvious since all the theory is based on representing algebraic numbers $\alpha \in K_2 \setminus K_1$ by assigning a polynomial $f(X) \in K_1[X]$ and building $K_1[\alpha]$ as $K_1[X]/f(X)$, but the latter is a field only if f is irreducible. In fact Kronecker himself proposed a factorization algorithm in $\mathbb{Q}[X]$, which, however, was unsatisfactory because of its complexity. For now, we just say that there exist factorization algorithms for each finite extension field K over its prime field and point the reader to Part II.
- The effectiveness of the computational model proposed by Kronecker is somehow limited by its inability to handle algorithms for solving real problems, such as deciding whether a root is real or complex, whether it is positive or negative, etc. Tools to do that were soon built (Sturm sequences) and new strong models have recently been built around Sturm sequences and Thom's Lemma. (see Chapter 13).
- Given a finite extension field K over its prime field, and a number $\alpha \in L \setminus K$ where L is any field extension $L \supseteq K$, we need to know whether α is algebraic over K (in which case, we need to have its monic polynomial), or transcendental. For the sake of the discussion, we will assume that we have an Oracle giving us this information; but honesty requires us to admit that the problem is still open, up to the point that there is not yet a proof of the conjecture that there is no algebraic relation between e and π.

8.2 Explicitly Given Fields

In order to discuss the Kronecker model, let us consider the class $\mathcal{K}$ of fields K to which it applies, i.e. finite extension fields over their prime field. To build any such field, we start with a prime field and repeatedly extend it either algebraically or transcendentally. The class $\mathcal{K}$ can then be defined recursively by:

$\mathbb{Q} \in \mathcal{K}$, $\mathbb{Z}_p \in \mathcal{K}$;
if $K \in \mathcal{K}$ and $f(X) \in K[X]$ is an irreducible polynomial then

$$K[X]/f(X) \in \mathcal{K};$$

if $K \in \mathcal{K}$ and $L \supseteq K$, $\alpha \in L \setminus K$ are such that α is transcendental over K and $L = K(\alpha)$, then $L \in \mathcal{K}$.

In order to understand better how Kronecker's model represents the fields in $\mathcal{K}$, let us first of all note that

Lemma 8.2.1. *Let $K_1 \subseteq K_2 \subseteq K_3$ be three fields; $\alpha_1, \ldots, \alpha_k, \beta \in K_3 \setminus K_1$ and let us assume that $K_2 := K_1(\alpha_1, \ldots, \alpha_k)$ is an algebraic extension. Then:*

(1) β is algebraic over $K_1 \iff$ it is algebraic over K_2.
(2) If β is transcendental over K_2, then $K_2(\beta) \cong K_1(\beta)[\alpha_1, \ldots, \alpha_k]$.

Proof Clearly it is sufficient to show that, if β is algebraic over K_2, then it is the same over K_1, but this is obvious, since $K_2[\beta]$ is an algebraic extension of K_1. ♄

As a consequence, if K is a finite extension over its prime field k, i.e.

there exists $\beta_1, \ldots, \beta_{d+r} \in K \setminus k : K = k(\beta_1, \ldots, \beta_{d+r})$,

then, up to reordering the βs we can assume that

for all $i \leq d$, β_i is transcendental over $k(\beta_1, \ldots, \beta_{i-1})$;
for all $i > d$, β_i is algebraic over $k(\beta_1, \ldots, \beta_{i-1})$;

so that

$k(\beta_1, \ldots, \beta_d) \cong k(Y_1, \ldots, Y_d) =: K_0$;
for all $i > d$, β_i is algebraic over $K_0(\beta_{d+1}, \ldots, \beta_{i-1})$.

We can then reinterpret $\mathcal{K}$ to be the class of all the finite algebraic extensions over any rational function field over a prime field, i.e. $\mathcal{K}$ is defined recursively by:

$\mathbb{Q}(Y_1, \ldots, Y_d) \in \mathcal{K}, \mathbb{Z}_p(Y_1, \ldots, Y_d) \in \mathcal{K}$;
if $K \in \mathcal{K}$ and $f(X) \in K[X]$ is an irreducible polynomial then

$$K[X]/f(X) \in \mathcal{K}.$$

On that basis, a field K, whose prime field is k, belongs to $\mathcal{K}$ if there is a tower of fields

$$k(Y_1, \ldots, Y_d) = K_0 \subseteq K_1 \subseteq \cdots \subseteq K_r = K;$$

for all $i \geq 1$, a monic irreducible polynomial $g_i(Z_i) \in K_{i-1}[Z_i]$;
for all $i \geq 1$, $\alpha_i \in K_i \setminus K_{i-1}$ such that $g_i(\alpha_i) = 0$,

8.2 Explicitly Given Fields

so that

$$\text{for all } i > 1, K_i = K_{i-1}[Z_i]/g_i(Z_i) \cong K_{i-1}[\alpha_i];$$

and it is quite clear that this representation allows us to perform the arithmetical operations on $K = K_r$ discussed in Section 5.1: while sums and subtractions only require us to do K_{r-1}-vector space algebra, products and divisions require the Euclidean algorithms in $K_{r-1}[Z]$ and therefore require us to perform operations in the field K_{r-1}, which, therefore, require Euclidean algorithms in $K_{r-2}[Z]\ldots$ and so on, recursively.

This iterative definition of K by quotient fields of univariate polynomials over a field can be collapsed to represent K as the quotient of a multivariate polynomial ring over K_0 by a suitable ideal:

$$K \cong k(Y_1, \ldots, Y_d)[Z_1, \ldots, Z_r]/(f_1, \ldots, f_r).$$

To prove this, let us put ourselves in a more general situation: let k be *any* field – not necessarily a prime one – and let **k** be the class of all its finite extension fields.

Let us also consider a sequence

$$\mathbf{f} := \{f_1, \ldots, f_r\}$$

of polynomials

$$f_i \in k(Y_1, \ldots, Y_d)[Z_1, \ldots, Z_i];$$

let us denote iteratively

$$L_j := k(Y_1, \ldots, Y_d)[Z_1, \ldots, Z_j]/(f_1, \ldots, f_j) \cong L_{j-1}[Z_j]/\pi_{j-1}(f_j),$$

by π_j both the canonical projection

$$\pi_j : k(Y_1, \ldots, Y_d)[Z_1, \ldots, Z_j] \mapsto L_j$$

and its polynomial extensions

$$\pi_j : k(Y_1, \ldots, Y_d)[Z_1, \ldots, Z_r] \mapsto L_j[Z_{j+1}, \ldots, Z_r];$$

$d_j := \deg_j(f_j)$, where, for a polynomial

$$f \in k(Y_1, \ldots, Y_d)[Z_1, \ldots, Z_r],$$

$\deg_j(f)$ denotes its degree in the variable Z_j;

and let us introduce the following

Definition 8.2.2. *We say* **f** *is an* admissible sequence *in*

$$k(Y_1, \ldots, Y_d)[Z_1, \ldots, Z_r]$$

if:

$f_1 \in k(Y_1, \ldots, Y_d)[Z_1]$ *is monic irreducible, so that L_1 is a field;*
for all i, $2 \leq i \leq r$, $\pi_{i-1}(f_i) \in L_{i-1}[Z_i]$ is monic irreducible, so that L_i is a field;
for all i, $2 \leq i \leq r$, for all $j < i$, $\deg_j(f_i) < d_j$.

which allows us to prove that

Proposition 8.2.3. *If $\mathbf{f} := \{f_1, \ldots, f_r\}$ is an admissible sequence in*

$$k(Y_1, \ldots, Y_d)[Z_1, \ldots, Z_r],$$

then $L_r \in \mathbf{k}$.

Conversely if $K \in \mathbf{k}$, then there are $d \geq 0$, $r \geq 0$, and an admissible sequence

$$\mathbf{f} := \{f_1, \ldots, f_r\} \subset k(Y_1, \ldots, Y_d)[Z_1, \ldots, Z_r],$$

such that

$$K \cong k(Y_1, \ldots, Y_d)[Z_1, \ldots, Z_r]/(f_1, \ldots, f_r).$$

Proof In fact, inductively, $L_0 := k(Y_1, \ldots, Y_d) \in \mathbf{k}$, and if $L_i \in \mathbf{k}$, then $L_{i+1} = L_i[Z_{i+1}]/\pi_i(f_{i+1}) \in \mathbf{k}$.

Then we conclude that

$$L_0 \subseteq L_1 \subseteq \cdots \subseteq L_r$$

is a tower of fields and, for all i, denoting $g_i := \pi_{i-1}(f_i)$, $\alpha_i := \pi_i(Z_i)$ we have

$$L_i = L_{i-1}[Z_i]/g_i(Z_i) \cong L_{i-1}[\alpha_i].$$

To prove the converse, let us remark that, if K is a finite extension of k, there are

a tower of fields

$$k(Y_1, \ldots, Y_d) = K_0 \subseteq K_1 \subseteq \cdots \subseteq K_r = K,$$

for all $i > 1$, a monic irreducible polynomial $g_i(Z_i) \in K_{i-1}[Z_i]$,
for all $i > 1$, $\alpha_i \in K_i \setminus K_{i-1}$ such that $g_i(\alpha_i) = 0$,

so that

for all $i > 1$, $K_i = K_{i-1}[Z_i]/g_i(Z_i) \cong K_{i-1}[\alpha_i]$.

8.2 Explicitly Given Fields

Let us begin by defining $f_1 := g_1 \in K_0[Z_1]$, so that (f_1) is admissible and

$$K_1 = K_0[Z_1]/f_1 =: L_1,$$

and assume inductively that we have defined

$$f_j \in K_0[Z_1, \ldots, Z_j], \text{ for all } j, 1 \le j \le i-1,$$

such that $\{f_1, \ldots, f_{i-1}\}$ is admissible and

$$L_{i-1} := K_0[Z_1, \ldots, Z_{i-1}]/(f_1, \ldots, f_{i-1}) \cong K_{i-1} = K_0[\alpha_1, \ldots, \alpha_{i-1}].$$

Then $g_i(Z_i) \in K_{i-1}[X_i] = K_0[\alpha_1, \ldots, \alpha_{i-1}][Z_i]$ can be interpreted as a polynomial over K_0 whose variables are the α_js and Z_i. It is then sufficient to substitute Z_j to α_j in order to obtain a polynomial $f_i \in K_0[Z_1, \ldots, Z_i]$ such that $\pi_{i-1}(f_i) = g_i$. Moreover, any coefficient of g_i is an element in K_{i-1} and so is represented by a polynomial in $K_0[\alpha_1, \ldots, \alpha_{i-1}]$ whose degree in the variable α_j is less than d_j; so we can conclude that $\deg_j(f_i) < d_j$, for all $j < i$. Finally, since g_i is irreducible, then so is f_i.

In conclusion, $\{f_1, \ldots, f_i\}$ is admissible and

$$K_i = L_{i-1}[Z_i]/g_i \cong L_{i-1}[Z_i]/\pi_{i-1}(f_i) \cong K_0[Z_1, \ldots, Z_i]/(f_1, \ldots, f_i).$$

♮

Corollary 8.2.4. *If $K \in \mathcal{K}$, then there are $d \ge 0$, $r \ge 0$, and an admissible sequence*

$$\mathbf{f} := \{f_1, \ldots, f_r\} \subset k(Y_1, \ldots, Y_d)[Z_1, \ldots, Z_r],$$

where k is the prime field of K, such that

$$K \cong k(Y_1, \ldots, Y_d)[Z_1, \ldots, Z_r]/(f_1, \ldots, f_r).$$

♮

Definition 8.2.5. *We then say that a finite extension field $K \supset k$ is explicitly given if $d \ge 0$, $r \ge 0$ are specified together with an admissible sequence*

$$\{f_1, \ldots, f_r\} \subset k(Y_1, \ldots, Y_d)[Z_1, \ldots, Z_r],$$

so that

$$K \cong k(Y_1, \ldots, Y_d)[Z_1, \ldots, Z_r]/(f_1, \ldots, f_r).$$

8.3 Representation and Arithmetics

8.3.1 Representation

Let k be a given effective field and let $K \supset k$ be a finite extension which is explicitly given by specifying an admissible sequence

$$(f_1, \ldots, f_r) \in k(Y_1, \ldots, Y_d)[Z_1, \ldots, Z_r]$$

so that

$$K \cong k(Y_1, \ldots, Y_d)[Z_1, \ldots, Z_r]/(f_1, \ldots, f_r).$$

Let us denote

$K_0 := k(Y_1, \ldots, Y_d)$,
$K_j := K_0[Z_1, \ldots, Z_j]/(f_1, \ldots, f_j) \cong K_{j-1}[Z_j]/\pi_{j-1}(f_j)$,
π_j both the canonical projection

$$\pi_j : K_0[Z_1, \ldots, Z_j] \mapsto K_j$$

and its polynomial extensions

$$\pi_j : K_0[Z_1, \ldots, Z_r] \mapsto K_j[Z_{j+1}, \ldots, Z_r],$$

$\alpha_j := \pi_j(Z_j)$,
$d_j := \deg_j(f_j)$.

Up to the end of this section we will keep the notation related to K fixed, and in the algorithms we will assume that K_0, and $(f_1, \ldots, f_r)$ are explicitly given.

Then K is a K_0-vector space of dimension $D := \prod_i d_i$ generated by the K_0-basis $\pi_r(\mathbf{B})$ where

$$\mathbf{B} := \{Z_1^{a_1} \cdots Z_r^{a_r} : a_i < d_i, \text{ for all } i\}.$$

Let $K_0[\mathbf{B}]$ be the K_0-subvector space of $K_0[Z_1, \ldots, Z_r]$ generated by $\mathbf{B}$; then the restriction of π_r to $K_0[\mathbf{B}]$ is a K_0-vector isomorphism from $K_0[\mathbf{B}]$ to K.

So, analogously to the simple extension case(Section 5.1), we have two representations of K as a K_0-vector space:

as K_0^D, so that each element is (identified with) a D-dimensional vector;
as a subset of the polynomial ring $K_0[Z_1, \ldots, Z_r]$, so each element is (identified with) a polynomial $g \in K_0[Z_1, \ldots, Z_r]$ such that $\deg_j(g) < d_j$, for all j.

8.3 Representation and Arithmetics

Let us order the semigroup **T** of terms of $K_0[Z_1, \ldots, Z_r]$ by[1] the lexicographical total semigroup ordering $<$, such that $Z_1 < Z_2 < \cdots < Z_n$, given by:

$$Z_1^{a_1} \cdots Z_r^{a_r} < Z_1^{b_1} \cdots Z_r^{b_r} \iff \text{there exists } j : a_j < b_j \text{ and } a_i = b_i,$$

for all $i > j$.

Let us index the D terms in **B** by increasing order so that

$$1 = b_1 < b_2 < \cdots < b_D = Z_1^{d_1-1} \cdots Z_r^{d_r-1}.$$

Switching from one representation to another is then very easy: if g is a polynomial such that $\deg_j(g) < d_j$, for all j, then $g = \sum_{i=1}^{D} c_i b_i$, $c_i \in K_0$, and is isomorphic to the vector $(c_1, \ldots, c_D) \in K_0^D$; conversely the polynomial $\sum_{i=1}^{D} c_i b_i$ corresponds to each vector $(c_1, \ldots, c_D) \in K_0^D$.

Moreover, in the second representation, elements of K_j are identified with polynomials in $K_0[\mathbf{B}] \cap K_0[Z_1, \ldots, Z_j]$.

8.3.2 Vector space arithmetics

Since K is explicitly represented as a K_0-vector space, the K_0-linear operations on K are immediately available; so it is possible to test for equality, add, subtract elements of K and test if an element of K is 0.

8.3.3 Canonical representation

Multiplication requires some more work; if we are given two polynomials $g_1, g_2 \in K_0[\mathbf{B}] \cong K$, their product will in general no longer be in $K_0[\mathbf{B}]$.

So we need to think about how to compute, given

$$g \in K_0[Z_1, \ldots, Z_r],$$

the unique $h \in K_0[\mathbf{B}]$ such that $\pi_r(h) = \pi_r(g)$, which will be called the *canonical representative* of g.

Since the algorithm is recursive denoting

$$\mathbf{B}_i := \mathbf{B} \cap K_0[Z_1, \ldots, Z_i],$$

we will assume the availability[2] of an algorithm which, given

$$g \in K_0[Z_1, \ldots, Z_i],$$

---

[1] Compare this with the identical definition and approach used in Remark 6.2.2.
[2] Such an algorithm is obviously available if $i = 0$, since then we simply to take $h := g \in K_0$.

computes $h \in K_0[\mathbf{B}_i]$ such that $\pi_i(h) = \pi_i(g)$ and we will show how this algorithm is used to solve the same problem at the $(i + 1)$th level.

First of all we note that, while we defined polynomial division only for polynomials with coefficients in a field, the algorithm obviously also applies when the coefficients are in a domain, provided that the divisor is monic, since then no inverse computation is needed.

So we are given $g \in K_0[Z_1, \ldots, Z_{i+1}]$; we perform polynomial division in $K_0[Z_1, \ldots, Z_i][Z_{i+1}]$ of g by the monic polynomial f_{i+1} to obtain

$$\mathbf{Rem}(g, f_{i+1}) = \sum_{j=0}^{d_{i+1}-1} q_j(Z_1, \ldots, Z_i) Z_{i+1}^j;$$

then, by recursive application, for each j we compute

$$p_j(Z_1, \ldots, Z_i) \in K_0[\mathbf{B}_i]$$

such that $\pi_i(p_j) = \pi_i(q_j)$. We obtain

$$\begin{aligned}
\pi_{i+1}(g) &= \pi_{i+1}(\mathbf{Rem}(g, f_{i+1})) \\
&= \pi_{i+1}\left(\sum_j q_j Z_{i+1}^j\right) \\
&= \sum_j \pi_i(q_j) \pi_{i+1}(Z_{i+1}^j) \\
&= \sum_j \pi_i(p_j) \pi_{i+1}(Z_{i+1}^j) \\
&= \pi_{i+1}\left(\sum_j p_j Z_{i+1}^j\right)
\end{aligned}$$

and $\sum_j p_j Z_{i+1}^j \in k[\mathbf{B}_{i+1}]$.

Algorithm 8.3.1. While this proposal is nothing more than a recursive application of the same algorithm in the univariate case, there is actually no need for recursion, since the same result, with essentially the same computations, can be obtained by the algorithm of Figure 8.1, where for $g \in K_0[Z_1, \ldots, Z_r]$, $g = \sum_{t \in \mathbf{T}} c_t t$, with $c_t \in K_0$, we denote

$$T(g) := \max_<\{t : c_t \neq 0\}, \quad \mathrm{lc}(g) := c_{T(g)} \in K_0.$$

Termination of this algorithm is guaranteed since at each step $T(f)$ is decreasing in the well-ordered semigroup $\mathbf{T}$.

Correctness is guaranteed since at each step $h \in K_0[\mathbf{B}]$ and

$$\pi_r(g) = \pi_r(f + h)$$

and, at termination, $f = 0$ and $\pi_r(g) = \pi_r(h)$.

Fig. 8.1. Canonical representations

$h := \mathbf{Reduction}(g; \{f_1, \ldots, f_r\})$
where
   $\{f_1, \ldots, f_r\}$ is an admissible sequence
   $g \in K_0[Z_1, \ldots, Z_r]$
   $h \in K_0[\mathbf{B}]$ is such that $\pi_r(g) = \pi_r(h)$
$f := g, h := 0$
While $f \neq 0$ **do**
   If $T(f) \in \mathbf{B}$ **then**
      $h := h + \mathrm{lc}(f)T(f), f := f - \mathrm{lc}(f)T(f)$
   else $- T(f) \notin \mathbf{B}$
      let f_i be such that $T(f_i)$ divides $T(f)$
      let $t \in \mathbf{T}$ be such that $T(f) = tT(f_i)$
      $f := f - \mathrm{lc}(f)\mathrm{lc}(f_i)^{-1}tf_i$

8.3.4 Multiplication

With an algorithm to compute canonical representatives, multiplication of $g_1, g_2 \in K_0[\mathbf{B}]$ can then be performed by computing $g_1 \cdot g_2$ in $K[Z_1, \ldots, Z_r]$, followed by the canonical representative h of $g_1 \cdot g_2$, because then

$$\pi_r(g_1)\pi_r(g_2) = \pi_r(g_1 \cdot g_2) = \pi_r(h).$$

8.3.5 Inverse and division

We have now turned K into an effective ring, and, as a consequence, the polynomial ring $K[Z]$ is an effective ring too. We have yet to produce an algorithm for computing inverses in K, that will turn K into an effective field, and so to have polynomial division in $K[Z]$ and, as a consequence, gcd computation, the Euclidean and the Extended Euclidean Algorithms, the squarefree test, the squarefree-associate and the distinct power factorization algorithms.

Unlike multiplication, where we could also use canonical representatives, we can only recursively use Kronecker's ideas, since we have to use the full power of polynomial algorithms in $K_i[Z_{i+1}]$ to get inverses in K_{i+1}.

So assuming recursively that K_i is an effective field (K_0 is such), let us show how to compute inverses in K_{i+1}. Let $g \in K[\mathbf{B}_{i+1}] \cong K_{i+1}, g \neq 0$,

$$g = \sum_{j=0}^{d_{i+1}-1} q_j(Z_1, \ldots, Z_i)Z_{i+1}^j.$$

Since $q_j \in K_0[\mathbf{B}_i] \cong K_i$ we can interpret them as field elements; the same argument applies to

$$f_{i+1} := X_{i+1}^{d_{i+1}} + \sum_{j=0}^{d_{i+1}-1} p_j(Z_1,\ldots,Z_i) Z_{i+1}^j,$$

with $p_j \in K_0[\mathbf{B}_i] \cong K_i$.

So both g and f_{i+1} can be interpreted as univariate polynomials in $K_i[Z_{i+1}]$. By the Extended Euclidean Algorithm we obtain $s \in K_i[Z_{i+1}]$ such that

$$sg + tf_{i+1} = 1, \quad \deg_{i+1}(s) < d_{i+1},$$

for a suitable $t \in K_i[Z_{i+1}]$, which we do not need to compute. By interpreting the coefficients of s in K_i as elements of $K_0[\mathbf{B}_i]$, we can interpret s as an element of $K[\mathbf{B}_{i+1}]$ and then $\pi_{i+1}(s)\pi_{i+1}(g) = 1$; so

$$s \in K[\mathbf{B}_{i+1}] \cong K_{i+1}$$

is the inverse of $g \in K[\mathbf{B}_{i+1}] \cong K_{i+1}$.

8.3.6 Polynomial factorization

As we remarked before, 'solving' requires a factorization algorithm.

Such an algorithm will be discussed in detail in the second Part. We simply give here a resumé of that part:

> Gauss studied the relation between the factorization algorithm over a domain and the one over its quotient field (Section. 6.1);
>
> an algorithm by Berlekamp allows us to factorize polynomials over any finite field and, in particular, over prime fields $\mathbb{Z}_p$;
>
> an algorithm by Hensel allows us, given a domain D and a principal ideal $(p) \subset D$, to use the existence of a polynomial factorization algorithm for $D/(p)$ in order to obtain one for $D/(p^n)$ for each n;
>
> Zassenhaus proposed how to apply Hensel algorithms to obtain, under some assumptions, a polynomial factorization algorithm for a domain D from that for their quotients by principal ideals; by Gauss such an algorithm gives a polynomial factorization algorithm for the quotient field of D;
>
> if $L \supseteq K$ is a single algebraic field extension, how to obtain a factorization algorithm for L from that of K is a 'classical' result.

On the basis of these results we obtain the following:

> we know how to factorize over the finite prime fields $\mathbb{Z}_p$, via Berlekamp;
>
> Zassenhaus, via Berlekamp and Hensel, gives a factorization algorithm over $\mathbb{Z}$;
>
> and, via Gauss, over $\mathbb{Q}$;

8.3 Representation and Arithmetics

as a consequence, Zassenhaus allows us to use Hensel and Gauss for obtaining a factorization algorithm over a single transcendental extension $K(X)$ of a field K which possesses a factorization algorithm;

and, by iteration, over $K(X_1, \ldots, X_r)$;

finally a factorization algorithm exists for any single algebraic field extension of a field K which possesses a factorization algorithm,

so that any field in $\mathcal{K}$ possesses a polynomial factorization algorithm.

8.3.7 Solving polynomial equations

Let $g(Z) \in K[Z]$ be a polynomial; it is now easy to find a larger field $L \in \mathcal{K}$ which contains a root α of g and to explicitly produce such a root.

We 'only' have to factorize $g(Z)$ in $K[Z]$ and choose a monic irreducible factor $h(Z) \in K[Z]$ (of least degree for gaining efficiency in subsequent arithmetical computations!). We then interpret the coefficients of h to be polynomials in $K_0[\mathbf{B}]$ and we obtain a monic polynomial in $K_0[Z_1, \ldots, Z_r][Z]$. We then consider the admissible sequence

$$(f_1, \ldots, f_r, h(Z_1, \ldots, Z_r, Z_{r+1}))$$

in $K_0[Z_1, \ldots, Z_{r+1}]$ defining a field $K_{r+1} \in \mathcal{K}$, which is a simple extension of K_r by the root $\pi_{r+1}(Z_{r+1})$ of $h(Z) \in K[Z]$.

Algorithm 8.3.2. Repeating this procedure, we obtain an algorithm which, given $g(Z) \in K[Z]$, computes a larger field $L \in \mathcal{K}$ which is the splitting field of g over K and explicitly produces all the roots of g in L. In presenting it, we will restrict ourselves to the case of a squarefree polynomial g, since reduction to this case and multiplicity count can be obtained in any effective field.

This algorithm is described in Figure 8.2 and it is just a formalization of the computations shown in Example 5.2.5.

8.3.8 Monic polynomials

Given $\alpha \in K$ there are algorithms for computing the minimal polynomial of α over K_0.

One approach[3] uses linear algebra; it iteratively checks whether (the vector representations of) $1, \alpha, \ldots, \alpha^r$ are linearly dependent over K_0. When a linear relation

$$\alpha^r - \sum_{i=0}^{r-1} a_i \alpha^i$$

---

[3] An alternative one is based on the concept of norm (cf. Section 10.5 and Section 16.3): since $Z - \alpha$ divides $N_{K/k}(Z - \alpha) =: h(X) \in k[Z]$, then $h(\alpha) = 0$; moreover $Z - \alpha$ is linear and so irreducible; therefore (by Proposition 16.3.1) h is a power of an irreducible polynomial, so that the minimal polynomial of α is $SQFR(h)$.

Fig. 8.2. Solving polynomial equations

$(L, \alpha_1, \ldots, \alpha_s) := \mathbf{Solve}(g(Z))$
where
    K is the explicitly given field $K_0[Z_1, \ldots, Z_r]/(f_1, \ldots, f_r)$
    $g(Z) \in K[Z]$ is a squarefree polynomial
    L is an explicitly given splitting field of g over K
    $\alpha_i \in L$
    $\{\alpha_1, \ldots, \alpha_s\}$ is the set of the roots of g
$t := r, L := K, \mathbf{Roots} := \emptyset, h := g$
While $\deg(h) > 0$ **do**
    Factor $h(Z) \in L[Z]$
    For each linear factor $(Z - \alpha)$ **do**
        $\mathbf{Roots} := \mathbf{Roots} \cup \{\alpha\}, h := h/(Z - \alpha)$
    If $\deg(h) > 0$ **then**
        choose a monic irreducible factor $p(Z) \in L[Z]$ of h
        $f_{t+1} := p(Z_{t+1})$
        $L := K_0[Z_1, \ldots, Z_{t+1}]/(f_1, \ldots, f_{t+1})$
        $\alpha := \pi_{t+1}(Z_{t+1})$
        $\mathbf{Roots} := \mathbf{Roots} \cup \{\alpha\}$
        $h := h/(Z - \alpha)$
        $t := t + 1$
---

has been found, then

$$g(Z) := Z^r - \sum_{i=0}^{r-1} a_i Z^i$$

is the minimal polynomial of α.

8.4 Primitive Element Theorems

It is possible to interpret Corollary 7.4.9 as a result proving that each finite algebraic extension of a finite field is simple.

In this section we intend to generalize this result to each finite algebraic extension of a perfect field, provided the extension involves at most a single inseparable element. We begin by noting:

Lemma 8.4.1. *Let* $K \supset k$ *be a finite algebraic extension of a perfect field k; then K is also perfect.*

Proof Either:

    $\mathrm{char}(k) = 0$ and so $\mathrm{char}(K) = 0$, or

    $\mathrm{char}(k) \neq 0$, k is finite, and so, setting $q := \mathrm{card}(k)$, $[K:k] := n$, we have $\mathrm{card}(K) = q^n$,   ♄

and then we introduce the following crucial

8.4 Primitive Element Theorems

Lemma 8.4.2. *Let k be an infinite field and let $K := k[\beta, \alpha]$ be a finite algebraic extension of k, where α is separable. Then there is $\xi \in K$ such that $K = k[\xi]$.*

Proof Let $f, m \in k[X]$ be the minimal polynomials over k of β and α respectively, and let L be a field where both polynomials split into linear polynomials; let $\beta = \beta_1, \ldots, \beta_r \in L$ be the roots of f and $\alpha = \alpha_1, \ldots, \alpha_s \in L$ the roots of m.

Since $\alpha_j \neq \alpha_1$, for $j \neq 1$, then for all i, $1 \leq i$, and all $j \neq 1$, there is at most one[4] $c_{ij} \in k$ such that

$$\beta_1 + c_{ij}\alpha_1 = \beta_i + c_{ij}\alpha_j. \tag{8.1}$$

Since k is infinite, there is $c \in k$ such that (cf. Proposition 6.5.7)

$$\beta_1 + c\alpha_1 \neq \beta_i + c\alpha_j, \text{ for all } i, \text{ for all } j \neq 1.$$

Let $\xi := \beta + c\alpha$; α is a root both of $m(X) \in k[X]$ and of

$$h(X) := f(\xi - cX) \in k[\xi][X],$$

because

$$h(\alpha) = f(\xi - c\alpha) = f(\beta) = 0;$$

therefore α is a root in L of $\gcd(m, h) \in k[\xi][X]$.

Any other root in L of $\gcd(m, h)$ is among $\alpha_2, \ldots, \alpha_s$ (since it must be a root of m). However, if $h(\alpha_j) = 0$ for some $j \neq 1$, then

$$f(\xi - c\alpha_j) = h(\alpha_j) = 0$$

and therefore there would be i such that $\xi - c\alpha_j = \beta_i$, i.e.

$$\beta_1 + c\alpha_1 = \beta_i + c\alpha_j,$$

giving a contradiction. Since $\gcd(m, h)$ is squarefree, then

$$\gcd(m, h) = (X - \alpha)$$

and so $\alpha \in k[\xi]$ and $\beta = \xi - c\alpha \in k[\xi]$ too.

Therefore $k[\beta, \alpha] = k[\xi]$. ♄

Example 8.4.3. Consider $k := \mathbb{Q}$ and $K = k[\beta, \alpha]$ where β is a root of $f(X) := X^2 - 2$ and α is a root of $m(X) := X^2 - 3$. The computations below will prove that $K = k[\xi]$ where $\xi = \beta + \alpha$, i.e. $c = 1$.

---

[4] Here we assume that α is separable.

In order to represent K as $k[\xi]$ we need of course to compute the minimal polynomial of ξ. Based on that, an algorithm to compute ξ could consist of:

choose c;

compute the minimal polynomial of $\xi = \beta + c\alpha$;

check whether K and $k[\xi]$ have the same dimension: if so the solution is found; otherwise the procedure has to be repeated for a different choice of c.

Finding the minimal polynomial of ξ is just linear algebra; in this example we have:

$$\begin{array}{rcl} \xi^0 &=& +1 \\ \xi^1 &=& +\beta +\alpha \\ \xi^2 &=& +5 \quad\quad\quad +2\beta\alpha \\ \xi^3 &=& +11\beta +9\alpha \\ \xi^4 &=& +49 \quad\quad\quad +20\beta\alpha \end{array}$$

from which we obtain $\xi^4 - 10\xi^2 + 1$ and so

$$K = k[\xi] = k[X]/p(X) \text{ where } p(X) = X^4 - 10X^2 + 1.$$

The computation of the gcd of

$$h(X) = (\xi - X)^2 - 2 = X^2 - 2\xi X + \xi^2 - 2$$

and $m(X) = X^2 - 3$ gives $X - \frac{11}{2}\xi + \frac{1}{2}\xi^3$, so that we get

$$\alpha = \frac{11}{2}\xi - \frac{1}{2}\xi^3, \quad \beta = -\frac{9}{2}\xi + \frac{1}{2}\xi^3.$$

Remark 8.4.4. For those who are expert in Gröbner basis technology, I note that the linear algebra computation can be applied via Gröbner bases. In fact

$$k[X, Y]/(M(X), F(X, Y)) \cong k[X, Y, T]/(M(X), F(X, Y), T - Y - cX),$$

where $M(X) \in k[X]$ is the minimal polynomial of α over k and $F(X, Y) \in k[X, Y]$ is such that $F(\alpha, Y) \in k(\alpha)[Y]$ is the minimal polynomial of β over $k(\alpha)$.

A computation of the Gröbner basis of the ideal

$$(M(X), F(X, Y), T - Y - cX)$$

with respect to a lexicographical term-ordering such that $T < X < Y$ will give a basis $(p(T), X - q_X(T), Y - q_Y(T))$ where $p(T)$ is the minimal polynomial of ξ, $\alpha = q_X(\xi)$ and $\beta = q_Y(\xi)$.

8.4 Primitive Element Theorems

Theorem 8.4.5 (Primitive Element Theorem). *Let $K \supset k$ be a finite algebraic extension of k, $K = k[\beta, \gamma_1, \ldots, \gamma_n]$ where each γ_i is separable over k. Then there is $\xi \in K$ such that $K = k[\xi]$.*

Proof The result follows by Corollary 7.4.9 if k is finite, and by iterative application of Lemma 8.4.2 if k is infinite. ♄

Corollary 8.4.6. *Let k, K be as above. Then K is a simple algebraic extension of k.* ♄

Corollary 8.4.7. *Let $K \supset k$ be a finite algebraic extension of a perfect field k. Then K is a simple algebraic extension of k.* ♄

Historical Remark 8.4.8. In Gauss' language a *primitive element of a finite field* (more exactly a *primitive root* of $X^{q-1} - 1$) meant an element such that any other element was a power of it; Gauss introduced this notion mainly because it *provided a computational tool in $GF(q)$ which behaves as do logarithms for the reals* (c.f. 7.4.8).

So we should be aware that *strictu sense* it is not entirely appropriate to describe as primitive the element ξ such that $K = k[\xi]$; the definition arises from a misunderstanding of the original meaning of the concept and from an obvious confusion arising from Corollary 7.4.11. In particular such a primitive element does not allow us to define a logarithmic function!

However, in one sense it does: it allows us to compute with an algebraic extension as if it were a *single* one and to represent elements by univariate polynomials.

To conclude this analysis we should show that the same result cannot be further generalized, by illustrating what happens when k, char$(k) \neq 0$, is infinite and $K = k[\beta, \gamma]$ is an extension by two inseparable elements. In order to do so we need to state the following

Lemma 8.4.9. *Let k be a field and let $K = k[\xi]$ be a simple algebraic extension. Then there are only finitely many fields L such that*

$$k \subseteq L \subseteq K.$$

Proof Let $f(X) \in k[X]$ be the minimal polynomial of ξ over k and, for a field L such that $k \subseteq L \subseteq K$, let $g(X) = \sum_{i=1}^{n} \alpha_i X^i \in L[X]$ be the minimal polynomial of ξ over L and $L' := k[\alpha_1, \ldots, \alpha_n] \subset L$.

Then ξ has the same degree over L and L'. Therefore $[L' : k] = [L : k]$, $[L : L'] = 1$ so that $L = k[\alpha_1, \ldots, \alpha_n]$.

On the other hand, f is a multiple of g.

As a consequence the fields L, such that $k \subseteq L \subseteq K$, are in biunivocal relation with the factors of f in $K[X]$. ♮

Example 8.4.10. We are now able to show that Corollary 8.4.7 cannot hold for an infinite field k of finite characteristic.

Let k be an infinite field, $\text{char}(k) = p \neq 0$, let $a, b \in k \setminus k^p$ and let $K = k[\alpha, \beta]$ where α and β satisfy the relations $\beta^p = b$ and $\alpha^p = a$.

We need to prove that K is not a simple algebraic extension of k.

Assuming the contrary, then there are at most finitely many fields L such that $k \subseteq L \subseteq K$; therefore there are finitely many elements among the fields $k[\beta + c\alpha]$ with $c \in k$.

In other words, there is a field L, $k \subseteq L \subseteq K$, and two different elements $c_1, c_2 \in k$ such that

$$L = k[\beta + c_1 \alpha] = k[\beta + c_2 \alpha].$$

Since both $\beta + c_1\alpha$ and $\beta + c_2\alpha$ are in L, their difference $(c_1 - c_2)\alpha$ is also in L; since $c_1 - c_2 \neq 0$ we conclude that both α and β are in L.

Therefore the assumption that K is a simple algebraic extension of k allows us to conclude that

$$L = k[\alpha, \beta]$$

but this is impossible since we get the absurd result that

$[k[\alpha, \beta] : k] = p^2$ and
$[L : k] = p$.

The last statement follows by the simple verification that, for each $c \in k$

$$(\beta + c\alpha)^p = \beta^p + c^p \alpha^p = b + c^p a,$$

so that $\beta + c\alpha$ satisfies the polynomial $X^p - b - c^p a \in k[X]$.

9
Steinitz

This chapter is mainly devoted to dealing with the deeper aspects of field extensions.

In Section 9.1 I prove the existence of a 'universal extension field' $\bar{k}$ of a field k, in which the polynomials in $k[X]$ and even those in $\bar{k}[X]$ split into linear factors: this notion of *algebraic closure* generalizes the property of $\mathbb{C}$ with respect to $\mathbb{R}$.

In Section 9.2, I discuss the argument which was only hinted at in Lemma 8.2.1, namely, the fact that a set of (not necessarily finite) generators of a field extension $K \supset k$ can be reordered and separated so that there is an intermediate field K_{trasc} such that K is an algebraic extension of K_{trasc}, which is a purely transcendental extension of k. In so doing I introduce the notions of algebraic dependence and transcendental bases and show that it is possible to introduce the concept of degree for transcendental extensions, as we did for algebraic ones.

In Section 9.3 I describe the structure of finite extensions based on the above analysis and on the result that algebraic extensions of a field k are a purely inseparable extension of a separable extension[1] of k.

In Section 9.4 I introduce another crucial concept, that of the *universal field* of a prime field k: this is a field which contains an isomorphic copy of any finite extension field K over k, i.e. a field in which all fields satisfying Kronecker's Model have a representation.

In Section 9.5 I discuss further the structure of simple transcendental extensions, proving the Lüroth Theorem, which states that every field K, such that $k \subset \mathsf{K} \subseteq k(X)$, is a simple transcendental extension by a non-constant rational function $\eta(X) \in k(X)$.

---

[1] This result is significant if k is not perfect, i.e. it is an infinite field of finite characteristic. If k is perfect all algebraic extensions are separable.

9.1 Algebraic Closure

Lemma 9.1.1. *Let k be a field. The following conditions are equivalent:*

(1) each non-constant polynomial $f(X) \in k[X]$ has a root in k;
(2) for each polynomial $f(X) \in k[X]$, k is a splitting field of f;
(3) each polynomial $f(X) \in k[X]$ factorizes into linear factors in $k[X]$;
(4) each non-constant irreducible polynomial $f(X) \in k[X] \setminus k$ is linear;
(5) for each algebraic extension $K \supset k$, $K = k$.

Proof Clearly conditions (1), (2), (3), (4) are equivalent.
(5) $\implies$ (1) follows from the fact that $K := k[X]/g(X)$, where $g(X)$ is a factor of $f(X)$ in $k[X]$, is an algebraic extension of k and contains a root of f.
Conversely, (4) $\implies$ (5) holds since, if $K \supset k$ is an algebraic extension of k and $\alpha \in K$, the minimal polynomial of α over k is linear, and so $\alpha \in k$. Therefore, $K = k$ holds for each algebraic extension $K \supset k$. ♮

Definition 9.1.2. *A field k is called* algebraically closed *if it satisfies the conditions of Lemma 9.1.1.*

Definition 9.1.3. *If k is a subfield of a field K, K is called an* algebraic closure *of k if it is algebraically closed and an algebraic extension of k.*

Lemma 9.1.4. *Let K be an algebraic extension of k. K is an algebraic closure of k iff each polynomial $f(X) \in k[X]$ splits in $K[X]$.*

Proof We have to prove that, if each polynomial $f(X) \in k[X]$ splits in $K[X]$, then each polynomial $g(X) \in K[X]$ has a root in K. Assuming this is false, let $g(X) \in K[X]$ be an irreducible polynomial which has no root in K and let us consider the algebraic extension

$$K[\alpha] := K[X]/g(X) \supset K \supset k;$$

since α is algebraic over K which is algebraic over k, there exists a polynomial $f(X) \in k[X]$ such that $f(\alpha) = 0$. As a consequence, f splits in K, $X - \alpha \in K[X]$, α is a root of g in K. ♮

Proposition 9.1.5. *Let k be a field. Then there is an algebraic closure of k.*

9.1 Algebraic Closure

Proof We can suppose that k is well-ordered. As a consequence, the polynomial ring $k[X]$ is also well-ordered by $\prec$ and in such a way that $k \subset k[X]$ is a section[2] of $k[X]$[3].

For each polynomial $f(X) \in k[X]$ we will build two well-ordered fields F_f, G_f which will satisfy the following properties

(1_f) k is a section of F_f;
(2_f) for all $g \prec f$, G_g is a section of F_f;
(3_f) $f(X)$ splits in G_f;
(4_f) F_f is a section of G_f.

The construction of these fields will be done by *transfinite induction*: we will assume that we are given fields F_g, G_g satisfying conditions (1_g), (2_g), (3_g), (4_g) for each $g \in k[X]$, $g \prec f$, and we will construct the required fields $\mathsf{F}_f, \mathsf{G}_f$.

Let us define

$$\mathsf{F}_f := k \bigcup \left(\bigcup_{g \prec f} \mathsf{G}_g \right)$$

and impose on it the unique well-ordering $<_f$ such that both k and each $\mathsf{G}_g, g \prec f$, are sections of it. It is then clear that F_f satisfies conditions (1_f), (2_f).

To construct G_f, we will build a splitting field of F_f as described in Theorem 5.2.3 – of which we will use the same notation – with a twist: in building the tower

$$\mathsf{F}_f = k_0 \subseteq k_1 \subseteq \cdots \subseteq k_{n-1} = \mathsf{G}_f,$$

we will

> impose an ordering on each k_r extending the one on k_{r-1} in such a way that k_{r-1} is a section[4] of k_r;
> extend it on $k_r[X]$ and
> in each step, choose as g_i the minimal irreducible factor of f_i with respect to $\prec$.

---

[2] If the set B is well-ordered by $\prec$, $A \subset B$ is called a *section* if

$$a \in A, b \in B \setminus A \implies a \prec b.$$

[3] If we are given a well-ordering $\prec$ on k, we generalize it to $k[X]$ as follows: given $f(X) = \sum_{i=0}^{n} a_i X_i$, $g = \sum_{i=0}^{m} b_i X_i$ in $k[X]$, then $f \prec g$ will hold iff either

$n < m$, or
$n = m$ and there is j such that $a_j \prec b_j$ and $a_i = b_i$, for all $i > j$.

[4] To establish this, given a well-ordering $\prec$ on k_{r-1}, we deduce a well-ordering on $k_{r-1}[X]$ such that k_{r-1} is a section of k_r, and we restrict it to the set of all the polynomials in $k_{r-1}[X]$ of degree lower than g_{r-1}, which is the classical representation of k_r.

It is then clear that G_f satisfies the conditions (3_f), (4_f).
Having in this way obtained the set

$$\mathfrak{S} := \{k\} \cup \{G_g : g \in k[X]\},$$

we impose on it an ordering induced by $\prec$ – and denoted as such – by

$G_f \prec G_g \iff f \prec g$, and
$k \prec G_g$, for all $g \in k[X]$.

Since, for each $f \in k[X]$,

k is a section of F_f which is a section of G_f, and
G_g is a section of F_f which is a section of G_f, for each $g \prec f$,

we conclude that

$$K := \bigcup_{F \in \mathfrak{S}} F$$

is a field[5].
Moreover, since K is

algebraic over k by its construction, and
algebraically closed by Lemma 9.1.4, since each polynomial $g(X) \in k[X]$ splits in G_g,

it is an algebraic closure of k.
♄

Proposition 9.1.6. *Let k be a field and let K be the algebraic closure defined in the proof of Proposition 9.1.5. Let K' be an algebraic closure of k; then there is a k-isomorphism $\Psi : K \mapsto K'$.*

---

[5] We use the fact that:

if, in an ordered set of fields, every preceding field is a subfield of its successor, then the union of this set of fields is a field too.

The proof is quite elementary: let $\mathfrak{S}$ be this set of fields ordered by $\prec$ and let $K := \cup_{F \in \mathfrak{S}} F$. For any two elements $\alpha_1, \alpha_2 \in K$, let $F_i \in \mathfrak{S}$ be such that $\alpha_i \in F_i$ for $i = 1, 2$. One among these two fields, let us call it F, contains the other and so contains α_1 and α_2. Therefore $\alpha_1 + \alpha_2$ and $\alpha_1 \alpha_2$ are defined in F and this definition coincides with that in each field $F \in \mathfrak{S}$ such that $F \succ \mathsf{F}$ and can so be used as the definition in K.
To prove the laws of field operations, e.g. the distributivity law, for any three elements $\alpha_1, \alpha_2, \alpha_3 \in K$ we consider $F_i \in \mathfrak{S}$ such that $\alpha_i \in F_i$ for $i = 1, 2, 3$. Then, in each field $F \in \mathfrak{S}$ such that $F \succ F_i$, for all i, the distributivity law holds, and so it holds in K.

9.1 Algebraic Closure

Proof The proof will again be obtained by transfinite induction: for each section $\mathsf{K} \subset K$ we will construct a section $\mathsf{K}' \subset K'$ and a k-isomorphism $\Psi_\mathsf{K} : \mathsf{K} \mapsto \mathsf{K}'$ such that

(1_K) For each section $\mathsf{H} \subset \mathsf{K}$ we have $\Psi_\mathsf{K}(a) = \Psi_\mathsf{H}(a)$, for all $a \in \mathsf{H}$.

(2_K) If K has a last element l, let us define $\mathsf{H} \subset \mathsf{K}$ to be the section such that $\mathsf{K} = \mathsf{H} \cup \{l\}$, and $f(X) \in \mathsf{H}[X]$ to be the irreducible polynomial whose root is l; then $\Psi_\mathsf{K}(l)$ is the first root – with respect to the well-ordering of K' – of $\Psi_\mathsf{H}(f(X)) \in \mathsf{H}'[X]$;

assuming that we already have, for each section $\mathsf{H} \subset \mathsf{K}$, the section H' and the k-isomorphism Ψ_H.

We have to consider two cases:

K has a last element l: in this case, with the notation of (2_K), let $l' \in K'$ be the first root of $\Psi_\mathsf{H}(f(X))$. Then defining

$\mathsf{K}' := \mathsf{H}' \cup \{l'\}$, and

$$\Psi_\mathsf{K}(a) := \begin{cases} l' & \text{iff } a = l \\ \Psi_\mathsf{H}(a) & \text{iff } a \in \mathsf{H}, \end{cases}$$

(1_K) and (2_K) hold.

K has no last element: in this case $\mathsf{K} = \bigcup_{\mathsf{H} \in \mathfrak{S}(\mathsf{K})} \mathsf{H}$, where $\mathfrak{S}(\mathsf{K})$ is the set of all the sections of K; then we define $\mathsf{K}' = \bigcup_{\mathsf{H} \in \mathfrak{S}} \mathsf{H}'$. For each $a \in \mathsf{K}$, (1_K) implies that there is a unique $a' \in \mathsf{K}'$ such that $\Psi_\mathsf{H}(a) = a'$, for all $\mathsf{H} \in \mathfrak{S}$; then setting

$$\Psi_\mathsf{K}(a) = a', \text{ for all } a \in \mathsf{K},$$

(1_K) and (2_K) hold.

Therefore, for each section $\mathsf{K} \subset K$, we have constructed a section $\mathsf{K}' \subset K'$ and a k-isomorphism $\Psi_\mathsf{K} : \mathsf{K} \mapsto \mathsf{K}'$ satisfying conditions (1_K) and (2_K). Let $K'' := \bigcup_{\mathsf{H} \in \mathfrak{S}(K)} \mathsf{H}$, where $\mathfrak{S}(K)$ is the set of all the sections of K, and let $\Psi : K \mapsto K''$ be such that it satisfies

$$\Psi(a) = \Psi_\mathsf{H}(a), \text{ for all } a \in K, \forall \mathsf{H} : a \in \mathsf{H}.$$

Therefore Ψ is a k-isomorphism and, since K is algebraically closed, so also is K''; since K' is an algebraic extension of K'' and K'' is algebraically closed, then $K' = K''$ and both K and K' are k-isomorphic. ♄

Theorem 9.1.7 (Steinitz). *Let k be a field. Then there is an algebraic closure K of k. Any two algebraic closures of k are k-isomorphic.*

Proof This follows from Propositions 9.1.5 and 9.1.6[6]. ♄

As we know and will prove in Section 12.1, $\mathbb{C}$ is algebraic closed and is the algebraic closure of any field $K: \mathbb{R} \subseteq K \subseteq \mathbb{C}$.

Proposition 9.1.8. *Let $K \subset \mathbb{C}$ and let*

$$K_{\mathrm{alg}} := \{\alpha \in \mathbb{C} : \alpha \text{ is algebraic over } K\};$$

then K_{alg} is the algebraic closure of K.

Proof The fact that K_{alg} is a field is a consequence of Corollary 6.7.5; it is then sufficient to show that it is algebraically closed: if

$$f(X) = \sum_{i=0}^{n} a_i X^i \in K_{\mathrm{alg}}[X],$$

then it has a root $\alpha \in \mathbb{C}$; therefore α is algebraic over $K[a_0, a_1, \ldots, a_n]$ which is algebraic over K, since each a_i is algebraic over K; as a consequence $\alpha \in K_{\mathrm{alg}}$, i.e. f has a root in K_{alg}. ♄

Remark 9.1.9. I intend now to describe the algebraic closure of $\mathbb{Z}_p$, for prime p.

Let us denote, for all $i \in \mathbb{N} \setminus \{0\}$, $e_i := i!$, $q_i := p^{e_i}$, $F_i := GF(q_i)$. Since, by Corollary 7.1.6, $F_i \subset F_{i+1}$ for each i, we can define the field

$$F := \bigcup_i F_i.$$

Since for all i, $GF(p^i) \subset F_i \subset F$, and is the splitting field of each irreducible polynomial of degree i, then F is the algebraic closure of $\mathbb{Z}_p$.

9.2 Algebraic Dependence and Transcendency Degree

In this section, we will discuss the argument which was only hinted at in Lemma 8.2.1, by introducing the notions of algebraic dependence, transcendental extension and transcendental degree. In the context which we are contemplating, we are given a field k (e.g. $\mathbb{Q}$ or a finite field) and a field extension $K \supset k$ (e.g. a field in Kronecker's Model) and a *finite* set of elements $\alpha_1, \ldots, \alpha_n \in K$ (e.g. the generators of the *finite* extension $K = k(\alpha_1, \ldots, \alpha_n)$ of k). In introducing the notion and in our first rudimentary results we will drop

---

[6] The more recent presentations substitute 'transfinite induction' by Zorn's Lemma, but, as Gordan said, 'das ist Theologie und keine Mathematik'.

9.2 Algebraic Dependence and Transcendency Degree

the restriction of finiteness; however, we will reintroduce it in the proof – not the statement – of the main result.

Definition 9.2.1. *Let $k \subset K$ be two fields and let $\mathsf{A}, \mathsf{B} \subset K \setminus k$ be sets.*

An element $v \in K$ is called

algebraically dependent *on A over k, if there is a polynomial*

$$f(X_1, \ldots, X_n, Y) \in k[X_1, \ldots, X_n, Y] \setminus k[X_1, \ldots, X_n]$$

and elements $\alpha_1, \ldots, \alpha_n \in \mathsf{A}$ such that $f(\alpha_1, \ldots, \alpha_n, v) = 0$.
algebraically independent *on A over k, if for each polynomial*

$$f(X_1, \ldots, X_n, Y) \in k[X_1, \ldots, X_n, Y] \setminus k[X_1, \ldots, X_n]$$

and for each element $\alpha_1, \ldots, \alpha_n \in \mathsf{A}$, $f(\alpha_1, \ldots, \alpha_n, v) \neq 0$.

The set B is said to be algebraically dependent *on A over k if all elements of B are algebraically dependent on A over k.*
The sets A and B are said to be equivalent *over k if they are mutually dependent.*
The set A is called

algebraically dependent *over k, if there is a polynomial*

$$f(X_1, \ldots, X_n) \in k[X_1, \ldots, X_n] \setminus \{0\}$$

and elements $\alpha_1, \ldots, \alpha_n \in \mathsf{A}$ such that

$$f(\alpha_1, \ldots, \alpha_n) = 0;$$

algebraically independent *over k, or a* transcendental set *of K over k, if for each polynomial $f(X_1, \ldots, X_n) \in k[X_1, \ldots, X_n] \setminus \{0\}$ and for each element $\alpha_1, \ldots, \alpha_n \in \mathsf{A}$, $f(\alpha_1, \ldots, \alpha_n) \neq 0$.*

If A is a transcendental set of K over k, and $K = k(\mathsf{A})$ then K is called a pure transcendental extension of k.
A transcendental set A over k is called a transcendental basis *of K over k if it is not a proper subset of another transcendental set of K over k.*

Remark 9.2.2. Clearly

if B is algebraically dependent on A and C is algebraically dependent on B, then C is algebraically dependent on A;
'equivalence' is an equivalence relation;

if A is algebraically independent over k and I is any set with the same cardinality as A then $k(A)$ is k-isomorphic with the field of the rational functions $k(X_i : i \in I)$.

Lemma 9.2.3. *Let* $A \subset K \setminus k$ *be a transcendental set of K over k; $z \in K \setminus k$ be such that $z \notin A$, and let $B = A \cup \{z\}$. Then the following are equivalent:*

(1) B *is a transcendental set of K over k;*
(2) z *is transcendental over* $k(A)$.

Proof

$(2) \implies (1)$ Assume there is a polynomial

$$f(X_1, \ldots, X_n, Z) \in k[X_1, \ldots, X_n, Z] \setminus \{0\}$$

and elements $\alpha_1, \ldots, \alpha_n, \beta \in B$ such that $f(\alpha_1, \ldots, \alpha_n, \beta) = 0$.
There are suitable polynomials $a_i \in k[X_1, \ldots, X_n]$ such that

$$f = \sum_i a_i(X_1, \ldots, X_n) Z^i$$

and at least one of them, say a_I, is not null.
Since A is a transcendental set, we can deduce that $a_I(\alpha_1, \ldots, \alpha_n) \neq 0$ and we can assume wlog that $\beta = z$.
Then z is a root of the polynomial $g(Z) := f(\alpha_1, \ldots, \alpha_n, Z) \in k(A)[Z]$; since z is transcendental, $g(Z) = 0$ in $k(A)[Z]$.
From $g(Z) = 0$ and $a_I(\alpha_1, \ldots, \alpha_n) \neq 0$ we obtain a contradiction.

$(1) \implies (2)$ Conversely, let $g(Z) = \sum \beta_i Z^i \in k(A)[Z]$. There are finitely many elements $\alpha_1, \ldots, \alpha_n \in A$ and polynomials $a_i \in k[X_1, \ldots, X_n]$ such that $a_i(\alpha_1, \ldots, \alpha_n) = \beta_i, \forall i$.
Let $f := \sum a_i(X_1, \ldots, X_n) Z^i \in k[X_1, \ldots, X_n, Z] \setminus \{0\}$; since B is a transcendental set of K over k we have

$$g(z) = \sum \beta_i z^i = \sum a_i(\alpha_1, \ldots, \alpha_n) z^i = f(\alpha_1, \ldots, \alpha_n, z) \neq 0.$$

♄

Corollary 9.2.4. *A transcendental set* A *of K over k is a transcendental basis of K over k iff K is an algebraic extension of $k(A)$.* ♄

Proposition 9.2.5. *Let $K \supset k$ be an extension field generated by a set* A. *Then there is a subset* $B \subseteq A$ *such that*

A *and* B *are equivalent;*

9.2 Algebraic Dependence and Transcendency Degree

B *is algebraically independent;*
A *depends algebraically over* B*;*
B *is a transcendental basis of K over k;*
and

$$k \subseteq k(\mathsf{B}) \subseteq k(\mathsf{A}) = K,$$

where $k(\mathsf{B})$ is a pure transcendental extension of k and K is an algebraic extension of $k(\mathsf{B})$.

Proof Let A be well-ordered by $\prec$ and let B consists of all elements $\alpha \in \mathsf{A}$ such that α is algebraically independent on $\{\beta \in \mathsf{A} : \beta \prec \alpha\}$.
With this construction, the required properties obviously hold. $\boxed{\natural}$

Lemma 9.2.6 (Steinitz). *Let $\mathsf{B}_n := \{\beta_1, \ldots, \beta_n\}$ be a finite subset of $K \setminus k$ which is algebraically independent and let A be a subset of $K \setminus k$ such that for all i, β_i is algebraically dependent on A over k.*
Then, there are $\alpha_1, \ldots, \alpha_n \in \mathsf{A}$ such that

$$\mathsf{C}_n := \{\alpha \in \mathsf{A} : \alpha \neq \alpha_i, 1 \leq i \leq n\} \cup \mathsf{B}_n \tag{9.1}$$

and A are equivalent over k.

Proof The proof is by induction since it obviously holds for $n = 0$.
Assume it holds for $n := \nu - 1$ and let us prove it for $n := \nu$.
We know that β_ν is algebraically dependent on A over k, and so is algebraically dependent on $\mathsf{C}_{\nu-1}$.
Therefore there is a smallest subset in $\mathsf{C}_{\nu-1}$ such that β_ν is algebraically dependent on it.
This subset cannot be contained in $\mathsf{B}_{\nu-1} = \{\beta_1, \ldots, \beta_{\nu-1}\}$, since B_n is algebraically independent; so it contains at least an element

$$\alpha_\nu \in \mathsf{C}_{\nu-1} \setminus \mathsf{B}_{\nu-1}.$$

Therefore there is a polynomial

$$f \in k[Y_1, \ldots, Y_{\nu-1}, Z_1, \ldots, Z_r][T, X] \setminus k[Y_1, \ldots, Y_{\nu-1}, Z_1, \ldots, Z_r][X]$$

such that

$$f(\beta_1, \ldots, \beta_{\nu-1}, \gamma_1, \ldots, \gamma_r, \alpha_\nu, \beta_\nu) = 0$$

where, for each i, $\gamma_i \in \mathsf{C}_{\nu-1} \setminus \mathsf{B}_{\nu-1}$ and $\gamma_i \neq \alpha_\nu$.

Clearly, f gives a dependence relation of α_ν over C_ν. The equivalence between C_n and A then follows. ♄

Corollary 9.2.7. *Let* A *and* B *be two transcendental bases of K over k. If A is finite, then A and B have the same cardinality.*

Proof Let $m := \operatorname{card} \mathsf{A}$. By Lemma 9.2.6, $\operatorname{card} \mathsf{B} \leq \operatorname{card} \mathsf{A}$ and B is finite. Since B is finite, Lemma 9.2.6 proves that $\operatorname{card} \mathsf{A} \leq \operatorname{card} \mathsf{B}$ and the claim. ♄

As a consequence, we can conclude that any transcendental bases of K over k have the same cardinality, if *at least one of them is finite*. The results hold without this restriction which we claim, without proof, in

Fact 9.2.8. *Any transcendental bases of K over k have the same cardinality.*
♄

Definition 9.2.9. *The common cardinality of any transcendental basis of K over k is called the* transcendency degree *of K over k.*

9.3 The Structure of Field Extensions

In order to analyse in more depth the structure of field extensions by considering the structure of algebraic extensions let us introduce

Definition 9.3.1. *Let $K \supset k$ be an algebraic extension of k. Then the set*

$K_{\text{sep}} := \{\beta \in K : \beta \text{ is separable over } k\}$

will be called the greatest separable extension of k in K,

and we claim

Fact 9.3.2. *Let $K \supset k$ be an algebraic extension of k. Then K_{sep} is a field,*
♄

which will be proved in Corollary 10.3.15.

Remark 9.3.3. If k is perfect, for each algebraic extension $K \supset k$ we know $K = K_{\text{sep}}$. Therefore what follows is significant when k is an infinite field of finite characteristic.

9.3 The Structure of Field Extensions

Lemma 9.3.4. *Let $K \supset k$ be an algebraic extension, $\mathrm{char}(k) = p$. Then each $\alpha \in K$ is purely inseparable over K_{sep}.*

Proof Let $f(X) \in k[X]$ be the minimal polynomial of α and e be their exponent of inseparability.
Then $f(X) = g(X^{p^e}) = g(X)^{p^e}$ for a suitable separable polynomial $g(X) \in k[X]$ and $\beta := \alpha^{p^e}$ is a root of g.
As a consequence, α^{p^e} is separable, and so $\alpha^{p^e} \in K_{\mathrm{sep}}$, i.e. α is purely inseparable over K_{sep} (cf. Remark 5.4.10). ♄

Corollary 9.3.5. *For any algebraic extension $K \supset k$, there is a field K_{sep} such that*

$K \supset K_{\mathrm{sep}} \supset k$;
K_{sep} is a separable extension of k;
K is a purely inseparable extension of K_{sep}.

♄

Let $K \supset k$ be a finite algebraic extension of k, $\mathrm{char}(k) = p$. Let $n_0 := [K_{\mathrm{sep}} : k]$ and let e be such that $[K : K_{\mathrm{sep}}] = p^e$ (cf. Remark 5.4.10) so that

$$[K : k] = [K : K_{\mathrm{sep}}][K_{\mathrm{sep}} : k] = n_0 p^e =: n.$$

It is evident that in the case of a simple algebraic extension $K := k(\alpha)$, this result can be read as follows:

Corollary 9.3.6. *Let $K \supset k$ be a simple algebraic extension $K := k(\alpha)$.*
Let $f(X) \in k[X]$ be the minimal polynomial of α, let e be its exponent of inseparability, n its degree and n_0 its reduced degree. Then

$$f(X) = g(X^{p^e}) = g(X)^{p^e}$$

for a suitable separable polynomial $g(X) \in k[X]$ and $\beta := \alpha^{p^e}$ is a root of g; moreover

- *$k(\beta)$ is the greatest separable extension of k in $k(\alpha)$ and $k(\alpha)$ is a purely inseparable extension of $k(\beta)$;*
- *$[k(\alpha) : k] = n$ is the degree of α and f;*
- *$[k(\beta) : k] = n_0$ is the reduced degree of α and f and it is the degree of β and g;*
- *$[k(\alpha) : k(\beta)] = p^e$ is the degree of inseparability of α and f.*

Proof We only have to prove that $k(\beta)$ is the greatest separable extension of k in $k(\alpha)$ and $k(\alpha)$ is a purely inseparable extension of $k(\beta)$.

Let $\gamma \in k(\alpha)$ and let $h(X) \in k[X]$ be such that $\gamma = h(\alpha)$; then

$$\gamma^{p^e} = h(\alpha)^{p^e} = h(\alpha^{p^e}) = h(\beta) \in k(\beta).$$

So, all elements in $k(\alpha)$ are purely inseparable over $k(\beta)$.

Moreover $k(\beta)$ is separable over k, since β is such.

Each $\gamma \in k(\alpha)$ which is separable, being purely inseparable over $k(\beta)$, is an element of $k(\beta)$ (Lemma 5.4.11). ♄

The results of Proposition. 9.2.5 and Corollary 9.3.5 allow us to describe the structure of field extensions by

Theorem 9.3.7. *Let $K \supset k$ be a field extension; then there is a tower*

$$K \supset K_{\text{sep}} \supset K_{\text{trasc}} \supset k$$

where

K_{trasc} is a purely transcendental extension of k;
K_{sep} is a separable algebraic extension of K_{trasc};
K is a purely inseparable algebraic extension of K_{sep}. ♄

9.4 Universal Field

Let k be a field and let us consider the polynomial ring over k with *infinitely* many variables $k[Y_1, Y_2, \ldots, Y_n, \ldots]$, its quotient field $k(Y_1, Y_2, \ldots, Y_n, \ldots)$ and its algebraic closure $\Omega(k)$.

Definition 9.4.1. $\Omega(k)$ *is the* universal field *over k.*

If k is a prime field and $\mathcal{K}(k)$ denotes the class of all the fields $K \supset k$ to which Kronecker's Model is applicable – i.e. all the finite extension fields K over their prime field k –, $\Omega(k)$ is the field which contains a representation of each field $K \in \mathcal{K}(k)$, in the sense that:

Proposition 9.4.2. *Any field $K \in \mathcal{K}(k)$ can be isomorphically embedded in $\Omega(k)$.*

Proof We know that K is explicitly given by specifying integers $d \geq 0$, $r \geq 0$ and an admissible sequence $f_1, \ldots, f_r$ with $f_i \in k(Y_1, \ldots, Y_d)[Z_1, \ldots, Z_i]$ so that

$$K \cong k(Y_1, \ldots, Y_d)[Z_1, \ldots, Z_r]/(f_1, \ldots, f_n).$$

Therefore, using freely the notation of Definition 8.2.2 and Proposition 8.2.3, we define isomorphic embeddings $\Pi_i : L_i \mapsto \Omega$, for all i.
We begin by noting that $L_0 = k(Y_1, \ldots, Y_d) \subset \Omega$ so that Π_0 is the immersion.
Assuming we have already defined Π_{i-1}, we extend it to

$$\Pi_i : L_i = L_{i-1}[Z_i]/g_{i-1}(Z_i) \cong L_{i-1}[\alpha_i] \mapsto \Omega$$

by sending α_i to any root of $\Pi_{i-1}(g_{i-1}(Z)) \in \Omega[Z]$. ♄

9.5 Lüroth's Theorem

Let k be a field and $K := k(\xi)$ a simple transcendental field; its elements are rational functions $\eta := f(\xi)/g(\xi)$ where $f, g \in k[X]$, $g \neq 0$, and, wlog we assume $\gcd(f, g) = 1$.

Definition 9.5.1. *The* degree *of η is* $\deg(\eta) := \max(\deg(f), \deg(g))$.

Lemma 9.5.2. *With this notation:*

(1) *The polynomial $P(X, Y) := g(X)Y - f(X) \in k[X, Y]$ is irreducible.*
(2) *The polynomial $Q(X, Y) := g(X)f(Y) - f(X)g(Y) \in k[X, Y]$ has no factor in $k[X]$ nor in $k[Y]$.*

Proof

(1) Assume P were reducible; since it is linear in Y, one of its factors must be independent of Y and so a polynomial in $k[X]$. The existence of such a factor is denied by the assumption that $\gcd(f, g) = 1$.
(2) The irreducibility of P implies that it has no factor in $k[X]$ and this holds if we substitute Y with $f(Y)/g(Y)$ and multiply by $g(Y)$.
Therefore Q has no factor in $k[X]$.
By symmetry we also deduce that Q has no factor in $k[Y]$. ♄

Proposition 9.5.3. *Let $\eta \in k(\xi) \setminus k$. Then:*

(1) *$k(\xi)$ is algebraic over $k(\eta)$;*
(2) *η is transcendental over k;*
(3) *$[k(\xi) : k(\eta)] = \deg(\eta)$.*

Proof Let $f := \sum a_i X^i$, $g := \sum b_i X^i$ and let
$$\Upsilon(X) := g(X)\eta - f(X) \in k(\eta)[X].$$
Then ξ satisfies the equation $0 = \Upsilon(\xi)$.
Also $\Upsilon(X) \neq 0$ in $k(\eta)[X]$: in fact, since $g \neq 0$, there is j such that $b_j \neq 0$; then, if $\Upsilon = 0$, we would have that $b_j\eta - a_j = 0$, $\eta = a_j/b_j \in k$, giving a contradiction with the assumption that η is non-constant.
Therefore

(1) ξ is algebraic over $k(\eta)$;
(2) if η were algebraic over k, then $k(\xi)$ would be an algebraic extension of k, contrary to the assumption that ξ is transcendental;
(3) by Lemma 9.5.2 we know that Υ is irreducible over $k(\eta)$; therefore it is the minimal polynomial of ξ over $k(\eta)$. ♄

The main significance of Proposition 9.5.3 is the case in which K is the rational function field $k(X)$ and η is a non-constant rational function $\eta(X) := f(X)/g(X)$; in this context we can interpret Proposition 9.5.3 as:

Corollary 9.5.4. *Let $K := k(X)$ and let $\rho(X) \in K \setminus k$ be any non-constant rational function. Then:*

$k(X)$ is an algebraic extension over $k(\rho)$, which is a transcendental extension over k;

$[k(X) : k(\rho)] = \deg(\rho)$;

$\deg(\rho)$ depends only on the two fields $k(\rho)$ and $k(X)$, since it is independent from our choice of the generator of $k(X)$ over k;

$k(X) = k(\rho) \iff \deg(\rho) = 1$;

the simple generators of the field $k(X)$ over k are all and only the linear rational functions;

the k-automorphisms of $k(X)$ are all and only the transformations
$$X \mapsto \frac{aX + b}{cX + d}, \ ad - bc \neq 0$$

Theorem 9.5.5 (Lüroth's Theorem). *Every field K such that $k \subset \mathsf{K} \subseteq k(X)$ is a simple transcendental extension, so that $\mathsf{K} = k(\eta)$ for some non-constant rational function $\eta(X)$.*

9.5 Lüroth's Theorem

Proof Let $\rho \in \mathsf{K} \setminus k$. By Corollary 9.5.4, X is algebraic over $k(\rho)$ and so, over K. Then let

$$h_1(Y) := \sum_{i=0}^n a_i Y^i \in \mathsf{K}[Y] \subset k(X)[Y]$$

be the minimal polynomial of X over K; in particular $a_n = 1$. Multiplying h_1 by $b_n(X)$, the least common multiple of the denominators of the a_is, we get the polynomial $H_1(X, Y) := \sum_{i=0}^n b_i(X) Y^i \in k[X, Y]$, where the coefficients satisfy $\gcd_i(b_i) = 1$.

If all the coefficients $a_i = b_i/b_n$ were independent from X, then X would be algebraic over k. Therefore one of them, say a_i, is in $k(X) \setminus k$ and we can represent it as

$$\eta(X) := a_i = \frac{b_i}{b_n} = \frac{f(X)}{g(X)},$$

where $f, g \in k[X]$, $g \neq 0$, and $\gcd(f, g) = 1$.
The polynomial

$$\Upsilon(Y) := g(Y)\eta - f(Y) = g(Y)\frac{f(X)}{g(X)} - f(Y) \in k(\eta)[Y] \subset k(X)[Y]$$

is such that $\Upsilon \neq 0$ and X is a root of it; therefore it is a multiple of $h_1(Y)$ over $k(X)$; i.e. there is $h_2(Y) \in k(X)[Y]$ such that

$$g(Y)\frac{f(X)}{g(X)} - f(Y) = h_1(Y)h_2(Y)$$

in $k(X)[Y]$.

As a consequence, by the Gauss Lemma (Corollary 6.1.6), there is $H_2(X, Y) \in k[X, Y]$ such that

$$Q(X, Y) := g(X)f(Y) - f(X)g(Y) = H_1(X, Y)H_2(X, Y).$$

Obviously, $\deg_X(H_1)$, the degree of H_1 in X, can be bounded by

$$\deg_X(H_1) = \max_i(\deg(b_i)) \geq \max(\deg(f), \deg(g)) = \deg_X(Q);$$

this implies that $H_2(X, Y) \in k[Y]$, which contradicts Lemma 9.5.2. Therefore, up to a constant, we have

$$g(X)f(Y) - f(X)g(Y) = Q(X, Y) = H_1(X, Y) = \sum_{i=0}^n b_i(X) Y^i \in k[X, Y].$$

Then, by symmetry

$$\deg(\eta) = \deg_X(Q) = \deg_Y(Q) = \deg_Y(H_1) = n,$$

and

$$[k(X) : k(\eta)][k(\eta) : \mathsf{K}] = [k(X) : \mathsf{K}] = n = \deg(\eta) = [k(X) : k(\eta)],$$

from which we obtain $[k(\eta) : \mathsf{K}] = 1$, i.e. $\mathsf{K} = k(\eta)$. ♄

10
Lagrange

In this chapter, I introduce some other concepts related to the solution of polynomial equations, and Lagrange's intuition of their rôle in solving root permutations and the analysis of expressions which are invariant under them, which led to Galois Theory.

Conjugate (Section 10.1) algebraic numbers over a field k are those which share the same minimal polynomial $f(X) \in k[X]$; this elementary definition could be formulated within Kronecker's Philosophy by saying that all arithmetical operations over algebraic expressions of a generic root behave in the same way when they are performed on conjugate algebraic numbers.

The reason for this behaviour is simply that, given two conjugate roots $\alpha_1, \alpha_2 \in K \supset k$ whose minimal polynomial is $f(X) \in k[X]$, there is the obvious isomorphism

$$L_1 := k[\alpha_1] \cong k[X]/f(X) \cong k[\alpha_2] =: L_2,$$

so that both the algebraic expressions and the algebraic operations in $k[X]/f(X)$ represent the corresponding ones in each L_i.

It is then natural to consider all the subfield structures[1] $L_i \subseteq K$ which represent $k[\alpha_1]$. This leads me to introduce the notion of *k-isomorphisms of L into K* (Section 10.3).

In many applications (mainly the ones related to Galois Theory) we need to consider *all* possible k-isomorphisms of $L = k[\alpha]$; since, in order to do so, it would be sufficient to consider all the conjugates of α, we must require that K – the setting in which we are considering the subfield structures k-isomorphic

---

[1] Be aware that this expression has been chosen in order to clarify that I am not only simply considering the subfields $L_i \subseteq K$ which are isomorphic to $k[\alpha_1]$: there is a single subfield $L \subset \mathbb{R}$ which is isomorphic to $\mathbb{Q}[\sqrt{2}]$, namely $\mathbb{Q}[\sqrt{2}]$ itself, but it has two different field structures, the one in which $\sqrt{2}$ is positive and the one in which it is negative. This crucial distinction is strictly related to the discussion of Examples 8.1.1 and 8.1.2.

to L – contains the splitting field of f. Kronecker's Philosophy gives us a tool for repeatedly solving polynomial equations by repeatedly adjoint roots; therefore the consideration of k-isomorphisms into K will be a success if K is such that it contains, together with a number, all its conjugates: such fields are called *normal* and their basic properties are discussed in Section 10.2.

In Section 10.4, using the notions of this chapter and the Primitive Element Theorem (8.4.5) I discuss the properties of the splitting field of an irreducible polynomial, both in the separable and inseparable cases.

If we consider a finite algebraic extension $K \supset k$ and an element $\tau \in K \setminus k$ then multiplication by τ is a k-linear function $\Psi_\tau : K \mapsto K$; the notions of the *trace* and *norm* of Ψ_τ, which can be associated to τ itself, are discussed in Section 10.5, where I present their essential properties and their basic relation to the notions discussed before; in particular, I show that, if $K = k(\tau)$, the trace (respectively norm) of τ is the sum (respectively product) of all the conjugates of τ and that in the general case it is the sum (respectively product) of the images of τ under all the k-isomorphisms of K into a normal extension.

Given a finite basis Ω of a finite algebraic extension $K \supset k$, via the trace we can associate a *discriminant*[2] to Ω which has different applications, from testing whether Ω is a k-basis when k is finite, to testing whether K is separable (Section 10.6).

I also show (Section 10.7) that if K is a finite separable normal extension then there is an element $\xi \in K$ such that the set of the conjugates of ξ is a k-basis of K (*normal basis*).

10.1 Conjugates

Definition 10.1.1. *Let $K \supset k$ be a field extension and let $\alpha, \beta \in K$ be algebraic over k. They are called* conjugate *over k if they have the same minimal polynomial over k.*

The conjugacy property does not depend on the larger field K, but certainly depends on k: for instance, the polynomial $X^4 - 2$ is irreducible over $\mathbb{Q}$ (so its four roots are conjugate over $\mathbb{Q}$) but factors as

$$X^4 - 2 = (X^2 - \sqrt{2})(X^2 + \sqrt{2})$$

over $\mathbb{Q}(\sqrt{2})$, so that the two real roots are conjugate over $\mathbb{Q}(\sqrt{2})$ and so are the two complex roots, but a real and a complex root are not conjugate over $\mathbb{Q}(\sqrt{2})$.

---

[2] The concept coincides with the discriminant of the polynomial $f(X)$ if K is separable and $K = k[X]/f(X)$.

Proposition 10.1.2. *Let $K \supset k$ be a field extension and let $\alpha, \beta \in K$ be algebraic over k. Then the following conditions are equivalent:*

α and β are conjugate over k;
there is a k-isomorphism $\Phi : k[\alpha] \mapsto k[\beta]$ such that $\Phi(\alpha) = \beta$.

Proof If α and β are conjugate, the existence of Φ is a consequence of Lemma 5.5.3.

Conversely, if there is a k-isomorphism $\Phi : k[\alpha] \mapsto k[\beta]$ such that $\Phi(\alpha) = \beta$, let $f(X) \in k[X]$ be the minimal polynomial of α over k; then

$$f(\beta) = f(\Phi(\alpha)) = \Phi(f(\alpha)) = \Phi(0) = 0,$$

so that f is the minimal polynomial of β over k. ♮

Let $K \supset k$ be a field extension and $\alpha \in K$ be an algebraic element over k of reduced degree n_0; let $f(X) \in k[X]$ be the minimal polynomial of α over k and let $\alpha =: \alpha_1, \ldots, \alpha_\nu$ be all the elements in K which are conjugate to α.
Then:

for each $i \leq \nu$ there is a k-isomorphism

$$k(\alpha) \cong k[X]/f(X) \cong k(\alpha_i)$$

under which α is carried into α_i;
$\nu \leq n_0$, i.e. there are at most n_0 elements in K which are conjugates to α over k,
and $\nu = n_0$ iff K contains the splitting field of $f(X)$,
in which case $f(X)$ splits over K as

$$f(X) = \prod_{i=1}^{n_0} (X - \alpha_i)^{p^e},$$

where $p = \text{char}(k)$ and e is the exponent of inseparability of α.

10.2 Normal Extension Fields

The above remark leads us to consider the question of whether a field K contains all the conjugates of its elements over k, and to characterize such fields:

Definition 10.2.1. *A field algebraic extension $K \supset k$ is called a* normal extension field *of k, if*

for all $\alpha \in K$, K contains all the d conjugates of α over k, where d denotes the reduced degree of α,

or, equivalently,

each irreducible polynomial $g(X) \in k[X]$ which has a root in K, factors into linear factors over K.

Proposition 10.2.2. *Let $K \supseteq k$ be a finite field extension.*
Then K is normal iff it is the splitting field of a (not necessarily irreducible) polynomial $f(X) \in k[X]$.

Proof

$\Rightarrow$ Assume K is normal. Since K is a finite field extension of k, let $\alpha_1, \ldots, \alpha_n \in K \setminus k$ be such that $K = k[\alpha_1, \ldots, \alpha_n]$.
For each i, let $f_i(X) \in k[X]$ be the minimal polynomial of α_i over k; since K is normal, it contains the splitting field of each $f_i(X)$ and, therefore, also the splitting field K of the product $f(X) := \prod_i f_i(X)$, i.e. $\mathsf{K} \subset K$.
Conversely, since K contains all the roots of $f(X)$, including the α_is, $K = k[\alpha_1, \ldots, \alpha_n] \subset \mathsf{K}$.
Therefore $K = \mathsf{K}$ is the splitting field of f.

$\Leftarrow$ Let $g(X) \in k[X]$ be an irreducible polynomial, $\alpha \in K$ be such that $g(\alpha) = 0$ in K and β a conjugate of α in the splitting field of g over K; we intend to prove that $\beta \in K$.
Note that the splitting field of f over $k[\alpha]$ is K, and that over $k[\beta]$ is $K[\beta]$; therefore by Proposition 5.5.4 the k-isomorphism $\Phi : k[\alpha] \mapsto k[\beta]$ such that $\Phi(\alpha) = \beta$ extends to a k-isomorphism $\Psi : K \mapsto K[\beta]$ such that $\Psi(\alpha) = \beta$.
Let $\mathcal{R} := \{\gamma_1, \ldots, \gamma_r\} \subset K$ be the set of the roots of f.
Since $f(X) \in k[X]$ is preserved by Ψ and splits linearly into K, we can deduce that Ψ produces a permutation on the set of the roots of f, i.e.

there exists $\pi : \{1, \ldots, r\} \mapsto \{1, \ldots, r\} : \Psi(\gamma_i) = \gamma_{\pi(i)}$.

But $K = k(\mathcal{R})$, so that Ψ is an automorphism of K.

10.2 Normal Extension Fields

Since, there is a rational function $\rho \in k(X_1, \ldots, X_r)$ such that $\alpha = \rho(\gamma_1, \ldots, \gamma_r)$, we have

$$\beta = \Psi(\alpha) = \rho(\Psi(\gamma_1), \ldots, \Psi(\gamma_r)) = \rho(\gamma_{\pi(1)}, \ldots, \gamma_{\pi(r)}) \in K.$$

♮

Theorem 10.2.3. *If $K \supset k$ is a finite algebraic extension of k, then there is a finite normal extension L of k containing K.*

If L and L' are two such normal extensions, and they are minimal, then they are K-isomorphic.

Proof Since $K = k[\alpha_1, \ldots, \alpha_n]$ and is algebraic over k, for each i let $f_i(X) \in k[X]$ be the minimal polynomial of α_i over k; let $f(X) := \prod_{i=1}^n f_i(X)$ and L be the splitting field of $f(X)$ over K.

Since L is generated over K by all the roots of f, and K is generated over k by some roots of f, it follows that L is generated over k by the roots of f, i.e. it is the splitting field of $f(X)$ over k and, therefore, a normal extension of k.

Moreover it is minimal over this property, because if L_0 is a normal extension of k such that $k \subseteq K \subseteq L_0 \subseteq L$, then, for all i, f_i, being irreducible in $k[X]$, splits linearly in L_0, so that the same is true for f; being L generated over K by all the roots of f, we conclude that $L = L_0$ is minimal.

To conclude the argument, we need to prove that any other minimal normal extension L_1 of k containing K is K-isomorphic to L. In fact, L_1 contains the root α_i (and so all the roots) of f_i, for all i; therefore it contains a splitting field of $f(X)$ over k, which is necessarily a normal extension of k containing K; we can conclude that if L_1 is minimal, it coincides with this splitting field. Being a splitting field of f over k, L_1 is also a splitting field of f over K, and so is K-isomorphic with L.

♮

The argument of Proposition 10.2.2 can be extended to infinite field extensions via

Proposition 10.2.4. *Let $K \supseteq k$ be a field extension.*

Then K is normal iff it is obtained from k by the adjunction of all the roots of a set of polynomials in $k[X]$.

Proof

⇐ Each element $\gamma \in K$ depends only on the roots of *finitely many* polynomials $\{f_1, \ldots, f_m\}$.

So, in order to prove that γ is algebraic over k and that its minimal polynomial splits, we can just consider the field K_1, $\gamma \in K_1 \subseteq K$, which is obtained by the adjunction of all the roots of the *single* polynomial $f(X) := \prod_{j=1}^{m} f_j(X) \in k[X]$.

That the minimal polynomial of γ splits in K_1 – and so also in K – is then a corollary of Proposition 10.2.2.

⇒ Let $K \supset k$ be normal and let $S \subset K$ be such that K is obtained from k by the adjunction of S, $K = k(S)$.

Every element $\alpha \in S$ has a minimal polynomial $f(X) \in k[X]$ over k, which therefore splits in K. Then, setting $\mathcal{R}$ to be the set of all the conjugates of the elements $\alpha \in S$, clearly $K = k(\mathcal{R})$. ♮

Remark 10.2.5. In the following sections we will often consider a field k and a normal extension field K of k and we will study algebraic field extensions L such that $k \subset L \subset K$. The results will become more understandable if the readers remember that there exists a 'universal normal extension field' $\bar{k}$ of k, such that $k \subset L \subset \bar{k}$ for each algebraic field extension $L \supset k$: the *algebraic closure* of k (Section 9.1).

10.3 Isomorphisms

Definition 10.3.1. *Let $k \subseteq L \subseteq K$ be three fields. A k-isomorphism of L into K is any assignment of a subfield $L' \subset K$ – not necessarily different from L – and a k-isomorphism $\psi : L \mapsto L'$.*

Let $K \supset k$ be a normal extension field of k, $\alpha \in K \setminus k$, $f(X) \in k[X]$ be its minimal polynomial over k, $L := k[\alpha]$.

Then we know that f has n_0 roots $\alpha =: \alpha_1, \ldots, \alpha_{n_0}$ in K, all of them having (by Proposition 4.5.2) the same multiplicity p^e where $p = \text{char}(k)$ and $\deg(f) = n = n_0 p^e$.

For each i, let $L_i := k[\alpha_i]$ and $\psi_i : L \mapsto L_i$ be the unique k-isomorphism such that $\psi_i(\alpha) = \alpha_i$; then

Lemma 10.3.2. *With the setting above, let $M \subset K$ be a subfield which is k-isomorphic to L under the isomorphism $\phi : L \mapsto M \subset K$.*

Then there exists $i \leq n_0 : M = L_i$, $\phi = \phi_i$.

10.3 Isomorphisms

Proof In fact, let $\gamma := \phi(\alpha) \in M \subset K$. Then, since ϕ is a k-isomorphism, we have $\phi(f) = f$, $f(\gamma) = 0$, there exists $i : \gamma = \alpha_i$, and $L_i = k[\alpha_i] \subseteq M$. Since

$$[L_i : k] = n = [L : k] = [M : k]$$

then $M = L_i$; since ψ_i is the unique k-isomorphism such that

$$\psi_i(\alpha) = \alpha_i = \gamma,$$

then $\phi = \psi_i$, ☐

from which we deduce

Corollary 10.3.3. *With the setting above, there are exactly n_0 k-isomorphisms of L into K.*

Proof They are the n_0 k-isomorphisms $\psi_i : L \mapsto L_i$. ☐

Remark 10.3.4. In the notation above, we have $\alpha = \alpha_1$ and so ψ_1 is the identity of $L = L_1$.

Also, note that, while the α_is are all different, this does not imply that the L_is are different too; it means only that there could be k-automorphisms of L_i.

This result can be generalized to

Proposition 10.3.5. *With the above setting, let K be any field extension of L; then there are at most n_0 k-isomorphisms of L into K.*

There are exactly n_0 of them if K contains a splitting field of L.

Proof If $\psi : L \mapsto M \subseteq \mathsf{K}$ is a k-isomorphism of L into K, the same argument as in Lemma 10.3.2 allows us to deduce that $\psi(\alpha)$ is a root of $f(X)$ in K. The claim is then obvious. ☐

Lemma 10.3.6. *Let $K \supset k$ be a finite normal extension field and let $\alpha, \beta \in K$ be conjugate over k. Then there is a k-isomorphism Ξ of K such that $\Xi(\alpha) = \beta$.*

Proof By definition there is a k-isomorphism $\Psi : k[\alpha] \mapsto k[\beta]$. Since, by definition, K is the splitting field of a polynomial $f(X) \in k[X]$, it is also the splitting field of f over both $k[\alpha]$ and $k[\beta]$; then the existence of Ξ follows from Proposition 5.5.4. ☐

Corollary 10.3.7. *Let $L \supset k$ be a finite extension. Then the following conditions are equivalent:*

L is normal.

If $K \supset L$ is any field extension, then each k-isomorphism

$$\phi : L \mapsto M \subset K$$

is an automorphism.

Under this assumption, L is the splitting field of a polynomial $f(X) \in k[X]$, and:

L has n_0 k-automorphisms, where n_0 is the reduced degree of f;
if $\alpha, \beta \in L$ are conjugate over k, then there is a k-automorphism Ξ of L such that $\Xi(\alpha) = \beta$.

Proof If L is normal, from Proposition 10.3.5, setting $L = \mathsf{K}$ we deduce that the n_0 k-isomorphisms of L are automorphisms.

Conversely, let us fix a normal extension $K \supset L$. We consider any element $\alpha \in L$; then K contains all its conjugates over L and we need to show that each of them lies in L.

Let $\beta \in K$ be any such conjugate. Then there is a k-automorphism $\Xi : K \mapsto K$ such that $\Xi(\alpha) = \beta$; then Ξ induces a k-isomorphism of L into K and, by assumption, a k-automorphism of L; therefore $\beta = \Xi(\alpha) \in L$. ♄

Example 10.3.8. Let us consider the following cases:

$k := \mathbb{R}$, $K := \mathbb{C}$, $\alpha := i$, $f(X) := X^2 + 1$.
Then we have $L_1 = L_2 = \mathbb{C}$, and $\psi_i : L_1 \mapsto L_i$ are respectively the identity $\psi_1(a+ib) = a+ib$ and the conjugation map $\psi_2(a+ib) = a-ib$.

$k := \mathbb{Q}$, $K := \mathbb{R}$, $\alpha := \sqrt{2}$, $f(X) := X^2 - 2$.
Thus $L_1 = L_2 = \mathbb{Q}(\sqrt{2})$, and $\psi_i : L_1 \mapsto L_i$ are respectively the identity $\psi_1(a + \sqrt{2}b) = a + \sqrt{2}b$ and the 'conjugation' map $\psi_2(a + \sqrt{2}b) = a - \sqrt{2}b$.

With the notations of Examples 5.2.5, 5.2.6 and 5.5.1 we can consider

$k := \mathbb{Q}$, $K := \mathbb{Q}(\sqrt[3]{2}, \sqrt{-3})$, $\alpha := \beta := \sqrt[3]{2}$, $f(X) := X^3 - 2$.

10.3 Isomorphisms

Then we have $L_1 = \mathbb{Q}[\beta]$, $L_2 = \mathbb{Q}[\gamma]$, $L_3 = \mathbb{Q}[-\beta - \gamma]$; $k := k_1 = \mathbb{Q}[\beta]$, $K := \mathbb{C}$, $\alpha := \gamma$, $f(X) := X^2 + \beta X + \beta^2$. Thus

$$L_1 = L_2 = k_2 = k_1[\gamma] = \mathbb{Q}[\beta, \gamma],$$

where, denoting by $\phi_i : k_2 \mapsto k_2$ the two k_1-automorphisms, we have ϕ_1 is the identity and ϕ_2 is defined by $\phi_2(\gamma) = -\beta - \gamma$ so that

$$\phi_2 \left(a_{00} + a_{10}\beta + a_{20}\beta^2 + a_{01}\gamma + a_{11}\beta\gamma + a_{21}\beta^2\gamma \right)$$
$$= (a_{00} - 2a_{21}) + (a_{10} - a_{01})\beta + (a_{20} - a_{11})\beta^2$$
$$- a_{01}\gamma - a_{11}\beta\gamma - a_{21}\beta^2\gamma;$$

$k := k_1 = \mathbb{Q}[\beta]$, $K := \mathbb{Q}(\sqrt[3]{2}, \sqrt{-3})$, $\alpha := \delta = \sqrt{-3}$, $f(X) := X^2 - 3$. Then we have $L_1 = L_2 = k_2' = k_1[\beta, \delta]$, where, denoting $\phi_i' : k_2' \mapsto k_2'$ the two k_1-automorphisms, we have ϕ_1' is the identity and ϕ_2' is defined by[3]

$$\phi_2' \left(a_{00} + a_{10}\beta + a_{20}\beta^2 + a_{01}\delta + a_{11}\beta\delta + a_{21}\beta^2\delta \right)$$
$$= a_{00} + a_{10}\beta + a_{20}\beta^2 - a_{01}\delta - a_{11}\beta\delta - a_{21}\beta^2\delta.$$

Example 10.3.9. If $k = \mathbb{Z}_p$, p prime, and $L = K = GF(p^n)$, we know (Corollary 7.2.6) that, for each irreducible polynomial $f(X) \in k[X]$, $\deg(f) = n$, K is the splitting field of f and, for each root $\alpha \in K \setminus k$ of f,

$GF(p^n) = K = L = k[\alpha] = k[X]/f(X)$, and
the conjugates of α are $\alpha_i := \alpha^{p^i}$, $0 \leq i < n$,

so that the n k-automorphisms of K are the Frobenius automorphisms Φ_i defined by $\Phi_i(\beta) = \beta^{p^i}$.

---

[3] It is easy and illuminating to check that under the isomorphism $\Psi : k_2 \mapsto k_2'$ we have $\Phi\phi_2 = \phi_2'\Phi$:

$$\Phi\phi_2 \left(a_{00} + a_{10}\beta + a_{20}\beta^2 + a_{01}\gamma + a_{11}\beta\gamma + a_{21}\beta^2\gamma \right)$$
$$= \Phi \left((a_{00} - 2a_{21}) + (a_{10} - a_{01})\beta + (a_{20} - a_{11})\beta^2 \right.$$
$$\left. - a_{01}\gamma - a_{11}\beta\gamma - a_{21}\beta^2\gamma \right)$$
$$= (a_{00} - a_{21}) + (a_{10} - \tfrac{1}{2}a_{01})\beta + (a_{20} - \tfrac{1}{2}a_{11})\beta^2$$
$$+ a_{21}\delta + \tfrac{1}{2}a_{01}\beta\delta + \tfrac{1}{2}a_{11}\beta^2\delta$$
$$= \phi_2' \left((a_{00} - a_{21}) + (a_{10} - \tfrac{1}{2}a_{01})\beta + (a_{20} - \tfrac{1}{2}a_{11})\beta^2 \right.$$
$$\left. - a_{21}\delta - \tfrac{1}{2}a_{01}\beta\delta - \tfrac{1}{2}a_{11}\beta^2\delta \right)$$
$$= \phi_2'\Phi \left(a_{00} + a_{10}\beta + a_{20}\beta^2 + a_{01}\gamma + a_{11}\beta\gamma + a_{21}\beta^2\gamma \right).$$

Lemma 10.3.10. *Let $k \subset L \subset M \subset K$ be field extensions such that*

K is a normal extension of k,
L is a finite algebraic extension of k,
$[L : k] = m$,
M is a finite algebraic extension of L,
$[M : L] = g$.

Then we know that L has m k-isomorphisms, which we will denote

$$\psi_i : L \mapsto L_i, \ 1 \leq i \leq m,$$

in K.

Then the gm k-isomorphisms of M into K can be partitioned into m subsets $\mathcal{G}_i$, $1 \leq i \leq m$, each consisting of those g k-isomorphisms ϕ which extend ψ_i, in the sense that they satisfy

$$\phi(\gamma) = \psi_i(\gamma), \ \text{for all } \gamma \in L.$$

Proof For each k-isomorphism ϕ of M into K, its restriction to L is a k-isomorphism ψ_i of L into K.

Then, the set $\mathcal{G}$ of the gm k-isomorphisms of M into K can be partitioned as $\mathcal{G} = \cup_i \mathcal{G}_i$ where, for each i, $\mathcal{G}_i$ contains those k-isomorphisms ϕ such that their restriction to L is ψ_i:

$$\phi(\gamma) = \psi_i(\gamma), \ \text{for all } \gamma \in L.$$

We can wlog assume that M is a simple extension of L by $\alpha \in M$ whose minimal polynomial $f(X) \in L[X]$ has reduced degree g:

$$M = L[X]/f(X) = L[\alpha].$$

For each i, $f_i := \psi_i(f)$ is an irreducible polynomial in $L_i[X]$ of reduced degree g whose roots in K will be denoted $\beta_{i1}, \ldots, \beta_{ig}$. Therefore for each k-isomorphism $\phi \in \mathcal{G}_i$ we have

$$f_i(\phi(\alpha)) = \psi_i(f)(\phi(\alpha)) = \phi(f)(\phi(\alpha)) = \phi(f(\alpha)) = \phi(0) = 0$$

so that $\phi(\alpha) = \beta_{ij}$ for some j.

Also, for all i, j, ψ_i can be extended to a k-isomorphism

$$\phi_{ij} : M = L[\alpha] \mapsto L_i[\beta_{ij}] \subset K$$

such that $\phi_{ij}(\gamma) = \psi_i(\gamma)$, for all $\gamma \in L$, and $\phi_{ij}(\alpha) = \beta_{ij}$, in a single way by defining

for all $\sum_h b_h \alpha^h \in L[\alpha] : \phi_{ij}\left(\sum_h b_h \alpha^h\right) := \sum_h \psi_i(b_h)\beta_{ij}^h.$

In conclusion, we have $\mathcal{G}_i = \{\phi_{ij} : 1 \leq j \leq g\}$. ♄

Example 10.3.11. As an example let us continue the computation developed in Examples 5.2.5, 5.2.6, 5.5.1 and 10.3.8 where

$$k = \mathbb{Q}, \quad M = K = \mathbb{Q}(\beta, \gamma), \quad L = \mathbb{Q}[\beta].$$

Let us first note that, with the notation of those examples, in $K := \mathbb{Q}(\beta, \gamma)$:

for ψ_2, for which $\psi_2(\beta) = \gamma$, we have:

$$\psi_2(f_1(X)) = X^2 + \gamma X + \gamma^2 \in L_2[X] = \mathbb{Q}[\gamma][X]$$

one of whose roots is obviously[4] β and the other is therefore $-\beta - \gamma$; and, for ψ_3, for which $\psi_3(\beta) = \beta + \gamma$, we have

$$\begin{aligned}\psi_3(f_1(X)) &= X^2 + (-\beta - \gamma)X + (-\beta - \gamma)^2 \\ &= X^2 - (\beta + \gamma)X + \beta\gamma \\ &= (X - \beta)(X - \gamma) \in L_3[X] = \mathbb{Q}[\beta + \gamma][X]\end{aligned}$$

whose roots are obviously β and γ.

This lets us describe all the k-isomorphisms of $K := \mathbb{Q}(\beta, \gamma)$. To set a notation which allows us to further discuss this example, let us freely set the identification (as Theorem 5.5.6 permits)

$$\alpha_1 := \beta, \quad \alpha_2 := \gamma, \quad \alpha_3 := -\beta - \gamma.$$

Lemma 10.3.10 gives us the following 6 k-automorphisms of K:

Φ_{123} : $\Phi_{123}(\alpha_1) = \alpha_1$ $\Phi_{123}(\alpha_2) = \alpha_2$ $\Phi_{123}(\alpha_3) = \alpha_3$
Φ_{132} : $\Phi_{132}(\alpha_1) = \alpha_1$ $\Phi_{132}(\alpha_2) = \alpha_3$ $\Phi_{132}(\alpha_3) = \alpha_2$
Φ_{213} : $\Phi_{213}(\alpha_1) = \alpha_2$ $\Phi_{213}(\alpha_2) = \alpha_1$ $\Phi_{213}(\alpha_3) = \alpha_3$
Φ_{231} : $\Phi_{231}(\alpha_1) = \alpha_2$ $\Phi_{231}(\alpha_2) = \alpha_3$ $\Phi_{231}(\alpha_3) = \alpha_1$
Φ_{312} : $\Phi_{312}(\alpha_1) = \alpha_3$ $\Phi_{312}(\alpha_2) = \alpha_1$ $\Phi_{312}(\alpha_3) = \alpha_2$
Φ_{321} : $\Phi_{321}(\alpha_1) = \alpha_3$ $\Phi_{321}(\alpha_2) = \alpha_2$ $\Phi_{321}(\alpha_3) = \alpha_1$

Lemma 10.3.10 can be expanded, by iteration, to yield

---

[4] Remember that $\gamma^2 + \gamma\beta + \beta^2 = 0$ in K.

Theorem 10.3.12. *Let $K \supset L \supset k$ be such that K is a normal extension of k and L is a finite algebraic extension of k, $L = k[\alpha_1, \ldots, \alpha_s]$, so that each α_i is algebraic over $L_{i-1} := k[\alpha_1, \ldots, \alpha_{i-1}]$ of reduced degree n_i.*
Then L has $\prod_{i=1}^{s} n_i$ k-isomorphisms into K.

Proof By induction. If $s = 1$ the claim follows from Proposition 10.3.5. For $s > 1$, then $M := L_{s-1}$ has, by induction, $m := \prod_{i=1}^{s-1} n_i$ k-isomorphisms into K. Then the result follows from Lemma 10.3.10. ♄

Corollary 10.3.13. *With the same notation L has $[L : k]$ k-isomorphisms into K iff each α_i is separable over L_{i-1}.*

Proof In fact α_i is separable over L_{i-1} iff $n_i = [L_i : L_{i-1}]$. ♄

As a consequence of Corollary 10.3.13 we obtain

Corollary 10.3.14. *Let $L = k[\alpha_1, \ldots, \alpha_s] \supset k$ be a finite algebraic extension of k, such that each α_i is separable over $L_{i-1} := k[\alpha_1, \ldots, \alpha_{i-1}]$.*
Then each $\beta \in L$ is separable over k.

Proof We can in fact obtain that L has an algebraic extension of $k(\beta)$: since, denoting by $K \supset L$ any normal extension of k, L has $[L : k] = [L : k(\beta)][k(\beta) : k]$ k-isomorphisms and $[L : k(\beta)]$ $k(\beta)$-isomorphisms into K, we conclude that $k(\beta)$ has $[k(\beta) : k]$ k-isomorphisms into K, i.e. β is separable. ♄

From which we can conclude immediately Fact 9.3.2:

Corollary 10.3.15. *Let $K \supset k$ be an algebraic extension of k. Then K_{sep} is a field.*

Proof For any $\alpha_1, \alpha_2 \in K_{\text{sep}}$, $\alpha_2 \neq 0$, $\alpha_1 \pm \alpha_2$, $\alpha_1 \alpha_2$, $\alpha_1 \alpha_2^{-1}$ are elements of $k(\alpha_1, \alpha_2)$ and so are separable. ♄

Finally we recall the Dedekind Theorem:

Theorem 10.3.16 (Dedekind). *Let $k \subset K \subset M$ be field extensions and let $\sigma_1, \ldots, \sigma_n$ be a finite family of distinct k-isomorphisms of K into M.*

10.4 Splitting Fields

Then they are linearly independent over M, i.e. if

$$\sum_{i=1}^n c_i\sigma_i(x) = 0, \text{ for all } x \in K, \text{ with } c_i \in M, \implies c_i = 0, \text{ for all } i.$$

Proof The result being obvious if $n = 1$, we will prove it by induction on n. For each $\alpha \in K$

$$0 = \sum_{i=1}^n c_i\sigma_i(\alpha x) = \sum_{i=1}^n c_i\sigma_i(\alpha)\sigma_i(x)$$

so that

$$\begin{aligned}
0 &= \sum_{i=1}^n c_i\sigma_i(\alpha)\sigma_i(x) - \sigma_n(\alpha)\sum_{i=1}^n c_i\sigma_i(x)\\
&= \sum_{i=1}^n c_i\left(\sigma_i(\alpha) - \sigma_n(\alpha)\right)\sigma_i(x)\\
&= \sum_{i=1}^{n-1} c_i\left(\sigma_i(\alpha) - \sigma_n(\alpha)\right)\sigma_i(x).
\end{aligned}$$

By induction we conclude that, for all $i < n$, we have $c_i\left(\sigma_i(\alpha) - \sigma_n(\alpha)\right) = 0$, and $c_i = 0$, since the σ_js are distinct. From $c_i = 0$, for all $i < n$, we also deduce $c_n = 0$. ♄

10.4 Splitting Fields

Let us now consider a finite normal extension $K \supset k$; by the results above we know that K is the splitting field of a polynomial $F(X) \in k[X]$. We intend to study its structure, by means of Theorem 8.4.5, both when F is separable and is not.

Corollary 10.4.1. *Let $G(X) \in k[X]$ be a separable irreducible polynomial and let $K \supset k$ be its splitting field. Then:*

there is a separable primitive element $\xi \in K$, and a separable polynomial $g(X) \in k[X]$ such that

$$K = k[X]/g[X] = k[\xi];$$

denoting by $n := [K : k]$, the degree of g and ξ, K contains n simple roots
$\xi := \xi_1, \ldots, \xi_n$ *of g;*
$g = \prod_{i=1}^n (X - \xi_i)$;

there are n polynomials $p_i(X) \in k[X]$, $\deg(p_i) < \deg(g)$, such that $p_i(\xi) = \xi_i$;

K has the n k-automorphisms $\phi_i : K \mapsto K$ such that $\phi_i(\xi) = \xi_i$;

there are $d := \deg(G)$ polynomials $q_i(X) \in k[X]$, $\deg(q_i) < \deg(g)$, such that the roots of G are $q_i(\xi)$.

Proof The existence of ξ and of the q_is is a consequence of Theorem 8.4.5 and of the fact that the finite extension K is separable, since G is.

Since K is normal and contains a root of g, we have g splits in it. ♄

Example 10.4.2. Let us continue the computations developed in Examples 5.2.5, 5.2.6, 5.5.1, 10.3.8 and 10.3.11 where $k = \mathbb{Q}$, $G(X) := X^3 - 2$, $K = \mathbb{Q}(\beta, \gamma)$.

Setting $\xi := \beta + 2\gamma$, by linear algebra we obtain that

$$
\begin{aligned}
\xi^0 &= +1 \\
\xi^1 &= \beta + 2\gamma \\
\xi^2 &= -3\beta^2 \\
\xi^3 &= -6 \qquad\qquad\qquad -6\beta^2\gamma \\
\xi^4 &= 18\beta \\
\xi^5 &= 18\beta^2 \quad -36\beta\gamma \\
\xi^6 &= -108
\end{aligned}
$$

and the minimal polynomial of ξ is $g(X) := X^6 + 108$, whose roots are

$$
\begin{aligned}
\Phi_{123}(\xi) &= \beta + 2\gamma &= \xi \\
\Phi_{132}(\xi) &= -\beta - 2\gamma &= -\xi \\
\Phi_{213}(\xi) &= 2\beta + \gamma &= \tfrac{1}{12}\xi^4 + \tfrac{1}{2}\xi \\
\Phi_{231}(\xi) &= -2\beta - \gamma &= -\tfrac{1}{12}\xi^4 - \tfrac{1}{2}\xi \\
\Phi_{312}(\xi) &= \beta - \gamma &= \tfrac{1}{12}\xi^4 - \tfrac{1}{2}\xi \\
\Phi_{321}(\xi) &= -\beta + \gamma &= -\tfrac{1}{12}\xi^4 + \tfrac{1}{2}\xi.
\end{aligned}
$$

Also, setting

$$
\begin{aligned}
q_1(X) &= \frac{1}{18}X^4, \\
q_2(X) &= -\frac{1}{36}X^4 + \frac{1}{2}X, \\
q_3(X) &= -\frac{1}{36}X^4 - \frac{1}{2}X,
\end{aligned}
$$

then

$$q_i(\xi) = \alpha_i, \text{ for all } i.$$

Corollary 10.4.3. *Let k be a field such that* $\text{char}(k) = p \neq 0$ *and let $F(X) \in k[X]$ be an inseparable irreducible polynomial. Let $\mathsf{K} \supset k$ be the splitting field of F.*

Then:

- *there is an $e > 0$ and a separable irreducible polynomial $G(X) \in k[X]$ such that $F(X) = G(X^{p^e})$;*
- *if K, $k \subset K$, is the splitting field of G, there is a separable primitive element $\xi \in K$, and a separable polynomial $g(X) \in k[X]$ such that*

$$K = k[X]/g[X] = k[\xi];$$

- *K contains $n_0 = [K : k] = \deg(g)$ simple roots $\xi := \xi_1, \ldots, \xi_{n_0}$ of g;*
- *which are represented by n_0 polynomials $p_i(X) \in k[X]$, $\deg(p_i) < \deg(g)$, such that $p_i(\xi) = \xi_i$;*
- *there are $d := \deg(G)$ polynomials $q_i(X) \in k[X]$, $\deg(q_i) < \deg(g)$, such that the roots of G are $q_i(\xi)$.*

Consider the polynomial $f(X) = g(X^{p^e})$ and let ζ be, in a suitable extension of $k[\xi]$, the element such that $\zeta^{p^e} = \xi$. Then:

- *f splits in $K[\zeta]$ as*

$$f(X) = \prod_{i=1}^{n_0}(X - p_i(\zeta))^{p^e};$$

- *$[k[\zeta] : k[\xi]] = p^e$;*
- *$k[\xi] = k[\zeta]_{\text{sep}}$.*

Moreover

- *$k[\zeta] = \mathsf{K}$, the splitting field of F;*
- *the roots of F are $q_i(\zeta) \in k[\zeta]$;*
- *$k[\zeta]$ has the n_0 k-automorphisms $\phi_i : k[\zeta] \mapsto k[\zeta]$ such that $\phi_i(\zeta) = p_i(\zeta)$;*
- *f is irreducible in $k[X]$.*

Proof The existence of e and G is a consequence of the inseparability of F; the existence of ξ, g, ξ_i, p_i, q_i follows from the above corollary. By the definition of ζ, for each i

$$\begin{aligned} p_i(\zeta)^{p^e} &= p_i(\zeta^{p^e}) = p_i(\xi), \\ f(p_i(\zeta)) &= g(p_i(\zeta)^{p^e}) = g(p_i(\xi)) = 0, \end{aligned}$$

from which follows the splitting of f in $K[\zeta]$.
From it we deduce

$$[k[\zeta] : k[\xi]][k[\xi] : k] = [k[\zeta] : k] = \deg(f) = p^e \deg(g) = p^e[k[\xi] : k]$$

and we conclude that $[k[\zeta] : k[\xi]] = p^e$.
Since ζ is purely inseparable over $k[\xi]$, the same is true for any element of $k[\zeta]$:

$$h(\zeta)^{p^e} = h(\zeta^{p^e}) = h(\xi) \in k[\xi].$$

The other statements are obvious. ♄

10.5 Trace and Norm

Let K be a finite algebraic extension of k of degree n. Therefore K has a finite k-vector space basis $\Omega := \{\omega_1, \ldots, \omega_n\}$.

For any element $\tau \in K$ the function $\Phi_\tau : K \mapsto K$ defined by

$$\Phi_\tau(\upsilon) := \tau\upsilon, \text{ for all } \upsilon \in K$$

is a k-linear function, therefore it can be represented by an $n \times n$ matrix $M_\tau := \left(t_{ij}\right)_{ij}$ where $t_{ij} \in k$ and

$$\tau\omega_j = \sum_i t_{ij}\omega_i, \text{ for all } j.$$

We recall that if $\Omega' := \{\omega'_1, \ldots, \omega'_n\}$ is a different basis, and $M'_\tau := \left(t'_{ij}\right)_{ij}$ is the matrix representing Φ_τ in terms of Ω', then the matrix $E := \left(e_{ij}\right)_{ij}$ defined by

$$\omega'_j = \sum_i e_{ij}\omega_i, \text{ for all } j$$

is invertible and

$$M'_\tau = E M_\tau E^{-1}.$$

We recall that for an $n \times n$ matrix $M := \left(t_{ij}\right)_{ij}$, its *norm* is its determinant and its *trace* is $\text{Tr}(M) := \sum_{i=1}^n t_{ii}$ and that

Lemma 10.5.1. *Let M, E be two $n \times n$ matrices, E being invertible, and define $M' := EME^{-1}$. Then $\det(M) = \det(M')$, $\text{Tr}(M) = \text{Tr}(M')$.*

10.5 Trace and Norm

Proof We only need to note that for two $n \times n$ matrices A, B we have, $\mathrm{Tr}(AB) = \mathrm{Tr}(BA)$, from which we can deduce that $\mathrm{Tr}(M) = \mathrm{Tr}(EME^{-1}) = \mathrm{Tr}(M')$. ♮

On the basis of this, we can introduce the following

Definition 10.5.2. *For an element $\tau \in K$,*

the norm *of τ in K over k is the norm of the matrix M_τ:*

$$N_{K/k}(\tau) = \det(M_\tau);$$

the trace *of τ in K over k is the trace of the matrix M_τ:*

$$\mathrm{Tr}_{K/k}(\tau) = \mathrm{Tr}(M_\tau).$$

Since K is an algebraic extension, τ satisfies a minimal polynomial

$$f(X) := X^m + \sum_{i=1}^{m} a_i X^{m-i} \in k[X].$$

We intend to express the norm and trace of τ in K over k in terms of the coefficients of $f(X)$, and therefore, by Section 6.2, in terms of the conjugates $\tau =: \tau_1, \ldots, \tau_m$ of τ.

Proposition 10.5.3. *We have*

$N_{k[\tau]/k}(\tau) = (-1)^m a_m = \prod_i \tau_i;$
$\mathrm{Tr}_{k[\tau]/k}(\tau) = -a_1 = \sum_i \tau_i.$

Proof As a basis of $k[\tau]$ we choose $\Omega := \{1, \tau, \tau^2, \ldots, \tau^{m-1}\}$ obtaining the matrix $M_\tau := (t_{ij})_{ij}$ defined by

$$t_{ij} = \begin{cases} 1 & \text{if } 1 \leq i = j - 1 \leq m - 1 \\ -a_{m-j+1} & \text{if } i = m, 1 \leq j \leq m \\ 0 & \text{otherwise,} \end{cases} \quad (10.1)$$

from which we easily get the expression of the trace, while that of the norm just requires us to expand the determinant along the last row. ♮

Remarking that

$$n = [K : k] = [K : k[\tau]][k[\tau] : k] = [K : k[\tau]]m$$

we denote $g := [K : k[\tau]]$ and deduce

Corollary 10.5.4.

$$N_{K/k}(\tau) = \left(N_{k[\tau]/k}(\tau)\right)^g = (-1)^n a_m^g = \prod_i \tau_i^g;$$

$$\operatorname{Tr}_{K/k}(\tau) = g \operatorname{Tr}_{k[\tau]/k}(\tau) = -g a_1 = g \sum_i \tau_i.$$

Proof Let us choose a basis $\{\omega_1, \ldots, \omega_g\}$ of K over $k[\tau]$; then the basis of K over k is

$$\Omega' := \{\omega_1, \omega_1\tau, \omega_1\tau^2, \ldots, \omega_1\tau_{m-1}, \omega_2, \ldots, \omega_{g-1}\tau_{m-1}, \omega_g, \ldots, \omega_g\tau_{m-1}\}$$

in terms of which we express the multiplication by τ via the $gm \times gm$ matrix

$$\begin{pmatrix} M_\tau & 0 & 0 & \cdots & 0 \\ 0 & M_\tau & 0 & \cdots & 0 \\ \vdots & \vdots & \vdots & \ddots & 0 \\ 0 & 0 & 0 & \cdots & M_\tau \end{pmatrix}$$

where M_τ is the matrix defined by Equation 10.1. The result then follows obviously. ♄

As a consequence of the above results, it follows easily that:

Lemma 10.5.5. *For $\tau, \upsilon \in K, c \in k$,*

$N_{K/k}(\tau) = 0 \iff \tau = 0;$

$N_{K/k}(\tau\upsilon) = N_{K/k}(\tau) N_{K/k}(\upsilon);$

$\operatorname{Tr}_{K/k}(\tau + \upsilon) = \operatorname{Tr}_{K/k}(\tau) + \operatorname{Tr}_{K/k}(\upsilon);$

$\operatorname{Tr}_{K/k}(c\tau) = c \operatorname{Tr}_{K/k}(\tau);$

$N_{K/k} : K \setminus \{0\} \mapsto k \setminus \{0\}$ *is a group homomorphism between these multiplicative groups;*

$\operatorname{Tr}_{K/k} : K \mapsto k$ *is a k-linear function;*

$N_{K/k}(c) = c^n;$

$\operatorname{Tr}_{K/k}(c) = nc;$

if τ and υ are conjugate, then $\operatorname{Tr}_{K/k}(\tau) = \operatorname{Tr}_{K/k}(\upsilon);$

if τ and υ are conjugate, then $N_{K/k}(\tau) = N_{K/k}(\upsilon).$ ♄

10.5 Trace and Norm

Lemma 10.5.6. *Let $k \subset K \subset L$ be finite algebraic extensions and let $\tau \in K$. Denoting $g := [L : K]$ we have*

$N_{L/k}(\tau) = \left(N_{K/k}(\tau)\right)^g = (-1)^n a_m^g = \prod_i \tau_i^g;$
$\mathrm{Tr}_{L/k}(\tau) = g\, \mathrm{Tr}_{K/k}(\tau) = -g a_1 = g \sum_i \tau_i.$

Proof An argument similar to the proof of Corollary 10.5.4 is sufficient. ♄

We intend now to interpret the norm and the trace in terms of the k-isomorphisms of K.

Therefore let us fix a normal extension field

$$\mathsf{K} \supset K \supset k[\tau] \supset k$$

and begin by considering the m k-isomorphisms of $k[\tau]$ in K: denoting them as $\psi_i : k[\tau] \mapsto k[\tau_i]$, we remark that

$$N_{k[\tau]/k}(\tau) = \prod_i \tau_i = \prod_{i=1}^n \psi_i(\tau); \quad \mathrm{Tr}_{k[\tau]/k}(\tau) = \sum_i \tau_i = \sum_{i=1}^n \psi_i(\tau).$$

Now let us consider the $[K : k] = n = gm$ k-isomorphisms of K in K: according to Lemma 10.3.10 we know that they can be partitioned into m subsets $\mathcal{G}_i$ each consisting of g isomorphisms, in such a way that for each k-isomorphism ϕ

$$\phi \in \mathcal{G}_i \iff \phi(\tau) = \psi_i(\tau) = \tau_i;$$

from this we deduce

Proposition 10.5.7. *Let $\mathcal{G} := \{\psi_i : 1 \leq i \leq n\}$ be the set of all the k-isomorphisms of K in a suitable normal extension; then*

$$N_{K/k}(\tau) = \prod_{i=1}^n \psi_i(\tau); \quad \mathrm{Tr}_{K/k}(\tau) = \sum_{i=1}^n \psi_i(\tau).$$

♄

On the basis of this definition of norm, we can generalize it to polynomials in $K[X]$ as follows: let, as above, $\mathcal{G} := \{\psi_i : 1 \leq i \leq n\}$ be the set of all the k-isomorphisms $\psi_i : K \mapsto K_i$ of K in a suitable normal extension and let $\psi_i : K \mapsto K_i$ denote also the polynomial extension $\psi_i : K[X] \mapsto K_i[X]$. Then the *norm* of a polynomial $f(X)$ in $K[X]$ over k is

$$N_{K/k}(f(X)) = \prod_{i=1}^n \psi_i(f(X)).$$

Under this definition we have:

Lemma 10.5.8. *The following three polynomials:*

the minimal polynomial $f(X)$ of τ;
the norm of $(X - \tau)$ in $k[\tau][X]$ over k;
the characteristic polynomial of the k-linear form $\Phi_\tau : k[\tau] \mapsto k[\tau]$;

coincide.

Proof The three polynomials have the same degree n, are monic and have τ as a root. Since one of them is the minimal polynomial of τ, they must coincide. ♄

Corollary 10.5.9. *The following three polynomials:*

$f(X)^g$;
the norm of $(X - \tau)$ in $K[X]$ over k;
the characteristic polynomial of the k-linear form $\Phi_\tau : K \mapsto K$;

coincide.

Proof The result follows from the proof of Corollary 10.5.4 and from 10.5.8. ♄

In the case of finite fields, the trace and norm have further relevant properties

Proposition 10.5.10. *Let $F := GF(q)$, $E := GF(q^n)$ and $\alpha \in E$. Then:*

(1) $\operatorname{Tr}_{E/F}(\alpha) = \sum_{i=0}^{n-1} \alpha^{q^i}$.
(2) *The F-linear function $\operatorname{Tr}_{E/F} : E \mapsto F$ is surjective.*
(3) $\operatorname{Tr}_{E/F}(\alpha) = 0 \iff$ *there exists $\beta \in E : \alpha = \beta^q - \beta$.*
(4) $N_{E/F}(\alpha) = \alpha^{q^n - 1/q - 1}$.
(5) *The multiplicative function $N_{E/F} : E \mapsto F$ is surjective.*
(6) $N_{E/F}(\alpha) = 1 \iff$ *there exists $\beta \in E : \alpha = \beta^{q-1}$.*

Proof

(1) The n F-automorphisms of E are the morphisms Φ_i, $i = 0, \ldots, n - 1$, defined by $\Phi_i(\alpha) = \alpha^{q^i}$.
(2) Setting, for all $c \in F$, $f_c(X) := X + X^q + X^{q^2} + \cdots + X^{q^{n-1}} - c \in F[X]$, for each $\alpha \in E$ we have:

$$\operatorname{Tr}_{E/F}(\alpha) = c \iff f_c(\alpha) = 0;$$

since f_c has degree q^{n-1}, for each of the q polynomials f_c, there are at most q^{n-1} elements $\alpha \in E$ such that $\mathrm{Tr}_{E/F}(\alpha) = c$. As a consequence the q^n elements of E are partitioned into q sets $E_c, c \in F$, of q^{n-1} elements whose trace is c.

(3) Obviously $\mathrm{Tr}_{E/F}(\beta) = \mathrm{Tr}_{E/F}(\beta^q)$, for all $\beta \in E$.
Conversely, let $\alpha \in E$ be an element such that $\mathrm{Tr}_{E/F}(\alpha) = 0$; in a suitable finite extension of E let β be a root of $X^q - X - \alpha$: if we are able to prove that $\beta \in E$ we are through; to do that we only have to show that $\beta = \beta^{q^n}$:

$$0 = \mathrm{Tr}_{E/F}(\alpha) = \sum_{i=0}^{n-1} \alpha^{q^i} = \sum_{i=0}^{n-1} (\beta^q - \beta)^{q^{n-1-i}}$$
$$= (\beta^{q^n} - \beta^{q^{n-1}}) + (\beta^{q^{n-1}} - \beta^{q^{n-2}})$$
$$+ \cdots + (\beta^q - \beta).$$

(4) It follows from $N_{E/F}(\alpha) = \prod_{i=0}^{n-1} \alpha^{q^i} = \alpha^{\sum_{i=0}^{n-1} q^i} = \alpha^{\frac{q^n-1}{q-1}}$.

(5) Setting, for all $c \in F$, $g_c(X) := X^{\frac{q^n-1}{q-1}} - c \in F[X]$, for each $\alpha \in E \setminus \{0\}$ we have:

$$N_{E/F}(\alpha) = c \iff g_c(\alpha) = 0;$$

for each of the $q-1$ polynomials f_c, there are at most $\frac{q^n-1}{q-1}$ elements $\alpha \in E$ such that $\mathrm{Tr}_{E/F}(\alpha) = c$.
As a consequence the $q^n - 1$ elements of $E \setminus \{0\}$ are partitioned into $q-1$ sets $E_c, c \in F \setminus \{0\}$, of $\frac{q^n-1}{q-1}$ elements whose norm is c.

(6) If β is a primitive (q^n-1)th root of unity, we have

$$N_{E/F}(\alpha) = \alpha^{\frac{q^n-1}{q-1}} = 1 \iff \text{there exists } r : \alpha = \beta^{r(q-1)}$$
$$= (\beta^r)^{(q-1)}. \qquad \square$$

This allows us to characterize the duality[5] of the $GF(q)$-vector space $GF(q^n)$:

Proposition 10.5.11. *Let $F := GF(q)$ and $E := GF(q^n)$. For each $\beta \in E$ let $L_\beta : E \mapsto F$ be the linear functional defined by*

$$L_\beta(\alpha) := \mathrm{Tr}_{E/F}(\alpha\beta).$$

---

[5] If F is a field and E is an F-vector space, we can consider the *linear functionals* of E, i.e. the F-morphisms $L := E \mapsto F$; their set, called the *dual space*, denoted by $E^* := \mathrm{Hom}_F(E, F)$, is an F-vector space.
When E is a finite vector space, the same holds for E^* and we have $\dim_F(E) = \dim_F(E^*) =: n$, so that they are isomorphic.

The mapping
$$\Lambda : E \mapsto \mathrm{Hom}_F(E, F)$$
defined by $\Lambda(\beta) = L_\beta$, *for all* β, *is an isomorphism.*

Proof Obviously, each L_β is F-linear, so that $L_\beta \in \mathrm{Hom}_F(E, F)$; moreover Λ is F-linear. To prove that it is an isomorphism, since E and $\mathrm{Hom}_F(E, F)$ have the same dimension over F, it is sufficient to show that it is injective: since $\mathrm{Tr}_{E/F}$ is surjective and
$$\beta \neq 0 \implies E = \{\alpha\beta : \alpha \in E\},$$
we conclude that
$$L_\beta(\alpha) := \mathrm{Tr}_{E/F}(\alpha\beta) = 0, \text{ for all } \alpha \in E \implies \{\alpha\beta : \alpha \in E\} \not\subseteq E$$
$$\implies \beta = 0,$$
i.e. that Λ is injective. ♄

10.6 Discriminant

Let $K \supset k$ be a finite algebraic extension of degree n.

If $\Omega := \{\omega_1, \ldots, \omega_n\} \subset K$ is any set of n elements of k, we can consider the matrix
$$\begin{pmatrix} \mathrm{Tr}_{K/k}(\omega_1\omega_1) & \mathrm{Tr}_{K/k}(\omega_1\omega_2) & \cdots & \mathrm{Tr}_{K/k}(\omega_1\omega_n) \\ \mathrm{Tr}_{K/k}(\omega_2\omega_1) & \mathrm{Tr}_{K/k}(\omega_2\omega_2) & \cdots & \mathrm{Tr}_{K/k}(\omega_2\omega_n) \\ \vdots & \vdots & \ddots & \vdots \\ \mathrm{Tr}_{K/k}(\omega_n\omega_1) & \mathrm{Tr}_{K/k}(\omega_n\omega_2) & \cdots & \mathrm{Tr}_{K/k}(\omega_n\omega_n) \end{pmatrix}$$
and its determinant $\Delta_{K/k}(\Omega)$.

Definition 10.6.1. *With the notation above,* $\Delta_{K/k}(\Omega)$ *is called the* discriminant *of* Ω.

Lemma 10.6.2. *Let* $\Omega := \{\omega_1, \ldots, \omega_n\}$ *be a k-basis of K. If* $\Delta_{K/k}(\Omega) = 0$ *then there is* $\beta \in K \setminus \{0\}$ *such that* $\mathrm{Tr}_{K/k}(\alpha\beta) = 0$, *for all* $\alpha \in K$.

Proof Since $\Delta_{K/k}(\Omega) = 0$, there are $c_j \in k$, not all zero, such that
$$\sum_j c_j \mathrm{Tr}_{K/k}(\omega_i \omega_j) = 0, \forall i.$$

10.6 Discriminant

Therefore, defining $\beta := \sum_j c_j \omega_j$, we have $\mathrm{Tr}_{K/k}(\omega_i \beta) = 0$, for all i, from which it follows that $\mathrm{Tr}_{K/k}(\alpha\beta) = 0$, for all $\alpha \in K$. ♮

Based on the lemma above, the discriminant allows us, when k is a finite field, to decide whether a set Ω is a k-basis of K:

Proposition 10.6.3. *Let $F := GF(q)$ and $E := GF(q^n)$ and let*

$$\Omega := \{\omega_1, \ldots, \omega_n\} \subset E;$$

then the following are equivalent

(1) Ω is an F-basis of E;
(2) $\Delta_{E/F}(\Omega) \neq 0$.

Proof

(1) $\Longrightarrow$ (2) From Lemma 10.6.2, we know that $\Delta_{E/F}(\Omega) = 0$ implies the existence of $\beta \in K \setminus \{0\}$ such that $\mathrm{Tr}_{E/F}(\alpha\beta) = 0$, for all $\alpha \in K$.
Using the notation and results of Proposition 10.5.11 we deduce that

$$L_\beta(\alpha) = 0, \text{ for all } \alpha \in K$$

and therefore that $\beta = 0$, giving the required contradiction.

(2) $\Longrightarrow$ (1) Let $c_j \in k$ be such that $\sum_j c_j \omega_j = 0$; then, for each i

$$0 = \mathrm{Tr}_{K/k}(\omega_i \sum_j c_j \omega_j) = \sum_j c_j \mathrm{Tr}_{K/k}(\omega_i \omega_j).$$

Since $\Delta_{E/F}(\Omega) \neq 0$, this implies $c_j = 0$, for all j. ♮

Let us now give an interpretation of the discriminant when K is a separable extension of k. In this case there are n k-isomorphisms of K into a normal extension $N \supset K$, which we will denote as $\phi_1, \ldots, \phi_n$, where ϕ_1 is the identity.

Lemma 10.6.4. *Under the above notation, let $\Omega := \{\omega_1, \ldots, \omega_n\} \subset K$ be a k-basis of the separable field K and let*

$$\omega_i^{(j)} := \phi_j(\omega_i).$$

Then

$$\Delta_{K/k}(\Omega) = \begin{vmatrix} \omega_1^{(1)} & \omega_1^{(2)} & \cdots & \omega_1^{(n)} \\ \omega_2^{(1)} & \omega_2^{(2)} & \cdots & \omega_2^{(n)} \\ \vdots & \vdots & \ddots & \vdots \\ \omega_n^{(1)} & \omega_n^{(2)} & \cdots & \omega_n^{(n)} \end{vmatrix}^2.$$

Proof Note that, by definition,

$$\mathrm{Tr}_{K/k}(\omega_i \omega_h) = \sum_j \phi_j(\omega_i \omega_h) = \sum_j \omega_i^{(j)} \omega_h^{(j)}$$

from which the result is obvious. ♄

With the same notation, if K is separable, by the Primitive Element Theorem we know that there is χ such that $K = k[\chi]$, so that $\Omega := \{1, \chi, \ldots, \chi^{n-1}\}$ is a k-basis of K.

If $\chi_j := \phi_j(\chi)$, for all j, denotes the conjugate of χ, in the above formula we have

$$\omega_i^{(j)} = \phi_j(\chi^{i-1}) = \phi_j(\chi)^{i-1} = \chi_j^{i-1},$$

from which we obtain:

Theorem 10.6.5. *Let $K \supset k$ be a separable algebraic extension, $[K : k] = n$. Let χ be a primitive element, $\chi =: \chi_1, \ldots, \chi_n$ be its conjugates and $f(X)$ its minimal polynomial.*
Then:

$$\begin{aligned}
\Delta_{K/k}(\Omega) &= \begin{vmatrix} 1 & 1 & \cdots & 1 \\ \chi_1 & \chi_2 & \cdots & \chi_n \\ \chi_1^2 & \chi_2^2 & \cdots & \chi_n^2 \\ \vdots & \vdots & \cdots & \vdots \\ \chi_1^{n-1} & \chi_2^{n-1} & \cdots & \chi_n^{n-1} \end{vmatrix}^2 \\
&= \prod_{i>j}(\chi_i - \chi_j)^2 \\
&= \mathrm{Disc}(f) \\
&= \mathrm{Res}(f, f') \\
&= N_{K/k}(f'(X)) \neq 0.
\end{aligned}$$

Proof The first equality follows from Lemma 10.6.4, the second one by the Vandermonde determinant, the third by Definition 6.5.3 and the fourth by Corollary 6.7.3.

10.6 Discriminant

The fifth equation follows from the fact that $f(X) = \prod_j (X - \chi_j)$, so that

$$f'(X) = \sum_i \prod_{j \neq i}(X - \chi_j),$$
$$f'(\chi_i) = \prod_{j \neq i}(\chi_i - \chi_j),$$
$$N_{K/k}(f'(X)) = \prod_i \prod_{j \neq i}(\chi_i - \chi_j).$$

The inequality is due to the fact that $\chi_i - \chi_j \neq 0$ if $i \neq j$. ♮

On the basis of the above result, the notion of discriminant also allows us to test the separability of finite extension; to prove this result we need, however, to introduce a further lemma:

Lemma 10.6.6. *Let* $K \supset k$ *be a finite algebraic extension of degree n. Let* $\Omega^1 := \{\omega_1^1, \ldots, \omega_n^1\}$ *and* $\Omega^2 := \{\omega_1^2, \ldots, \omega_n^2\}$ *be two k-bases of K. Then*

$$\Delta_{K/k}(\Omega^1) = 0 \iff \Delta_{K/k}(\Omega^2) = 0.$$

Proof Let $A := (a_{ij})_{ij}$ be the invertible matrix such that

$$\omega_i^2 = \sum_j a_{ij} \omega_i^1;$$

so that

$$\mathrm{Tr}_{K/k}(\omega_I^2 \omega_J^2) = \sum_{ij} a_{Ii} a_{jJ} \, \mathrm{Tr}_{K/k}(\omega_i^1 \omega_j^1)$$

and

$$\Delta_{K/k}(\Omega^2) = \det(A)^2 \Delta_{K/k}(\Omega^1).$$ ♮

Theorem 10.6.7. *Let* $K \supset k$ *be a finite algebraic extension of degree n. The following conditions are equivalent:*

(1) for each k-basis Ω of K, $\Delta_{K/k}(\Omega) = 0$;
(2) for each $\alpha \in K$, $\mathrm{Tr}_{K/k}(\alpha) = 0$;
(3) K is an inseparable extension of k.

Proof

(1) $\implies$ (2) By Lemma 10.6.2 we conclude that there is $\beta \in K \setminus \{0\}$ such that $\mathrm{Tr}_{K/k}(\alpha\beta) = 0$, for all $\alpha \in K$. Therefore for each $\alpha \in K$, setting $\gamma := \alpha\beta^{-1}$ we have $\mathrm{Tr}_{K/k}(\alpha) = \mathrm{Tr}_{K/k}(\gamma\beta) = 0$.

(2) $\implies$ (1) Trivial.

(1) $\implies$ (3) If K were separable, by Theorem 10.6.5, would we conclude that $\Delta_{K/k}(\Omega) \neq 0$.

(3) $\implies$ (2) Let K_{sep} be the greatest separable extension of k in K. Using freely the notation of Corollary 9.3.6, we note that
if $\alpha \in K_{\text{sep}}$ then $\text{Tr}_{K/k}(\alpha) = p^e \text{Tr}_{K_{\text{sep}}/k}(\alpha) = 0$.
if $\alpha \in K \setminus K_{\text{sep}}$, then α is inseparable and so its minimal polynomial $g(X)$ is such that $g(X) = \prod_{i=1}^{n_0} X - \alpha_i^{p^e}$ so that $\text{Tr}_{K/k}(\alpha) = p^e \sum_i \alpha_i = 0$. ♄

Corollary 10.6.8. *Let $K \supset k$ be a finite algebraic extension. K is a separable extension of k iff there is $\alpha \in K$ such that $\text{Tr}_{K/k}(\alpha) \neq 0$.* ♄

10.7 Normal Bases

Let $K \supset k$ be a finite separable normal extension and let $n := [K : k]$; by Corollary 10.4.1 we know the existence of n k-automorphisms $\phi_i : K \mapsto K$.

If $\xi \in K$ is a primitive element, so that $K = k[\xi]$, we will denote its conjugates as $\xi_i := \psi_i(\xi)$, for all i, and we will consider the set

$$\Omega(\xi) := \{\xi_1, \ldots, \xi_n\};$$

as in Lemma 10.6.4 we will use the notation

$$\xi_i^{(j)} := \phi_j(\xi_i) = \phi_j(\psi_i(\xi)).$$

Definition 10.7.1. *With the above notation, a basis $\Omega := \{\omega_1, \ldots, \omega_n\}$ of K as a k-vector space is called a* normal basis *of K over k if each two elements of Ω are conjugate, or, equivalently, if there is an element $\xi \in K$ such that $\Omega = \Omega(\xi)$.*

Lemma 10.7.2. *With the above notation, if $\xi \in K$ is a primitive element, $\Omega(\xi)$ is a basis – and so a normal one – iff $\det(\xi_i^{(j)}) \neq 0$.*

Proof Let us assume that there are $c_i \in k$ such that $\sum_{i=1}^{n} c_i \xi_i = 0$; then

$$\sum_{i=1}^{n} c_i \xi_i^{(j)} = \phi_j \left(\sum_{i=1}^{n} c_i \xi_i \right) = 0, \text{ for all } j,$$

since $\phi_j(c_i) = c_i$, for all i. As a consequence, if $\det(\xi_i^{(j)}) \neq 0$, we deduce $c_i = 0$, for all i, and $\Omega(\xi)$ is a basis.

10.7 Normal Bases

Conversely, let us assume that $\det(\xi_i^{(j)}) = 0$, so that there are elements $\gamma_i \in K$, not all of them null, such that

$$\sum_{i=1}^{n} \gamma_i \xi_i^{(j)} = 0, \text{ for all } j; \tag{10.2}$$

let us assume wlog $\gamma_1 \neq 0$.

Since K is separable, by Theorem 10.6.7, we know the existence of $\alpha \in K$ such that $\text{Tr}_{K/k}(\alpha) \neq 0$; therefore, multiplying Equation 10.2 by $\alpha \gamma_1^{-1}$, we can assume that $\text{Tr}_{K/k}(\gamma_1) \neq 0$. From Equation 10.2 we obtain

$$\sum_{i=1}^{n} \phi_j^{-1}(\gamma_i) \xi_i = 0, \text{ for all } j,$$

and

$$\sum_{i=1}^{n} \text{Tr}_{K/k}(\gamma_i) \xi_i = \sum_{j=1}^{n} \sum_{i=1}^{n} \phi_j^{-1}(\gamma_i) \xi_i = 0;$$

since $\text{Tr}_{K/k}(\gamma_i) \in k$ and $\text{Tr}_{K/k}(\gamma_1) \neq 0$, we conclude that $\Omega = \{\xi_1, \ldots, \xi_n\}$ is not a linear basis. $\boxed{\text{h}}$

Lemma 10.7.3. *Let k be an infinite field and $K \supset k$ be a finite separable normal extension; let $d(X_1, \ldots, X_n) \in K[X_1, \ldots, X_n]$ be a polynomial such that*

$$d(\phi_1(\xi), \ldots, \phi_n(\xi)) = 0, \text{ for all } \xi \in K;$$

then $d = 0$.

Proof Let $\{\omega_1, \ldots, \omega_n\}$ be a k-basis of K so that each element $\xi \in K$ can be represented as $\xi = \sum_{i=1}^{n} c_i \omega_i$; therefore, we have

$$\psi_j(\xi) = \sum_{i=1}^{n} c_i \psi_j(\omega_i) = \sum_{i=1}^{n} c_i \omega_i^{(j)}, \text{ for all } j.$$

Consider the polynomial

$$D(Y_1, \ldots, Y_n) := d\left(\sum_{i=1}^{n} Y_i \omega_i^{(1)}, \ldots, \sum_{i=1}^{n} Y_i \omega_i^{(n)}\right) \in K[Y_1, \ldots, Y_n],$$

which, by assumption satisfies

$$D(c_1, \ldots, c_n) = 0, \text{ for all } (c_1, \ldots, c_n) \in k^n;$$

therefore the infiniteness of k allows us to conclude[6] that $D = 0$. By Theorem 10.6.7, the separability of K implies that the matrix $\left(\omega_i^{(j)}\right)$ is invertible; therefore denoting (γ_{ij}) to be its inverse we have

$$d(X_1, \ldots, X_n) := D\left(\sum_{j=1}^n \gamma_{1j} X_j, \ldots, \sum_{j=1}^n \gamma_{nj} X_j\right) = 0.$$

♄

Proposition 10.7.4. *Let k be an infinite field and $K \supset k$ be a finite separable normal extension; then there exists a normal basis of K over k.*

Proof For each $i, j, 1 \leq i, j \leq n$, let us denote $i \star j$, $1 \leq i \star j \leq n$, the element such that $\phi_{i \star j} = \phi_i \phi_j$.

Let us consider the matrix in the variables $X_1, \ldots, X_n$ whose (i, j) entry is $X_{i \star j}$, and the polynomial

$$d(X_1, \ldots, X_n) := \det(X_{i \star j}).$$

Since we have

$$i \star j = i \star h \implies j = h,$$
$$i \star j = h \star j \implies j = h,$$

we deduce that each row and column contains exactly one entry X_1; therefore $d(1, 0, \ldots, 0) = \pm 1$ and $d(X_1, \ldots, X_n) \neq 0$. Therefore, Lemma 10.7.3 allows us to deduce the existence of $\xi \in K$ such that

$$\det(\xi_i^{(j)}) = d(\phi_1(\xi), \ldots, \phi_n(\xi)) \neq 0$$

and so that of a normal basis $\Omega(\xi)$ of K over k.

♄

---

[6] In fact, if $g(Y_1, \ldots, Y_n) \in K[X_1, \ldots, X_n]$ is such that $g \neq 0$, and $k \subset K$ is an infinite set, we can prove the existence of $c_i \in k$, for all i, such that $g(c_1, \ldots, c_n) \neq 0$. The proof is by induction on n.
If $n = 1$ it is clear that $g(Y_1)$ has at most $\deg(g)$ roots in k, so we can choose $c_1 \in k$ such that $g(c_1) \neq 0$.
Iteratively, we can write $g(Y_1, \ldots, Y_n)$ as $g = \sum_j f_j(Y_1, \ldots, Y_{n-1}) Y_n^j$; since $g \neq 0$, there is at least an index J such that $f_J \in K[Y_1, \ldots, Y_{n-1}]$ is not zero, and, inductively, there are $c_i \in k, 1 \leq i < n$, such that $f_J(c_1, \ldots, c_{n-1}) \neq 0$ and $G(Y_n) = g(c_1, \ldots, c_{n-1}, Y_n) \in K[Y_n]$ is not zero and has at most $\deg(G)$ roots in k, so we can choose $c_n \in k$ such that $g(c_1, \ldots, c_{n-1}, c_n) = G(c_n) \neq 0$.

10.7 Normal Bases

Let k be finite, card$(k) = q$ and K be an algebraic extension such that $[K : k] = n$.

Then, denoting i, by $\Phi_i : K \mapsto K$ for each i, the morphism defined by $\Phi_i(\chi) = \chi^{q^i}$, the k-isomorphisms of K are $\{\Phi_i : 1 \leq i \leq n\}$ and we have

$$\Phi_i = \Phi_j \iff i \cong j \pmod{n}.$$

Proposition 10.7.5. *With the above assumptions, K has a normal basis over k.*

Proof We need to prove that there is $\xi \in K$ such that

$$\{\xi, \xi^q, \xi^{q^2}, \ldots, \xi^{q^{n-1}}\}$$

is a basis of K as a k-vector space.

To each polynomial $g(X) := \sum_k c_k X^k \in K(X)$ and to each element $\chi \in K$ let us denote

$$g \diamond \chi := \sum_k c_k \Phi_k(\chi) = \sum_k c_k \chi^{q^k} \in K.$$

To each $\chi \in K$ let us associate the unique polynomial $g_\chi \in K(X)$ of minimal degree such that

$$g_\chi \diamond \chi = 0.$$

Obviously, for each χ, g_χ divides $X^n - 1$ since Φ_n is the identity.
Let

$$X^n - 1 = p_1(X)^{e_1} \cdots p_h^{e_h}(X)$$

be the factorization in $k[X]$ and let

$$q_i(X) := \frac{X^n - 1}{p_i(X)}, \; r_i(X) := \frac{X^n - 1}{p_i^{e_i}(X)}, \; \text{for all } i.$$

For each i, there is β_i such that $q_i \diamond \beta_i \neq 0$: in fact $\deg(q_i) \leq n - 1$ and the set $\{\Phi_i : 1 \leq i \leq n\}$ is linearly independent, so that there is β_i such that

$$q_i \diamond \beta_i := \sum_k c_k \Phi_k(\beta_i) \neq 0.$$

Let $\gamma_i := r_i \diamond \beta_i$; then

$$p_i^{e_i} \diamond \gamma_i = p_i^{e_i} \diamond (r_i \diamond \beta_i) = (r_i p_i^{e_i}) \diamond \beta_i = (X^n - 1) \diamond \beta_i = 0,$$

while

$$p_i^{e_i - 1} \diamond \gamma_i = p_i^{e_i - 1} \diamond (r_i \diamond \beta_i) = q_i \diamond \beta_i \neq 0,$$

so that $p_i^{e_i} = g_{\gamma_i}$.

Denoting $\xi := \sum_i \gamma_i$ it is clear that $g_\xi = X^n - 1$.
As a consequence

$$\sum_{k=0}^{n-1} c_k \Phi_k(\xi) = \sum_{k=0}^{n-1} c_k \xi^{q^k} \neq 0, \text{ for all } c_k \in K,$$

so that

$$\{\xi, \xi^q, \xi^{q^2}, \ldots, \xi^{q^{n-1}}\}$$

is a basis of K as a k-vector space.

From the two previous propositions we deduce

Theorem 10.7.6. *Let $K \subset k$ be a finite separable normal extension; then there is a normal basis of K over k.*

11
Duval

> Clever triviality is the essence of geniality.
> E.B. Gebstadter, *Copper, Silver, Gold: an Indestructible Metallic Alloy*

Kronecker's Model gives a powerful tool for computing, at least within the field of the algebraic complex numbers, and for solving polynomial equations there, provided we have an algorithm for factorizing polynomials over a given algebraic extension of the rationals. Such an algorithm exists, but its practical complexity is so unsatisfactory, that the solution of polynomial equations provided by Kronecker's ideas has no practical impact and the state of the art on Solving Polynomial Equation Systems was again in an impasse: as Macaulay put it[1], 'the solution is only a theoretical one'...

... until in 1987, more than one hundred years after Kronecker's *Grundzüge*, Duval[2] added an unexpected twist to Kronecker's proposal, showing how factorization can be easily avoided. Her proposal threw light on Kronecker's ideas, clarifying the philosophy behind them.

I will introduce Duval's idea by discussing how to represent rings explicitly.

11.1 Explicit Representation of Rings

In all the cases we have seen up to now, a ring A is effectively given by taking a set R, whose elements are in biunivocal correspondence with the elements of A, and defining in R those operations which turn R into a ring isomorphic to A.

For instance, for $A = \mathbb{Z}_n$, we can put $R := \{z \in \mathbb{Z} : -\frac{n}{2} < z \leq \frac{n}{2}\}$, whose operations are those in $\mathbb{Z}$, followed by computation of the least absolute value remainder of their division by n.

---

[1] F. S. Macaulay, *The Algebraic Theory of Modular Systems*, Cambridge University Press (1916).
[2] D. Duval, *Diverses questions relatives au calcul formel avec les nombres algébriques*, Thèse d'Etat, Grenoble (1987).

In the same manner, for $A = k[Z_1, \ldots, Z_r]/(f_1, \ldots, f_r)$, we can use

$$R = \{g \in k[Z_1, \ldots, Z_r] : \deg_i(g) < d_i\},$$

whose operations are those in $k[Z_1, \ldots, Z_r]$, followed (in the case of multiplication) by a reduction procedure.

The advantages of such a representation are that:

- each element of A has a unique representative in R, so, in particular, testing the equality of two elements (and so deciding if an element is zero) is easy: the two representations must be equal!
- arithmetical operations have bounded complexity; so every arithmetical operation in the given representation of $\mathbb{Z}_n$ has polynomial complexity in $\log(n)$, and every arithmetical operation in the given representation of $k[Z_1, \ldots, Z_r]/(f_1, \ldots, f_r)$ has polynomial complexity[3] in $D = \prod_i d_i$.

However, biunivocal correspondence between R and A is not at all necessary: we could choose, for instance, to represent an element $a \in \mathbb{Z}_n$ by any element of $\mathbb{Z}$ whose residue class is a; arithmetical operations are then just performed in $\mathbb{Z}$, so that they are conceptually simpler, albeit with worse theoretical complexity. In the same way, elements of $k[Z_1, \ldots, Z_r]/(f_1, \ldots, f_r)$ can be represented just by polynomials in $k[Z_1, \ldots, Z_r]$.

Clearly, in such a representation, elements of A have in general more than one representative in R. As a consequence, 0-testing and equality testing, which are trivial in the case of unique representation, become more involved. In our example, if $a, b \in \mathbb{Z}_n$ are represented respectively by $\alpha, \beta \in \mathbb{Z}$, then

$a = b \iff \alpha - \beta$ is a multiple of n;
$a = 0 \iff \alpha$ is a multiple of n.

If we give up uniqueness of representation, we can then effectively obtain a ring A, by way of

an effective ring B, such that there exists a projection $\pi : B \mapsto A$;
a subset $R \subseteq B$ such that $\pi(R) = A$;
an algorithm which, given $b \in B$, computes $r \in R$ such that $\pi(b) = \pi(r)$;
an algorithm which, given $r_1, r_2 \in R$, decide whether $\pi(r_1) = \pi(r_2)$,

so that

addition, subtraction, multiplication in R are simply done in B, obtaining some b, and then computing $r \in R$ such that $\pi(b) = \pi(r)$;

---

[3] In a model, of course, in which we assume that each operation in k is constant.

equality testing and zero-testing are performed by means of the specific algorithms which, given $r_1, r_2 \in R$, decide whether $\pi(r_1) = \pi(r_2)$;

uniqueness of representation is achieved when the restriction of π to R is one-to-one.

Note that we have not discussed invertibility tests or algorithms for computing inverses, through this representation of the *ring* A. In fact, in all the examples of algorithms in a *ring* that we have discussed, these operations are not needed.

11.2 Ring Operations in a Non-unique Representation

In Chapter 8 I discussed the problem of how to explicitly obtain a field containing either one or all of the roots of a polynomial, in order to be able to explicitly perform arithmetical operations over arithmetical expressions in it and in its polynomial extensions, and showed how Kronecker's Model solved that problem. Let us now turn to the following variation of that problem:

let us be given a monic squarefree polynomial $f(X) \in k[X]$, and let $\alpha_1, \ldots, \alpha_n$ be all the roots of f; we want to perform one and the same specific arithmetical computation in each $k[\alpha_i][Z]$, whose result is some polynomial $g_i(\alpha_i, Z)$.

In other words we want to perform the same computation but separately for each root of f.

If f is irreducible, then the fields $k[\alpha_i]$ are all isomorphic to

$$k[X]/f(X) =: k[x],$$

so it is sufficient to perform the computation once in $k[X, Z]/f(X)$ and then interpret the solution $g(x, Z)$ in each $k[\alpha_i][Z]$, by substituting x with α_i, or (if you like) by interpreting x as α_i. In fact all the roots of f are conjugates, and we know that all arithmetical operations over algebraic expressions of a generic root x behave in the same way when performed on conjugate algebraic numbers; in particular, 0-testing and equality testing give the same solution whenever α_i is substituted for x.

If f is reducible, in Kronecker's Model we should first factorize $f(X)$ into irreducible factors in $k[X]$,

$$f(X) = f_1(X) \ldots f_r(X),$$

and then perform the same computation in each ring $K[X, Z]/f_i(X)$.

Let us assume that the arithmetical computation we are performing requires only the ring arithmetics of $k[\alpha_i]$; in particular this means that we do not need to test whether some polynomial expression in α_i is zero (which obviously depends on α_i) nor to compute inverses in $k[\alpha_i]$.

In this case, we do not need to perform 0-testing, equality testing and inverse computation, so that all computations in each $k[\alpha_i]$ are, in some sense, 'conjugate': the computations are exactly the same in each ring $K[X, Z]/f_i(X)$, except for the need to compute remainders mod f_i, which again gives different results for the different f_is.

We could, however, give up the unique representation implied by Kronecker's Model and, following the ideas sketched in the previous section, perform our computation only once in $k[X, Z]/f(X)$, getting a polynomial $g(X, Z) \in k[X, Z]$; in other words we can explicitly get all rings $k[\alpha_i][Z]$ by assigning the same effective ring $k[X, Z]/f(X)$ and just specifying the proper projection

$$\pi_i : k[X, Z]/f(X) \mapsto k[\alpha_i][Z] = k[X, Z]/f_i(X),$$

which is nothing more than

$$\pi_i \left(\sum_j g_j(X) Z^j \right) = \sum_j \mathbf{Rem}\left(g_i(X), f_i(X)\right) Z^j.$$

Clearly, for each i, $\pi_i(g(X, Z))$ is the canonical representative of the solution $g_i(\alpha_i, Z) \in k[\alpha_i][Z]$. Moreover the last step, the valuation of π_i, is simply required for converting from the common non-unique representation of each $k[\alpha_i][Z]$ by $k[X, Z]/f(X)$ to the Kronecker representation $k[X, Z]/f_i(X)$, and if we choose to represent each $k[\alpha_i][Z]$ by $k[X, Z]/f(X)$ we can dispose of it.

If we know the factorization of $f(X)$, the computational amount is essentially the same in both ways, as a consequence of the Chinese Remainder Theorem. More precisely, the k-vector space operations have exactly the same cost when performed in $k[X]/f$ or in parallel in each $k[X]/f_i$, while there is some advantage for multiplication in the 'parallel' computation since the total cost is $\sum_i \mathcal{O}(\deg(f_i)^2) \leq \mathcal{O}(\deg(f)^2)$.

If, instead, we do not know the factorization of f, we have a large advantage if we compute in $k[X]/f$: there is absolutely no need to factorize f !

11.3 Duval Representation

While representing each $k[\alpha_i]$ by the same $k[X]/f(X)$, instead of by each $k[X]/f_i(X)$, allows us to avoid factorization when we only have to perform the ring arithmetics of each $k[\alpha_i]$, in most computations we will need to test

11.3 Duval Representation

if an element is 0 or to compute the inverse of a non-zero element – this will obviously occur, for instance, if we are going to perform gcd computations in each $k[\alpha_i][Z]$.

Clearly then the different roots are no longer 'conjugate' for such computations, since the answer to a 0-test will obviously be different for the different roots: for instance, if α_1 and α_2 are roots of $f(X)$, and f_i is the minimal polynomial of α_i, for all i, then $f_1(\alpha_1) = 0$, while $f_1(\alpha_2) \neq 0$. Let us see in a concrete example what happens with polynomial algorithms.

Example 11.3.1. So let us choose

$$f(X) = X^4 - 13X^2 + 36 \in \mathbb{Q}[X],$$

and let us denote its four roots by $\alpha_1, \ldots, \alpha_4$ and assume that the computation we need to perform is to decide whether

$$g_i(Z) := Z^3 + 3\alpha_i Z^2 + 12Z + 4\alpha_i \in \mathbb{Q}[\alpha_i][Z]$$

is squarefree. So we should compute in each $\mathbb{Q}[\alpha_i][Z]$, $\gcd(g_i(Z), h_i(Z))$ where

$$h_i := \frac{1}{3}g_i' = Z^2 + 2\alpha_i Z + 4.$$

By representing each $\mathbb{Q}[\alpha_i]$ by $\mathbb{Q}[X]/f(X)$, each g_i is represented by

$$g(X, Z) := Z^3 + 3XZ^2 + 12Z + 4X,$$

and each h_i by

$$h(X, Z) := Z^2 + 2XZ + 4.$$

Since h is monic in Z, polynomial division by h is also possible in the ring $\mathbb{Q}[X, Z]$ without the need to compute inverses. By standard arithmetic we obtain:

$$g(X, Z) = (Z + X)h(X, Z) + 2(4 - X^2)Z,$$

which can be interpreted as

$$g_i(\alpha_i, Z) = (Z + \alpha_i)h_i(\alpha_i, Z) + 2(4 - \alpha_i^2)Z.$$

Can we reach some conclusion? Well:

if $\alpha_i^2 = 4$, then h_i divides g_i, $h_i = \gcd(g_i, h_i)$, $SQFR(g_i) = (Z + \alpha_i)$, $g_i = (Z + \alpha_i)^3$;

if, instead, $\alpha_i^2 \neq 4$, then we must go on and divide h_i by Z (since in this case, we are lucky and we do not need to compute the inverse of $4 - \alpha_i^2$).

The result is then:

$$h(X, Z) = (Z + 2X)Z + 4, \quad h_i(\alpha_i, Z) = (Z + 2\alpha_i)Z + 4,$$

and so, since $4 \neq 0$ (we are very lucky this time!) we can conclude that g_i is squarefree.

Which one is the correct answer? The bad news is that both are correct! In fact the example is so trivial that we can perform factorization:

$$f(X) = (X - 2)(X + 2)(X - 3)(X + 3),$$

so the four roots are $\alpha_1 = 2, \alpha_2 = -2, \alpha_3 = 3, \alpha_4 = -3$. Then, clearly if
$i = 1, 2$, then $\alpha_i^2 = 4$, and $g_i = (Z + \alpha_i)^3$,
$i = 3, 4$, then $\alpha_i^2 = 9 \neq 4$, and g_i is squarefree.

Therefore, apparently, we should factorize any time we need to perform a non-trivial task. However, the good news is that factorization is not needed at all.

In fact, even if we are unable to factorize the polynomial f, we can easily obtain the partial factorization

$$f(X) = (X^2 - 4)(X^2 - 9),$$

by the following argument: the roots α_i satisfy the polynomial $f(X)$ and therefore

satisfying the polynomial $p(X) := X^2 - 4 \iff$ satisfying the two polynomials f and $p \iff$ satisfying $f_1(X) := \gcd(f, p) = X^2 - 4$;
not satisfying $p \iff$ not satisfying f_1; satisfying $f \iff$ satisfying $f_2(X) := \frac{f(X)}{f_1(X)} = X^2 - 9$.

In conclusion, we see that we can answer whether or not a root of f satisfies a polynomial p and get a partial factorization of f as a consequence of the following (trivial):

Theorem 11.3.2 (Lazard). *Let $f(X) \in k[X]$ be a squarefree polynomial. Let $p(X) \in k[X]$ and let*

$$f_1(X) := \gcd(f, p), \quad f_2 := \frac{f}{f_1}, \quad p_1 := \frac{p}{f_1}.$$

Also let $s(X), t(X)$ satisfy

$$sp + tf = f_1,$$

11.3 Duval Representation

and $u(X), v(X)$ satisfy

$$uf_1 + vf_2 = 1.$$

Therefore:

if $f_1 = 1$, then for each α such that $f(\alpha) = 0$, we have $p(\alpha) \neq 0$, $p(\alpha)^{-1} = s(\alpha)$;

if $f_1 = f$, then for each α such that $f(\alpha) = 0$, we have $p(\alpha) = 0$;

otherwise, if α is such that $f(\alpha) = 0$, then:

$$p(\alpha) = 0 \iff f_1(\alpha) = 0,$$
$$p(\alpha) \neq 0 \iff f_2(\alpha) = 0, \text{ in which case } p(\alpha)^{-1} = u(\alpha)s(\alpha).$$

Proof If $f(\alpha) = 0$, either $f_1(\alpha) = 0$ or $f_2(\alpha) = 0$, but not both since f is squarefree.

If $f_1(\alpha) = 0$, then $p(\alpha) = f_1(\alpha)p_1(\alpha) = 0$.

If $f_2(\alpha) = 0$, then

$$\begin{aligned} u(\alpha)s(\alpha)p(\alpha) &= u(\alpha)s(\alpha)p(\alpha) + u(\alpha)t(\alpha)f(\alpha) \\ &= u(\alpha)f_1(\alpha) \\ &= u(\alpha)f_1(\alpha) + v(\alpha)f_2(\alpha) = 1. \end{aligned}$$

♃

In conclusion, any time we need to perform a zero-test on each $k[\alpha_i]$ or to compute inverses there, we have no need to factorize. In Duval's computational model, each $k[\alpha_i]$ is represented by $k[X]/f(X)$ and each element $\beta_i \in k[\alpha_i]$ is represented by a polynomial $p(X), \deg(p) < \deg(f)$. So let

$$\beta_j := p(\alpha_j) \in k[\alpha_j], \text{ for all } j,$$

and assume we need to test, for all j, whether $\beta_j \neq 0$, in which case we need to compute $\beta_j^{-1} \in k[\alpha_j]$. We compute $f_1(X) := \gcd(f(X), p(X))$ and $s(X)$ such that

$$sp \equiv f_1 \pmod{f}, \deg(s) < \deg(f).$$

If $f_1 = 1$, then β_j is invertible for all j, and $s(\alpha_j)$ is its inverse whose representation is $s(X)$.

If $f_1 \neq 1$, then we compute $f_2 := \frac{f}{f_1}$ – and f_2 is not constant since $\deg(p) < \deg(f)$ –, obtaining a partial factorization of f; we split our computation, representing those $k[\alpha_j]$ such that $\beta_j = 0$ by $k[X]/f_1(X)$ and those $k[\alpha_j]$ such that $\beta_j \neq 0$ by $k[X]/f_2(X)$.

We then compute $u(X)$ such that

$$uf_1 \equiv 1 (\mathrm{mod}\ f_2)$$

and then $w(X)$ such that

$$w \equiv us\ (\mathrm{mod}\ f_2), \deg(w) < \deg(f_2).$$

Then in those $k[\alpha_j]$ such that $\beta_j \neq 0$, we find $\beta_j^{-1} = w(\alpha_j)$ and their common representation is $w(X)$.

Remark 11.3.3. It is quite clear that we can interpret the Duval's result in terms of Theorem 2.3.2.

In fact, if f is squarefree and $f = g_1 \ldots g_n$ is its factorization, let us denote $R = k[X]/f(X)$, $R_i = k[X]/g_i(X)$, for all i. Up to a reordering of the g_is, we can assume that g_i is the minimal polynomial of a root α of f such that

$$p(\alpha) = 0 \iff i \leq s,$$
$$p(\alpha) \neq 0 \iff i > s,$$

so that $f_1 = \prod_{i=1}^{s} g_i$ and $f_2 = \prod_{i=s+1}^{n} g_i$. If we set $S_1 := k[X]/f_1(X)$ and $S_2 := k[X]/f_2(X)$, it obviously holds that

$$S_1 \cong R_1 \oplus \cdots \oplus R_s, \quad S_2 \cong R_{s+1} \oplus \cdots \oplus R_n,$$

and $R = S_1 \oplus S_2$.

11.4 Duval's Model

In Section 8.2 we showed that Kronecker's Model allowed us to perform arithmetical operations and solve polynomial equations in each finite extension field over its prime field. We denoted by $\mathcal{K}$ the class to which Kronecker's Model applied, we described its recursive definition and we showed that the fields in $\mathcal{K}$ were representable as a quotient by an admissible sequence of polynomial rings over a rational function field over a prime field.

In what follows we will restrict ourselves to fields of characteristic 0 and $\mathcal{K}_0$ will denote the class of the fields $K \in \mathcal{K}$ such that $\mathrm{char}(K) = 0$.

From our discussion above, it is clear that Duval's Model applies to a larger class of rings $\mathcal{D} \supset \mathcal{K}_0$, which is a subset of the class of all the rings that are direct sums of fields in $\mathcal{K}_0$.

Definition 11.4.1. *We say that*

$$\mathbf{f} = \{f_1, \ldots, f_r\}$$

11.4 Duval's Model

is an admissible Duval sequence *in*

$$\mathbb{Q}(Y_1, \ldots, Y_d)[Z_1, \ldots, Z_r]$$

if, denoting

$L_j := \mathbb{Q}(Y_1, \ldots, Y_d)[Z_1, \ldots, Z_j]/(f_1, \ldots, f_j),$
π_j *the canonical projections*

$$\pi_j : \mathbb{Q}(Y_1, \ldots, Y_d)[Z_1, \ldots, Z_r] \mapsto L_j[Z_{j+1}, \ldots, Z_r],$$

$d_j := \deg_j(f_j),$

then

$f_1 \in \mathbb{Q}(Y_1, \ldots, Y_d)[Z_1]$ *is squarefree, so that* L_1 *is a direct sum of fields*

$$L_1 := L_{11} \oplus \cdots \oplus L_{1r_1}$$

and there exist ring projections

$$\pi_{1j} : \mathbb{Q}(Y_1, \ldots, Y_d)[Z_1, \ldots, Z_r] \mapsto L_{1j}[Z_2, \ldots, Z_r];$$

for all $i, 2 \leq i \leq r,$ *for all* $j, 1 \leq j \leq r_{i-1},$ $\pi_{i-1\,j}(f_i) \in L_{i-1\,j}[Z_i]$ *is squarefree, so that* L_i *is a direct sum of fields*

$$L_i := L_{i1} \oplus \cdots \oplus L_{ir_i}$$

and there exist ring projections

$$\pi_{ij} : \mathbb{Q}(Y_1, \ldots, Y_d)[Z_1, \ldots, Z_r] \mapsto L_{ij}[Z_{i+1}, \ldots, Z_r];$$

for all $i, 2 \leq i \leq r,$ *for all* $j < i,$ $\deg_j(f_i) < d_j.$

Definition 11.4.2. *A* Duval field *is a ring* D *such that there exist* $d \geq 0,$ $r \geq 0$ *and an admissible Duval sequence*

$$\mathbf{f} = \{f_1, \ldots, f_r\} \subset \mathbb{Q}(Y_1, \ldots, Y_d)[Z_1, \ldots, Z_r]$$

so that

$$D = \mathbb{Q}(Y_1, \ldots, Y_d)[Z_1, \ldots, Z_r]/(f_1, \ldots, f_r).$$

Obviously, $\mathcal{D}$ is exactly the class of all Duval fields[4].

In Kronecker's Model, the basic tool used to recursively build fields in $\mathcal{K}$ consists of constructing the field $K[Z]/f(Z)$, where $K \in \mathcal{K}$ and $f(Z) \in$

---

[4] Yes, from the algebraist's point of view, they are rings and not fields!... but they behave computationally more like a field than a ring. Since in this context the computation is much more important than the algebra, I prefer the name Duval field rather than Duval ring.

$K[Z]$ is irreducible; analogously, in Duval's Model, the basic tool used to recursively build rings in $\mathcal{D}$ consists of constructing the ring $D[Z]/f(Z)$, where $D \in \mathcal{D}$ and $f(Z) \in D[Z]$ is squarefree.

In Kronecker's Model, when we were given a field $K \in \mathcal{K}$ and a polynomial $g(Z) \in K[Z]$, we needed to factorize g in order to construct the fields $K[Z]/f(Z)$ where f runs over the irreducible factors of g; analogously, in Duval's Model, when we are given a ring $D \in \mathcal{D}$ and a polynomial $g(Z) \in D[Z]$, we need to compute its squarefree associate in each $R_i[Z]$ where the R_is are fields such that $D \cong R_1 \oplus \cdots \oplus R_r$.

To discuss this algorithm let us assume that

$$D = \mathbb{Q}(Y_1, \ldots, Y_d)[Z_1, \ldots, Z_r]/(f_1, \ldots, f_r)$$

and use the same notation as in the previous definitions.

The algorithm is the one we discussed in Example 11.3.1: we have to compute $g/\gcd(g, g')$ in $D[Z]$; to do that, we need to perform invertibility tests and algorithms in $D = L_r$ which require us to compute gcds in $L_{r-1}[Z_r]$ and so invertibility algorithms in L_{r-1} and so gcds in $L_{r-2}[Z_{r-1}]$ and so ... *und so weiter*.

The result will be

a splitting of D, i.e. admissible Duval sequences

$$\mathbf{f}_\lambda = \{f_{\lambda 1}, \ldots, f_{\lambda r}\} \subset \mathbb{Q}(Y_1, \ldots, Y_d)[Z_1, \ldots, Z_r], 1 \leq \lambda \leq m$$

so that denoting

$$D_\lambda = \mathbb{Q}(Y_1, \ldots, Y_d)[Z_1, \ldots, Z_r]/(f_{\lambda 1}, \ldots, f_{\lambda r})$$

we have

$$D \cong D_1 \oplus \cdots \oplus D_m;$$

polynomials $g_\lambda(Z) \in D_\lambda[Z]$ such that denoting
$D_\lambda \cong D_{\lambda 1} \oplus \cdots \oplus D_{\lambda r_\lambda}$ to be the direct sum decompositions, and
$\Pi_{\lambda j} : D[Z] \mapsto D_{\lambda j}[Z], \pi_{\lambda j} : D_\lambda[Z] \mapsto D_{\lambda j}[Z]$ to be the canonical projections,

we have for each λ $\pi_{\lambda j}(g_\lambda)$ is the squarefree associate of $\Pi_{\lambda j}(g)$ in $D_{\lambda j}[Z]$,

so that the 'extension' of D by g is a set of λ Duval fields $\mathbf{D}_\lambda$ defined as

$$\mathbf{D}_\lambda = \mathbb{Q}(Y_1, \ldots, Y_d)[Z_1, \ldots, Z_r, Z]/(f_{\lambda 1}, \ldots, f_{\lambda r}, g_\lambda).$$

Example 11.4.3. Completing Example 11.3.1, we only have to note that the extension of
$$D := \mathbb{Q}[X]/(X^4 - 13X^2 + 36)$$
by the polynomial
$$g := Z^3 + 3XZ^2 + 12Z + 4X \in D[Z]$$
is the assignment of

$D_1 = \mathbb{Q}[X, Z]/(X^2 - 4, Z + X),$
$D_2 = \mathbb{Q}[X, Z]/(X^2 - 9, Z^3 + 3XZ^2 + 12Z).$

12
Gauss

> Aequationes [...] solvere oportebit
> C.F. Gauss, *Disquisitiones arithmeticae*

This chapter is devoted to two of Gauss' important contributions to solving:

- Section 12.1 is devoted to a proof of the Fundamental Theorem of Algebra: I use the second proof by Gauss, which is the most algebraic of his four proofs;
- Section 12.2 presents a résumé of the *Disquisitiones Arithmeticae*'s section devoted to the solution of the cyclotomic equation: I consider these results to be the best pages of Computational Algebra, and I hope to be able to transmit my feeling to the reader.

These two sections also play the rôle of introducing the arguments discussed in the last two chapters: the generalization of Kronecker's Method to real algebraic numbers and Galois Theory.

12.1 The Fundamental Theorem of Algebra

In order to present a proof of the Fundamental Theorem of Algebra, and, mainly, to give a statement and a proof which can be easily generalized to an interesting setting (real closed fields), I must start by discussing the elementary and well-known difference between $\mathbb{R}$ and $\mathbb{C}$, i.e. that one is 'ordered' and the other not:

Definition 12.1.1. *A field K is said to be ordered if there is a subset $P \subset K$, the positive cone, which satisfies the following conditions:*

for all $a, b \in P$, $a + b \in P$, $ab \in P$,
for all $a \in P$, just one of the following three cases holds:

$a \in P$,
$a = 0$,

12.1 The Fundamental Theorem of Algebra

$-a \in P$.

Obviously the definition generalizes the trivial property of the 'positiveness' relation over $\mathbb{R}$ where P is the set of the positive numbers,

$$P := \{a \in \mathbb{R} : a > 0\}.$$

In this generalization, it is clearer if we work in the other way: a positive cone $P \subset K$ induces on K the total ordering $<_P$ defined by:

$$\text{for all } a, b \in K : a >_P b \iff a - b \in P.$$

From now on I will write $<$ omitting the dependence on P.

Lemma 12.1.2. *If K is ordered then*

(1) for all $a \in K : a^2 = (-a)^2 \geq 0$;
(2) char$(K) = 0$.

♄

Proof

(1) In fact we have $a = 0$ and $a^2 = 0$; or $a > 0$ and $a^2 > 0$; or $a < 0$ and $(-a)^2 > 0$.
(2) Since $1 = 1^2 > 0$ and, for each $n \in \mathbb{N}, n > 1$, we have by induction $\chi(n) = \chi(n-1) + 1 > 0$.

Corollary 12.1.3. *If K has a root of $X^2 + 1 \in K[X]$ – e.g. $K := \mathbb{C}$ – then K is not ordered.*

Proof In fact, assume K is ordered and denote by $i \in K$ the root of $X^2 + 1$; we have $-1 = i^2 > 0$, while $1 = 1^2 > 0$ giving an obvious contradiction. ♄

Therefore, we deduce that, if K is an ordered field, then it is not algebraically closed, therefore we can construct the algebraic extension

$$K \subset K[X]/(X^2 - 1) = K[i].$$

Let us consider the complex conjugation ϕ which is the K-isomorphism of $K[i]$ defined by $\phi(i) = -i$ and its extension to $K[i][X]$; as usual we will denote $\phi(f)$ by $\bar{f}$ for each $f \in K[i][X]$.

Lemma 12.1.4. *Let K be an ordered field such that, for each positive $a \in K$ there is $b \in K : b^2 = a$. Then*

for all $\alpha \in K[i]$, there exists $\beta \in K[i] : \beta^2 = \alpha$.

Proof Let $\alpha = a_1 + ia_2 \neq 0$ with $a_i \in K$. By the assumption there is $c \in K$ such that $c^2 = a_1^2 + a_2^2, c > 0$.
Moreover, since $a_1^2 \leq c^2$, we deduce that $|a_1| \leq c$ so that $\pm a_1 + c \geq 0$ and there are $b_1, b_2 \in K$ such that

$$b_1^2 = \frac{a_1 + c}{2}, \quad b_2^2 = \frac{-a_1 + c}{2},$$

which satisfy

$$b_1^2 - b_2^2 = a_1,$$
$$4b_1^2 b_2^2 = -a_1^2 + a_1^2 + a_2^2 = a_2^2;$$

therefore, choosing properly the signs, they satisfy

$$2b_1 b_2 = a_2.$$

Then we have

$$(b_1 + ib_2)^2 = b_1^2 - b_2^2 + i2b_1 b_2 = a_1 + ia_2,$$

so that $\beta := b_1 + ib_2$ satisfies $\beta^2 = \alpha$. ♄

Corollary 12.1.5. *Let K be an ordered field such that, for each positive $a \in K$, there is $b \in K : b^2 = a$. Then each quadratic equation*

$$aX^2 + bX + c \in K[i][X]$$

has the solutions

$$\frac{-b \pm \sqrt{b^2 - 4c}}{2a}$$

where, with an obvious abuse of notation, $\pm\sqrt{b^2 - 4c}$ denotes the two elements $\beta \in K[i]$ such that $\beta^2 = b^2 - 4c$. ♄

Theorem 12.1.6. *Let K be an ordered field. Then the following conditions are equivalent:*

(1) K is such that

for each positive $a \in K$, there is $b \in K : b^2 = a$;
each odd polynomial $f(X) \in K[X]$ has a root in K;

12.1 The Fundamental Theorem of Algebra

(2) each irreducible polynomial $f(X) \in K[X]$ has a root in $K[i]$;
(3) $K[i]$ is algebraically closed.

Proof

(3) $\implies$ (1) If $a \in K$ is positive and $b = c + id \in K[i]$ is the root of $X^2 - a$, then $b^2 = (c^2 - d^2) + 2cdi = a$ and $2cd = 0$; if $c = 0$ then we would get the contradiction $0 < a = b^2 = -d^2 < 0$; therefore $d = 0$ and $b \in K$.

If $\alpha \in K[i] \setminus K$ is a root of $f \in K[X]$, then $\bar{\alpha}$ is a root of $\bar{f} = f$ and $X^2 - \alpha^2 \in K[X]$ divides f in $K[X]$. Therefore, if f is odd, it has a root in K.

(2) $\implies$ (3) We need to prove that each $h \in K[i][X]$ has a root in $K[i]$.

So let then $g(X) := N_{K[i]/K}(h) = h\bar{h} \in K[X]$ and let $f(X)$ be an irreducible factor of g; by the assumption, g has a root $\alpha \in K[i]$; either α is a root of h or it is a root of $\bar{h}$ and so $\bar{\alpha}$ is a root of h.

(1) $\implies$ (2) Let $f(X) \in K[X]$ be irreducible and let us represent its degree as $n := \deg(f) = 2^m k$ where k is odd; we will do our induction in terms of m.

If $m = 0$, i.e. f is odd, the existence of a root follows from (1)

If $m > 0$, let $\alpha_1, \ldots, \alpha_n \in F$ be the roots of f in its splitting field $F \supset K$, and let $\mathfrak{LG}(t, X) \in K[t][X]$ be the Laplace–Gauss resolvent of f (cf. Definition 6.5.6) for which we know:

$\deg_X(\mathfrak{LG}) = \binom{2^m k}{2} = 2^{m-1} k'$ with $k' = k(2^m k - 1)$ odd;
for infinitely many $\lambda \in K$ the discriminant of $\mathfrak{LG}(\lambda, X) \in K[X]$ is not zero.

By this and the induction assumption, we deduce that, for infinitely many $\lambda \in K$,

its roots $\alpha_i + \alpha_j + \lambda \alpha_i \alpha_j$, for all i, j, are all simple, by Proposition 6.5.7,
and one of them is in $K[i]$, by assumption.

Since K is infinite and $\mathfrak{LG}(t, X)$ has at most $\binom{2^m k}{2}$ solutions, we can produce pairs $\lambda_1, \lambda_2 \in K \setminus \{0\}$, $\lambda_1 \neq \lambda_2$, and $\mu_1, \mu_2 \in K[i]$ such that there exist i, j, $1 \leq i < j \leq 2^m k$:

$$\mu_1 := \alpha_i + \alpha_j + \lambda_1 \alpha_i \alpha_j, \quad \mu_2 := \alpha_i + \alpha_j + \lambda_2 \alpha_i \alpha_j;$$

then computing

$$v_1 := \frac{\mu_1 - \mu_2}{\lambda_1 - \lambda_2} = \alpha_i \alpha_j \in K[i],$$
$$v_2 := \mu_2 - v_1 \lambda_2 = \alpha_i + \alpha_j \in K[i],$$

the equation

$$X^2 - v_2 X + v_1 = (X - \alpha_i)(X - \alpha_2) \in K[i]$$

is solvable in $K[i]$ by Corollary 12.1.5, so that $\alpha_i, \alpha_j \in K[i]$, proving the inductive assumption[1]. ♄

Historical Remark 12.1.7. As remarked by van der Waerden, in *A History of Algebra*, Springer (1985) 97–99:

the proof works if it is known that[2] the equation $[f = 0]$ has $[n]$ roots [...] in some extension of the field of real numbers. The existence of such an extension can be proved by Kronecker's method of 'symbolic adjunction' [...]. However, Gauss does not follow this road. He constructs his auxiliary equations without assuming the existence of the roots.[...]

The proof of this theorem [Proposition 6.5.7] covers four pages in Netto's translation. Right at the beginning of the proof Gauss says: 'The proof of this theorem would be extremely simple if we could presuppose that $[f]$ is a product of linear factors.'

Corollary 12.1.8. *Let K be an ordered field satisfying the conditions of Theorem 12.1.1, then*

each polynomial $f \in K[i][X]$ splits over $K[i]$;

each polynomial $f \in K[X]$ factorizes over K in linear and quadratic factors.

♄

---

[1] Note that this proof is essentially a procedure for 'solving' any irreducible polynomial $f(X) \in K[X]$, if we can assume the knowledge of procedures for computing a root in K of an odd polynomial, and for extracting roots of any positive element in K.
In fact, under this assumption, in order to obtain a root of an equation f of degree $2^m k$, with k odd, what we have to do is

choose elements $\lambda_1, \lambda_2 \in K \setminus \{0\}, \lambda_1 \neq \lambda_2$ such that the discriminant of $\mathfrak{LG}(\lambda_i, X), i = 1, 2$, is not zero,
find a root μ_i of $\mathfrak{LG}(\lambda_i, X), i = 1, 2$,
compute v_1 and v_2 as defined above,
solve the equation $X^2 - v_2 X + v_1 = 0$,
check whether its solutions satisfy the polynomial f, until we have found a root of f.

The two assumptions were elementary at Gauss' time in the case $K := \mathbb{R}$, since there were numerical procedures.

[2] And in the presentation above I of course assumed that we knew that!

Lemma 12.1.9. *Each odd polynomial $f(X) \in \mathbb{R}[X]$ has a root in $\mathbb{R}[X]$.*

Proof Let
$$f(X) = X^n + a_1 X^{n-1} + \cdots + a_i X^{n-i} + \cdots + a_{n-1} X + a_n$$
and let
$$M := \max\{1, \sum_{i=1}^{n} |a_i|\}.$$
Then
$$f(s) > 0, \ f(-s) < 0, \text{ for all } s \in \mathbb{R}, M < s.$$

Therefore by the Weierstrass Theorem of continous functions, we can deduce there is $c \in \mathbb{R}, -s < c < s$, such that $f(c) = 0$. ♄

Corollary 12.1.10 (Fundamental Theorem of Algebra).

Each polynomial $f(X) \in \mathbb{C}[X]$ splits into linear factors in $\mathbb{C}$. ♄

12.2 Cyclotomic Equations

In this section, I present Gauss' Algorithm for 'solving' the cyclotomic equation
$$X^n - 1 = 0$$
where $n \in \mathbb{N}$ is an odd prime[3], following verbatim Gauss' presentation in *Disquisitiones Arithmeticae* Artt. 342–354.

The aim of Gauss' Algorithm is presented by him in his introduction (Art. 342):

Propositum disquisitionum sequentium, quod pauci declaravisse haud inutile erit, eo tendit, ut $\mathbf{X}^4$ in factores continuo plures GRADATIM resolvatur, et quidem ita, ut horum coëfficientes per aequationes ordinis quam infimi determinentur, usque dum hoc

---

[3] The solution of the cyclotomic equation is interesting due to the fact that if γ is an nth root of α, $\gamma = \alpha^n$, all the nth roots of α are $\beta\gamma$ where β runs among the nth roots of unity.
The restriction on solving the cyclotomic equations when n is prime, is justified by the fact that

if $n = jk$, $\gcd(j, k) = 1$, α is a primitive jth root of unity, β is a primitive kth root of unity, then $\gamma := \alpha\beta$ is a primitive nth root of unity;
if p is prime, α is a primitive (p^e)th root of unity, and γ is a (p)th root of α, $\gamma^p = \alpha$, then γ is a primitive (p^{e+1})th root of unity,

so that knowledge of a primitive pth root of unity, for each prime p, and the ability to extract an nth root of any number, is sufficient to compute all the roots of any number.
[4] $= \sum_{i=0}^{n-1} X^i$.

modo ad factores simplices sive ad radices Ω^5 ipsas perveniatur. Scilicet ostendemus, si numero $n - 1$ quomodocunque in factores integros α, β, γ etc. resolvatur (pro quibus singulis numeros primos accipere licet), $\mathbf{X}$ in α factores $n - 1/\alpha$ dimensionum resolvi posse, quorum coëfficientes per aequationem α^{ti} gradus determinentur; singulos hoc factores iterum in β alios $n - 1/\alpha\beta$ dimensionum adiumento aequationis β^{ti} gradus etc., ita ut designante ν multitudinem factorum α, β, γ etc. inventio radicum Ω ad resolutionum ν aequationum $\alpha^{ti}, \beta^{ti}, \gamma^{ti}$ etc. gradus reducatur. E.g. pro $n = 17$, ubi $n - 1 = 2 \cdot 2 \cdot 2 \cdot 2$, quatuor aequationes quadraticas solvere oportebit; pro $n = 73$ tres quadraticas duasque cubicas.

The aim of the following results, which it is not useless to declare here in a few words, is in GRADUALLY decomposing $\mathbf{X}$ into more and more factors, in such a way that their coefficients can be determined by equations of minimal degree, until in this way one achieves simple factors or the roots Ω. We will show that if a number $n - 1$ can be factorized into integer factors α, β, γ etc. (for which one can take the prime factors), $\mathbf{X}$ can be factorized in α factors of degree $\frac{(n-1)}{\alpha}$, whose coefficients will be determined by an equation of degree α; each of these factors can be factorized into β factors of degree $\frac{(n-1)}{\alpha\beta}$ with the help of an equation of degree β, etc; therefore, denoting by ν the number of the factors α, β, γ etc., finding the roots Ω is reduced to the solution of ν equations of degree α, β, γ etc. For example for $n = 17$, where $n - 1 = 2 \cdot 2 \cdot 2 \cdot 2$, four quadratic equations must be solved; for $n = 73$ three quadratics and two cubics.

In Kronecker's language, if $n - 1 = p_1 p_2 \cdots p_\nu$, with p_i prime, Gauss' Algorithm produces a set of ν polynomials $\mathfrak{f}_1, \ldots, \mathfrak{f}_\nu \in \mathbb{C}[X]$ so that, denoting

$$\mathbf{X}(X) = \sum_{i=0}^{n-1} X^i = \frac{X^n - 1}{X - 1}$$

and, for each $i \leq \nu$,

$\mathfrak{r}_i \in \mathbb{C}$ a root of $\mathfrak{f}_i$,
$\mathbb{K}_i := \mathbb{K}_{i-1}(\mathfrak{r}_i)$ where $\mathbb{K}_0 := \mathbb{Q}$,

we have

$\deg(\mathfrak{f}_i) = p_i$, for all i,
$\mathbb{Q} \subset \mathbb{K}_1 \subset \cdots \subset \mathbb{K}_\nu \subset \mathbb{C}$,
$\mathfrak{f}_i(X) \in \mathbb{K}_{i-1}[X]$, for all i, so that[6] $\mathbb{K}_i = \mathbb{K}_{i-1}[X]/\mathfrak{f}_i(X)$,
$\mathbb{K}_\nu$ is the splitting field of $\mathbf{X}(X)$,
$\mathfrak{r}_\nu$ is a root of $\mathbf{X}(X)$,

---

[5] Ω denotes the set of all the roots of $\mathbf{X}$.
[6] Since $\mathfrak{f}_i$ is irreducible because p_i is prime.

12.2 Cyclotomic Equations

so that $\mathfrak{r}_\nu$ can be computed by iteratively 'solving'[7] the equation

$$f_i(X) = 0, \text{ for all } i,$$

and all the roots of $\mathbf{X}$ can be obtained by squaring.

Moreover Gauss' Algorithm provides – via $\mathbb{Q}$-automorphisms of $\mathbb{K}_i$ in $\mathbb{C}$ – the factorization of $\mathbf{X}(X)$ in each $\mathbb{K}_i$. But Gauss' theory gives much much more.

Gauss began by introducing his notation: having fixed a generic root of $\mathbf{X} = \frac{X^n - 1}{X - 1}$

$$r \in \mathbb{C}, \ r^n = 1, \ r \neq 1,$$

and remarking that all the roots of $\mathbf{X}$ are then

$$\{r, r^2, \ldots, r^{n-1}\},$$

he denoted the roots r^λ with the notation $[\lambda]$, thus indexing the roots by their logarithms in $\mathbb{Z}_{n-1}$ modulo a generic root $r = [1]$; he also used the notation $[0] := 1$.

Then he remarks that, according to Corollary 7.4.7, there is a generator

$$g \in \mathbb{Z}_n \setminus \{0\}$$

such that

$$\Omega := \{r, r^2, \ldots, r^{n-1}\} = \{[\lambda] : \lambda \in \mathbb{Z}_n \setminus \{0\}\} = \{[g^\mu] : \mu \in \mathbb{Z}_{n-1}\}$$

and, moreover, for all $\lambda \in \mathbb{Z}_n \setminus \{0\}$, we have[8]

$$\{[g^\mu] : \mu \in \mathbb{Z}_{n-1}\} = \{[\lambda g^\mu] : \mu \in \mathbb{Z}_{n-1}\}.$$

In conclusion Corollary 7.4.7 allows us to introduce logarithms to transform multiplication in the set of all the roots of $\mathbf{X}$ to addition in $\mathbb{Z}_{n-1}$ (cf. Historical Remark 7.4.8).

Lemma 12.2.1 (Art. 343). *Let g and G be two generators; let e be a divisor of $n - 1$, $ef = n - 1$, and let $h := g^e$, $H := G^e$.*
Then

$$\{[h^\mu] : 0 \leq \mu \leq f - 1\} = \{[H^\mu] : 0 \leq \mu \leq f - 1\}.$$

---

[7] Obviously Gauss' notion of 'solving' is that of the eighteenth century: producing the roots by a straight line program which applies the five operations having as inputs the coefficients of the polynomial.

[8] Note that, with the present notation, $[g^0] = [g^{n-1}] = [1] = r$. In this representation 1 is still represented by $[0]$.

Table 12.1. *Logarithm table for* $\mathbb{Z}_{19}$

i	2^i	i	2^i
1	2	10	17
2	4	11	15
3	8	12	11
4	16	13	3
5	13	14	6
6	7	15	12
7	14	16	5
8	9	17	10
9	18	18	1

Proof There is an ω such that $G = g^\omega$; let $\mathsf{m} : \mathbb{Z}_f \mapsto \mathbb{Z}_f$ be the isomorphism defined by $\mathsf{m}(\mu) := \mu\omega$. Then $\mathsf{m}(\mu)e \equiv \mu\omega e \pmod{n-1}$ so that, for each $\mu \in \mathbb{Z}_f$, we have

$$H^\mu = g^{e\omega\mu} = g^{e\mathsf{m}(\mu)} = h^{\mathsf{m}(\mu)}$$

whence the conclusion. ♄

This lemma guarantees that the set $\{[\lambda g^{ej}] : 0 \leq j < f\}$ and the sum

$$\sum_{j=0}^{f-1} [\lambda g^{ej}]$$

do not depend on the choice of generator g; both the set and the sum are denoted by (f, λ) and labelled a *period*.

Example 12.2.2. If we take $n = 19$, then $g = 2$ is a generator, as is easy to check by constructing the Table 12.1.

Then the period $(6, 1)$ is

$$\begin{aligned}
(6, 1) &= [1] + [8] + [8^2] + [8^3] + [8^4] + [8^5] \\
&= [1] + [2^3] + [2^6] + [2^9] + [2^{12}] + [2^{15}] \\
&= [1] + [8] + [7] + [18] + [11] + [12];
\end{aligned}$$

and $(6, 2)$ is

$$\begin{aligned}
(6, 2) &= [2] + [2 \cdot 8] + [2 \cdot 8^2] + [2 \cdot 8^3] + [2 \cdot 8^4] + [2 \cdot 8^5] \\
&= [2] + [2^4] + [2^7] + [2^{10}] + [2^{13}] + [2^{16}] \\
&= [2] + [16] + [14] + [17] + [3] + [5];
\end{aligned}$$

for (6, 3) we have (6, 2) = (6, 3); and for (6, 4),

$$\begin{align} (6, 4) &= [4] + [4 \cdot 8] + [4 \cdot 8^2] + [4 \cdot 8^3] + [4 \cdot 8^4] + [4 \cdot 8^5] \\ &= [2^2] + [2^5] + [2^8] + [2^{11}] + [2^{14}] + [2^{17}] \\ &= [4] + [13] + [9] + [15] + [6] + [10]. \end{align}$$

Obviously

$$(6, 1) = (6, 7) = (6, 8) = (6, 11) = (6, 12) = (6, 18);$$

$$(6, 2) = (6, 3) = (6, 5) = (6, 14) = (6, 16) = (6, 17);$$

$$(6, 4) = (6, 6) = (6, 9) = (6, 10) = (6, 13) = (6, 15).$$

From now on we will fix a factorization $fe = n - 1$ and a generator g and we will denote $h := g^e$.

Remark 12.2.3 (Art. 344). It is clear that

two periods (f, λ), (f, μ), having a root in common are identical;
if $e = 1$, $f = n - 1$, then $(n - 1, 1)$ coincides with Ω, while, when $e \neq 1$, then Ω is the union of the e disjoint periods $(f, 1), (f, g), (f, g^2), \ldots, (f, g^{e-1})$;
for all $\lambda \neq 0$, Ω is the union of the e disjoint periods[9]

$$(f, \lambda), (f, \lambda g), (f, \lambda g^2), \ldots, (f, \lambda g^{e-1})$$

while
$(f, 0) = f$.
More generally, if $n - 1 = \alpha\beta\gamma$, then the period $(\beta\gamma, \lambda)$ is the union of the β disjoint periods $(\gamma, \lambda), (\gamma, \lambda g^\alpha), (\gamma, \lambda g^{2\alpha}), \ldots, (\gamma, \lambda g^{\alpha(\beta-1)})$.

Example 12.2.4. For instance

$$\begin{align} (6, 1) &= (2, 1) + (2, 2^3) + (2, 2^6) \\ &= (2, 1) + (2, 8) + (2, 7) \\ &= ([1] + [18]) + ([8] + [11]) + ([7] + [12]). \end{align}$$

Theorem 12.2.5 (Art. 345). *Let (f, λ) and (f, μ) be two periods. Then*

$$(f, \lambda)(f, \mu) = \sum_{j=0}^{f-1}(f, \lambda + \mu h^j);$$

---

[9] In most of the statements, we will implicitly assume that $\lambda \not\equiv 0 \pmod{n}$.

and, more generally,
$$(f, k\lambda)(f, k\mu) = \sum_{j=0}^{f-1}(f, k(\lambda + \mu h^j)).$$

Proof With $(f, \lambda) = \sum_{j=0}^{f-1}[\lambda h^j]$ and $(f, \mu) = \sum_{j=0}^{f-1}[\mu h^j]$, we obtain

$$\begin{aligned}
(f, \lambda)(f, \mu) &= \sum_{j=0}^{f-1}\sum_{i=0}^{f-1}[\lambda h^j][\mu h^i] \\
&= \sum_{j=0}^{f-1}\sum_{i=0}^{f-1}[\lambda h^j + \mu h^i] \\
&= \sum_{j=0}^{f-1}\sum_{i=0}^{f-1}[(\lambda + \mu h^j)h^{i-j}] \\
&= \sum_{j=0}^{f-1}\sum_{k=0}^{f-1}[(\lambda + \mu h^j)h^k] \\
&= \sum_{j=0}^{f-1}(f, \lambda + \mu h^j).
\end{aligned}$$

♄

Example 12.2.6. For instance we have

$$\begin{aligned}
(6, 1)(6, 2) &= \sum_{j=0}^{5}(6, 1 + 2 \times 2^{3j}) \\
&= (6, 3) + (6, 17) + (6, 15) + (6, 18) + (6, 4) + (6, 6) \\
&= (6, 2) + (6, 2) + (6, 4) + (6, 1) + (6, 4) + (6, 4) \\
&= (6, 1) + 2(6, 2) + 3(6, 4);
\end{aligned}$$

and

$$\begin{aligned}
(6, 1)(6, 1) &= \sum_{j=0}^{5}(6, 1 + 1 \times 2^{3j}) \\
&= (6, 2) + (6, 9) + (6, 8) + (6, 0) + (6, 12) + (6, 13) \\
&= (6, 2) + (6, 4) + (6, 1) + (6, 0) + (6, 1) + (6, 4) \\
&= 6 + 2(6, 1) + (6, 2) + 2(6, 4).
\end{aligned}$$

12.2 Cyclotomic Equations

Remark 12.2.7. It is then clear that, given any polynomial

$$F(Y_1, \ldots, Y_s) \in \mathbb{Z}[Y_1, \ldots, Y_s],$$

if we substitute a period (f, λ_i) for each indeterminate Y_i then there exists $a, b_0, b_1, \ldots, b_{e-1} \in \mathbb{Z}$:

$$F\big((f, \lambda_1), \ldots, (f, \lambda_s)\big) = a + \sum_{i=0}^{e-1} b_i (f, g^i).$$

Moreover, we also have, for all $k \neq 0$,

$$F\big((f, k\lambda_1), \ldots, (f, k\lambda_s)\big) = a + \sum_{i=0}^{e-1} b_i (f, kg^i).$$

As a consequence we can deduce:

Corollary 12.2.8. *Let $F(Y) \in \mathbb{Z}[Y]$. Then there exists λ:*

$$F\big((f, \lambda)\big) = 0 \implies F\big((f, g^i)\big) = 0, \quad \text{for all } i,\ 0 \leq i < e.$$

Proof $\{(f, k\lambda), 1 \leq k < n\} = \{((f, g^i), 0 \leq i < e\}.$ ♄

Theorem 12.2.9 (Art. 346). *Let us fix a period $\mathsf{p} := (f, \lambda)$. Then for any other period (f, μ) there is a polynomial*

$$F_\mu(X) := a + \sum_{i=1}^{e-1} b_i X^i \in \mathbb{Q}(X)$$

such that $(f, \mu) = F_\mu(\mathsf{p})$.

Proof Since the trace of any root of $X^n - 1$ is zero, we know that

$$0 = 1 + \sum_{\rho \in \Omega} \rho = 1 + \sum_{i=0}^{e-1} (f, g^i).$$

By the previous results we know that there are $a_i, b_{ij} \in \mathbb{Z}$ such that

$$\mathsf{p}^j = a_j + \sum_{i=0}^{e-1} b_{ij}(f, g^i),\ 2 \leq j \leq e-1;$$

these equations, together with the relation

$$0 = 1 + \sum_{i=0}^{e-1} (f, g^i)$$

give $e-1$ linear equations relating the $e-1$ elements $(f, g^i) \neq \mathsf{p}$ and whose known term is in $\mathbb{Z}[\mathsf{p}]$.

Therefore by Gaussian elimination for each $(f, g^j) \neq \mathsf{p}$ we can obtain an equation

$$a + \sum_{i=1}^{e-1} b_i \mathsf{p}^i + \beta(f, g^j) = 0.$$

To complete the proof we only have to show that $\beta \neq 0$: otherwise, the polynomial

$$F(X) := a + \sum_{i=1}^{e-1} b_i X^i$$

would have p as a root and, therefore, by the Corollary above, would have the e roots (f, g^i), which contradicts the fact that $\deg(F) < e$. ♮

Example 12.2.10. The proof of the Theorem above also gives an algorithm for producing the polynomials $F_\mu(X)$. Let us consider as before the case $n = 19$, $f = 6$. Then, setting $\mathsf{p} = (6, 1)$ we have from Example 12.2.6

$$\mathsf{p}^2 = 6 + 2(6, 1) + (6, 2) + 2(6, 4).$$

The two equations

$$\begin{aligned} -\mathsf{p} - 1 &= (6, 2) + (6, 4) \\ \mathsf{p}^2 - 2\mathsf{p} - 6 &= (6, 2) + 2(6, 4) \end{aligned}$$

allow us to deduce

$$\begin{aligned} (6, 2) &= -\mathsf{p}^2 + 4, \\ (6, 4) &= \mathsf{p}^2 - \mathsf{p} - 5. \end{aligned}$$

It is also easy to check that, as claimed by the Corollary above,

$$\begin{aligned} (6, 4) &= -(6, 2)^2 + 4, \\ (6, 1) = (6, 8) &= (6, 2)^2 - (6, 2) - 5, \\ (6, 1) = (6, 8) &= -(6, 4)^2 + 4, \\ (6, 2) = (6, 16) &= (6, 4)^2 - (6, 4) - 5. \end{aligned}$$

Theorem 12.2.11 (Art. 347). *Let*

$$F(Y_0, Y_1, \ldots, Y_j, \ldots, Y_{f-1}) \in \mathbb{Z}[Y_0, \ldots, Y_{f-1}]$$

be a symmetric function. Then

$$F\bigl([\mu], [\mu g^e], \ldots, [\mu g^{ej}], \ldots, [\mu g^{e(f-1)}]\bigr) = a + \sum_{i=0}^{e-1} b_i(f, g^i).$$

Proof Clearly, we have

$$F\bigl([\mu], [\mu g^e], \ldots, [\mu g^{ej}], \ldots, [\mu g^{e(f-1)}]\bigr) = a + \sum_{i=1}^{n-1} b_i[i].$$

Since the function is symmetric it is clear that

$$a + \sum_{i=1}^{n-1} b_i[i] = a + \sum_{i=1}^{n-1} b_i[ig^e],$$

so that $b_i = b_j$ iff $[i]$ and $[j]$ belong to the same period. ♄

Remark 12.2.12. It is worthwhile interpreting the results obtained so far with Kronecker's language: Corollary 12.2.8 states that the periods

$$(f, g^i),\ 0 \le i < e,$$

are conjugate and Theorem 12.2.9 states that each conjugate is in the field extension $\mathbb{K} := \mathbb{Q}(\mathsf{p}) \subset \mathbb{C}$, where we recall that $\mathsf{p} = (f, \lambda)$ for a fixed root $[\lambda] \in \mathbb{C}$ of **X**.

In this setting, of course, we know that the splitting field $\mathbb{F}$ of **X**, satisfies

$$\mathbb{Q} \subset \mathbb{K} \subset \mathbb{F} \subset \mathbb{C}$$

and is endowed with the $\mathbb{Q}$-isomorphisms Φ_k defined by $\Phi_k([\lambda]) = [k\lambda]$, for each root $[\lambda] \in \Omega$.

The effect of these isomorphisms is discussed in Remark 12.2.7 and they are directly applied in Theorem 12.2.11, in order to prove that any symmetric function of the roots contained in a period $\mathsf{q} = (f, \mu)$ can be expressed in terms of all the periods (f, g^i) and therefore as elements in $\mathbb{K}$.

In particular, the symmetric functions of the roots contained in

$$\mathsf{q} = (f, \mu)$$

can be expressed in $\mathbb{K}$.

Therefore setting, for each root $[\mu]$, $\mathsf{q} := (f, \mu)$ and denoting $P_\mu(X) \in \mathbb{C}[X]$ the polynomial whose roots are the elements of q, we can deduce that:

Corollary 12.2.13 (Art. 348). *With the notation above*

$$P_\mu(X) \in \mathbb{Q}(\mathsf{p})[X];$$

$P_\lambda(X)$ can be 'computed' in the sense that we can compute a polynomial $Q(Y, X) \in \mathbb{Q}[Y, X]$ such that
$$P_\lambda(X) = Q(\mathsf{p}, X);$$
$P_{k\lambda} = \Phi_k(P_\lambda)$.

Example 12.2.14. Consider
$$(6, 1) = \{[1], [7], [8], [11], [12], [18]\};$$
the coefficients of the corresponding polynomial
$$P_1(X) = \sum_{i=0}^{5}(-1)^i a_{6-i} X^i + X^6$$
are
$$a_i = \sigma_i([1], [7], [8], [11], [12], [18]),$$
which Gauss computed as
$$\begin{aligned} a_1 &= (6, 1), \\ a_2 &= 3 + (6, 1) + (6, 4), \\ a_3 &= 2 + 2(6, 1) + (6, 2), \\ a_4 &= 3 + (6, 1) + (6, 4), \\ a_5 &= (6, 1), \\ a_6 &= 1, \end{aligned}$$
which give us
$$\begin{aligned} P_1(x) &= (1 + 3x^2 + 2x^3 + 3x^4 + x^6) \\ &+ (x + x^2 + 2x^3 + x^4 + x^5)(6, 1) \\ &+ x^3(6, 2) \\ &+ (x^2 + x^4)(6, 4), \end{aligned}$$
from which we deduce that the period $(6, 2)$ corresponds to the polynomial
$$\begin{aligned} P_2(x) &= (1 + 3x^2 + 2x^3 + 3x^4 + x^6) \\ &+ (x + x^2 + 2x^3 + x^4 + x^5)(6, 2) \\ &+ x^3(6, 4) \\ &+ (x^2 + x^4)(6, 1), \end{aligned}$$

12.2 Cyclotomic Equations

and in the same way we get the polynomial P_4 corresponding to the period $(6, 4)$.

Remark 12.2.15 (Art. 349). Gauss discusses algorithms for computing the coefficients above and remarks that direct computations like

$$\begin{aligned}
a_2 = \ & [1+1] + [7+1] + [8+1] + [11+1] + [12+1] + [18+1] \\
& + [1+7] + [7+7] + [8+7] + [11+7] + [12+7] + [18+7] \\
& + [1+8] + [7+8] + [8+8] + [11+8] + [12+8] + [18+8] \\
& + [1+11] + [7+11] + [8+11] + [11+11] + [12+11] \\
& + [18+11] + [1+12] + [7+12] + [8+12] + [11+12] \\
& + [12+12] + [18+12] + [1+18] + [7+18] + [8+18] \\
& + [11+18] + [12+18] + [18+18] \\
= \ & \ldots
\end{aligned}$$

can be avoided since σ_i can be expressed via the Waring functions s_i, which can be easily computed, since we obviously have

$$\sum_{r \in (f, \lambda)} r^i = (f, i\lambda).$$

Using the Newton formula, and denoting

$$\mathfrak{s}_i := \mathsf{s}_i([1], [7], [8], [11], [12], [18])$$

we get

$$\begin{aligned}
a_1 &= \mathfrak{s}_1 &&= (6, 1), \\
a_2 &= \tfrac{1}{2}(a_1\mathfrak{s}_1 - \mathfrak{s}_2) &&= 3 + (6, 1) + (6, 4), \\
&\ldots
\end{aligned}$$

using the already proven formula

$$(6, 1)^2 = 6 + 2(6, 1) + (6, 2) + 2(6, 4).$$

Let me just add that such computations become elementary if we pre-compute all products

$$(6, a)(6, b), \quad a, b \in \{1, 2, 4\},$$

and, moreover, such products are obtained immediately by applying Remark 12.2.7 to the already computed formulas:

$$(6, 1)^2 = 6 + 2(6, 1) + (6, 2) + 2(6, 4),$$
$$(6, 1)(6, 2) = (6, 1) + 2(6, 2) + 3(6, 4).$$

Remark 12.2.16 (Art. 350). The results summarized in Corollary 12.2.13 allowed us to express the polynomial $P_\mu(X)$ whose roots are the elements of a period (f, μ) in terms of a specific period $\mathsf{p} := (f, \lambda) \in \mathbb{C}$ and therefore to compute these roots by 'solving' the equations $P_\mu(X) = 0$, *provided that we already have computed the value of* p.

The crucial remark is to consider the case $n - 1 = \alpha\beta\gamma$ and the period $\mathsf{P} := (\beta\gamma, \lambda)$ which 'consists' of the β disjoint periods $\mathsf{p}_i := (\gamma, \lambda g^{\alpha i}), 0 \leq i < \beta$, in the same sense that a period 'consists' of its roots; i.e. P is both the union and the sum of the p_is.

Therefore, the required computation of p can be undertaken if we generalize Corollary 12.2.13 to this setting, substituting roots with periods p_i and the period p with P. This is possible and is the aim of the following discussion.

If we consider a symmetric function

$$F(Y_0, \ldots, Y_{\beta-1}) \in \mathbb{Z}[Y_0, \ldots, Y_{\beta-1}]$$

we know that

$$F(\mathsf{p}_0, \ldots, \mathsf{p}_{\beta-1}) = a + \sum_{i=0}^{\alpha\beta-1} b_i(\gamma, g^i) = a + \sum_{j=0}^{\alpha-1} \sum_{i=0}^{\beta-1} b_{ij}(\gamma, g^{\alpha i+j}).$$

Since the periods $(\beta\gamma, \lambda)$ and $(\beta\gamma, \lambda g^\alpha)$ coincide and consist of the union of the periods p_is, the argument of the proof of Theorem 12.2.11 allows us to prove that for all $i, k, j, b_{ij} = b_{kj}$, from which we deduce that there exists $a, b_1, \ldots, b_{\alpha-1} \in \mathbb{Z}$:

$$F(\mathsf{p}_0, \ldots, \mathsf{p}_{\beta-1}) = a + \sum_{j=0}^{\alpha-1} b_j(\beta\gamma, g^j) \tag{12.1}$$

and that

$$\Phi_k\Big(F(\mathsf{p}_0, \ldots, \mathsf{p}_{\beta-1})\Big) = a + \sum_{j=0}^{\alpha-1} b_j(\beta\gamma, kg^j).$$

Therefore denoting $\mathsf{P}_i := (\beta\gamma, i)$ and by $G_i(X) \in \mathbb{K}[X]$ the polynomial whose roots are the β periods (γ, λ) contained in P_i, we can resume this discussion as:

Corollary 12.2.17 (Art. 351). *With the above notation:*

$G_1(X) \in \mathbb{Q}(\mathsf{P}_1)[X]$;
we can compute a polynomial $Q(Y, X) \in \mathbb{Q}[Y, X]$ such that

$$G_1(X) = Q(\mathsf{P}_1, X);$$

$G_k = \Phi_k(F_1)$. ♄

Example 12.2.18. Continuing the computation when $n = 19$ we can produce the polynomial $F_1(X) = X^3 + \sum_{i=1}^{3}(-1)^i c_i X^{3-i} \in \mathbb{Q}[X]$ whose roots are $(6, 1), (6, 2), (6, 4)$.

We know that its coefficients are

$$\begin{aligned}
c_1 &= (6, 1) + (6, 2) + (6, 4) = (18, 1) \\
&= -1; \\
c_2 &= (6, 1)(6, 2) + (6, 2)(6, 4) + (6, 1)(6, 4) \\
&= \Big((6, 1) + 2(6, 2) + 3(6, 4)\Big) \\
&\quad + \Big((6, 2) + 2(6, 4) + 3(6, 1)\Big) \\
&\quad + \Big((6, 4) + 2(6, 1) + 3(6, 2)\Big) \\
&= 6\Big((6, 1) + (6, 2) + (6, 4)\Big) \\
&= -6; \\
c_3 &= (6, 1)(6, 2)(6, 4) \\
&= (6, 1)(6, 4) + 2(6, 2)(6, 4) + 3(6, 4)(6, 4) \\
&= \Big((6, 4) + 2(6, 1) + 3(6, 2)\Big) \\
&\quad + 2\Big((6, 2) + 2(6, 4) + 3(6, 1)\Big) \\
&\quad + 3\Big(6 + 2(6, 4) + (6, 1) + 2(6, 2)\Big) \\
&= 18 + 11\Big((6, 1) + (6, 2) + (6, 4)\Big) \\
&= 7;
\end{aligned}$$

therefore

$$F_1(X) = X^3 + X^2 - 6X - 7.$$

Example 12.2.19. With the same technique, denoting $P_1 := (6, 1)$, we will now compute the polynomial $G_1(X) = X^3 + \sum_{i=1}^{3}(-1)^i c_i X^{3-i} \in \mathbb{Q}(P_1)[X]$ whose roots are $(2, 1), (2, 7), (2, 8)$.

First let us remark that

$(2, 1)$	$=$	$(2, 18)$	$(2, 2)$	$=$	$(2, 17)$	$(2, 4)$	$=$	$(2, 15)$
$(2, 7)$	$=$	$(2, 12)$	$(2, 5)$	$=$	$(2, 14)$	$(2, 10)$	$=$	$(2, 9)$
$(2, 8)$	$=$	$(2, 11)$	$(2, 16)$	$=$	$(2, 3)$	$(2, 13)$	$=$	$(2, 6)$

and

$$(2, 1)(2, 7) = (2, 8) + (2, 6),$$
$$(2, 1)^2 = (2, 2) + (2, 19)$$
$$= (2, 2) + 2,$$
$$(2, 6)(2, 8) = (2, 2) + (2, 5),$$

so that

$$(2, k1)(2, k7) = (2, k8) + (2, k6),$$
$$(2, k)^2 = (2, k2) + 2,$$
$$(2, k6)(2, k8) = (2, k2) + (2, k5),$$

and let us recall that

$$(6, 2) = -(6, 1)^2 + 4,$$
$$(6, 4) = (6, 1)^2 - (6, 1) - 5.$$

Therefore we obtain:

$$\begin{aligned}
c_1 &= (2, 1) + (2, 7) + (2, 8) = (6, 1) \\
&= P_1; \\
c_2 &= (2, 1)(2, 7) + (2, 7)(2, 8) + (2, 8)(2, 1) \\
&= \big((2, 8) + (2, 6)\big) + \big((2, 1) + (2, 4)\big) + \big((2, 7) + (2, 9)\big) \\
&= (6, 1) + (6, 4) \\
&= (6, 1) + (6, 1)^2 - (6, 1) - 5 \\
&= P_1^2 - 5; \\
c_3 &= (2, 1)(2, 7)(2, 8) \\
&= (2, 8)^2 + (2, 6)(2, 8) \\
&= \big((2, 3) + 2\big) + \big((2, 2) + (2, 5)\big) \\
&= 2 + (6, 2) \\
&= -P_1^2 + 6,
\end{aligned}$$

i.e.
$$G_1(X) = X^3 - \mathsf{P}_1 X^2 + (\mathsf{P}_1^2 - 5)X + \mathsf{P}_1^2 - 6.$$

Therefore the polynomial whose roots are $(2, 2), (2, 3), (2, 5)$ is[10]

$$\begin{aligned}G_2(X) &= X^3 - (6,2)X^2 + ((6,2)^2 - 5)X + (6,2)^2 - 6 \\ &= X^3 + \mathsf{P}_1^2 X^2 - 4X^2 - \mathsf{P}_1^2 X + \mathsf{P}_1 X + 4X - \mathsf{P}_1^2 + \mathsf{P}_1 + 3,\end{aligned}$$

and the one whose roots are $(2, 4), (2, 6), (2, 9)$ is

$$\begin{aligned}G_4(X) &= X^3 - (6,4)X^2 + ((6,4)^2 - 5)X + (6,4)^2 - 6, \\ &= X^3 - \mathsf{P}_1^2 X^2 + \mathsf{P}_1 X^2 + 5X^2 - 8\mathsf{P}_1 X + 6X - 8\mathsf{P}_1 + 5.\end{aligned}$$

Remark 12.2.20. The computations above allow us to 'solve' the equation

$$\mathbf{X}(X) := \frac{X^{19} - 1}{X - 1} :$$

what we have to do is

- 'solve' the equation

$$0 = F_1(X) = X^3 + X^2 - 6X - 7 \in \mathbb{Q}[X];$$

- for each solution ξ_i, $i = 1, \ldots$, 'solve' the three equations[11]

$$0 = G_1(X, \xi)$$
$$0 = G_2(X, \xi)$$
$$0 = G_4(X, \xi)$$

---

[10] Note that we use the fact that

$$\begin{aligned}(6, 2)^2 &= (-(6, 1)^2 + 4)^2 \\ &= (6, 1)^4 - 8(6, 1)^2 + 16 \\ &= -(6, 1)^2 + (6, 1) + 9\end{aligned}$$

where the last equation is obtained by division since

$$F_1((6, 1)) = (6, 1)^3 + (6, 1)^2 - 6(6, 1) - 7 = 0.$$

[11] This approach poses a minor problem: if we solve the three equations

$$G(X, \xi_1) = G(X, \xi_2) = G(X, \xi_4) = 0,$$

among the nine roots *which one is which?* That is, once we have fixed – and we know that we can do it freely – one among the roots in Ω and labelled it as [1], we can freely decide which among the roots of $F_1(X)$ is the one corresponding to the period $(6, 1)$ and, among the roots of $G(X, \xi_1)$, which represents $(2,1)$; but then we *must correctly associate the eight other roots with the symbols $(2, i)$.*

Gauss discusses this question and solves it with a numerical analysis approach: not only does he present roughly *varia alia artificia* but he proposes a general solution: since the tables allow us to approximate the roots via the de Moivre formula, all the periods can be approximated; it then becomes easy to decide which one is which.

I would also like to point to another trick proposed by Gauss: when the roots of F_1 are approximated and they are used to find the three solutions of the equation $G(X, \xi_i) = 0$, by checking that their sum gives approximately ξ_i we *calculi confirmationem* obtain.

where

$$G_1(X, Y) = X^3 - YX^2 + Y^2X - 5X + Y^2 - 6,$$
$$G_2(X, Y) = X^3 + Y^2X^2 - 4X^2 - Y^2X + YX + 4X - Y^2 + Y + 3,$$
$$G_4(X, Y) = X^3 - Y^2X^2 + YX^2 + 5X^2 - 8YX + 6X - 8Y + 5;$$

- since for each $(2,i) = \{[i],[n-i]\}$

$$[i][n - i] = 1, [i] + [n - i] = (2, i),$$

all the roots can then be derived by 'solving' the quadratic equations

$$H_i(X) := H(X, \theta_i) = 0, i = 1, \ldots, 9,$$

where $\theta_i, i = 1, \ldots, 9$, are the solutions obtained in the previous step, and

$$H(X, Y) := X^2 - YX + 1.$$

Alternatively, we can

- 'solve' the equation $0 = F_1(X)$ computing a single root $\mathfrak{r}_1$;
- 'solve' the equation $0 = G_1(X, \mathfrak{r}_1)$ computing a single root $\mathfrak{r}_2$;
- 'solve' the equation $0 = H(X, \mathfrak{r}_2)$ computing a single root $\mathfrak{r}_3$;
- setting $[1] := \mathfrak{r}_3$, compute all the other roots $[\lambda] = \mathfrak{r}_3^\lambda$ by repeated squaring.

'Solving' should be understood here in the sense of Gauss and his era: producing the roots by performing the five operations over the coefficients – note that we are required to 'solve' cubic and quadratic equations, for which Gauss' era had sufficient tools.

It is, however, significant to understand what we would get by 'solving' the equations above in the sense of Kronecker.

Let us note that we have a tower

$$\mathbb{Q} \subset \mathbb{K}_1 \subset \mathbb{K}_2 \subset \mathbb{F} \subset \mathbb{C}$$

where $\mathbb{F}$ is the splitting field of $\mathbf{X}(X)$ so that $\mathbb{F} = \mathbb{Q}([1])$, and $\mathbb{K}_1 := \mathbb{Q}(\mathsf{P}_1)$, $\mathbb{K}_2 := \mathbb{Q}(\mathsf{p}_1)$

We therefore have

$$[\mathbb{F} : \mathbb{K}_2] = 2, [\mathbb{K}_2 : \mathbb{K}_1] = [\mathbb{K}_1 : \mathbb{Q}] = 3,$$

12.2 Cyclotomic Equations

and
$$\mathbb{K}_1 = \mathbb{Q}(\mathsf{P}_1) = \mathbb{Q}[X]/F_1(X),$$

$$\mathbb{K}_2 = \mathbb{K}_1(\mathsf{p}_1) = \mathbb{K}_1[X]/G_1(X, \mathsf{P}_1),$$

$$\mathbb{F} = \mathbb{K}_2([1]) = \mathbb{K}_2[X]/H(X, \mathsf{p}_1),$$

so that, 'solving' here means to build the above tower and then

- split $F_1(X)$ in $\mathbb{K}_1$;
- for each solution ξ, split in $\mathbb{K}_1$ the three polynomials $G_1(X, \xi)$, $G_2(X, \xi)$, $G_4(X, \xi)$;
- for each solution θ, split in $\mathbb{K}_1$ the nine polynomials $H_1(X, \theta)$.

Again, the better approach is to build the above tower and repeatedly square [1].

Algorithm 12.2.21. On the bases of the results discussed upto here, Gauss presented his algorithm for 'solving' $\mathbf{X}(X) := \frac{X^n-1}{X-1}$ as follows (Art. 352):

Theoremata praecedentia cum consectariis annexis praecipua totius theoriae momenta continent, modusque valores radicum Ω inveniendi paucis iam tradi poterit.
Ante omnia accipiendus est numerus g, qui por modulo n sit radix primitiva, residuaque minima potestatum ipsius g usque ad g^{n-2} secundum modulum n eruenda. Resolvatur $n-1$ in factores, et quidem, si problema ad aequationes gradus quam infima reducere lubet, in factores primos; sint hi (ordine prorsus arbitrario) $\alpha, \beta, \gamma, \ldots, \zeta$, ponaturque

$$\frac{n-1}{\alpha} = \beta\gamma\ldots\zeta = a, \quad \frac{n-1}{\alpha\beta} = \gamma\ldots\zeta = b, \text{ etc.}$$

Distribuantur omnes radices Ω in α periodos a terminorum; hae singulae rursus in β periodos b terminorum; hae singulae denuo in γ periodos etc. Quaeratur per art praec. [Remank 12.2.16] aequatio α^{ti} gradus (A), cuius radices sint illa α aggregata a terminorum, quorum itaque valores per resolutionem huius aequationis innotescent.
At hic difficultas oritur, quum incertum videatur, cuinam radici aequationis (A) quodvis aggregatum aequale statuendum sit, puta quaenam radix per $(a, 1)$, quaenam per (a, g) etc. denotari debeat: huic rei sequenti modo remedium afferri poterit. Per $(a, 1)$ designari potest radix quaecunque aequationis (A); quum enim quaevis radix huius aequ. sit aggregatum a radicum ex Ω omninoque arbitrarium sit, quaenam radix ex Ω per [1] denotetur, manifesto supponere licebit, aliquam ex iis radicibus, e quibus radix quaecunque data aequ. (A) constat, per [1] exprimi, unde illa radix aequ. (A) fiet $(a, 1)$; radix [1] vero hinc nondum penitus determinatur, sed etiamnum prorsus arbitrarium seu indefinitum manet, quamnam radicem ex iis, quae $(a, 1)$ constituunt, pro [1]

adoptare velimus. Simulac vero $(a, 1)$ determinatum est, etiam omnia reliqua aggregata a terminorum rationaliter inde deduci poterunt (art. 346) [Theorem 12.2.9]. Hinc simul patet, unicam tantummodo radicem per huius resolutionem eruere oportere. – Potest etiam methodus sequens, minus directa, ad hunc finem adhiberi. Accipiatur pro [1] radix determinata, i.e. ponatur

$$[1] = \cos\frac{kP}{n} + i\sin\frac{kP}{n},$$

integro k ad lubitum electo, ita tamen ut per n non sit divisibilis; quo facto etiam [2], [3], etc. radices determinatas indicabunt, unde etiam aggreata $(a, 1)$, (a, g) etc. quantitates determinatas designabunt. Quibus e tabulis sinuum levi tantum calamo computatis, puta ea precisione, ut quae maiora quaeve minora sint decidi possit, nullum dubium superesse poterit, quibusnam signis singulae radices aequ. (A) sint distinguendae.

Quando hoc modo omnia α aggregata a terminorum inventa sunt, investigetur per art. praec. [Remark 12.2.16] aequatio (B) β^{ti} gradus, cuius radices sint β aggregata b terminorum sub $(a, 1)$ contenta; coëfficientes huius aequationis omnes erunt quantitates cognitae. Quum adhuc arbitrarium sit, quaenam ex $a = \beta b$ radicibus sub $(a, 1)$ contentis per [1] denotetur, quaelibet radix data aequ. (B) per $(b, 1)$ exprimi poterit, quia manifesto supponere licet, aliquam b radicum, e quibus composita est, per [1] denotari. Investigetur itaque una radix quaecunque aequationis (B) per eius resolutionem, statuatur $= (b, 1)$, deriventurque inde per art. 346 [Theorem 12.2.9] omnia reliqua aggregata b terminorum. Hoc modo simul calculi confirmationem nanciscimur, quum semper ea aggregata b terminorum, quae ad easdem periodos a terminorum pertinent, summas notas conficere debeant. – In quibusdam casibus aeque expeditum esse potest, $\alpha - 1$ alias aequationes β^{ti} gradus eruere, quarum radices sint resp. singula β aggregata b terminorum in reliquis periodis a terminorum, (a, g), (a, gg) etc. contenta, atque *omnes* radices tum harum aequationum tum aequationis B per resolutionem investigare: tunc vero simili modo ut supra adiumento tabulae sinuum decidere oportebit, quibusnam periodis b terminorum singulae radices hoc modo prodeuntes aequales statui debeant. Ceterum ad hocce iudicium varia alia artificia adhiberi possunt, quae hoc loco complete explicare non licet; unum tamen, pro eo casu ubi $\beta = 2$, quod imprimis utile est, ac per exempla brevius quam per praecepta declarari poterit, in exemplis sequentibus cognoscere licebit.

Postquam hoc modo valores omnium $\alpha\beta$ aggregatorum b terminorum inventi sunt, prorsus simili modo hinc per aequationes γ^{ti} gradus omnia $\alpha\beta\gamma$ aggregata c terminorum determinari poterunt. Scilicet *vel unam* aequationem γ^{ti} gradus, cuius radices sint γ aggregata c terminorum sub $(b, 1)$ contenta, per art. 350 [Remark 12.2.16] eruere; per eius resolutionem unam radicem quamcunque elicere et $= (c, 1)$ statuere, tandemque hinc per art. 346 [Theorem 12.2.9] omnia reliqua similia agregata deducere oportebit; *vel* simili modo omnino $\alpha\beta$ aequationes γ^{ti} gradus evolvere, quarum radices sint resp. γ aggregata c terminorum in singulis periodis b terminorum contenta, valores omnium radicum omnium harum aequationum per resolutionem extrahere, tandemque ordinem harum radicum perinde ut supra adiumento tabulae sinuum, vel, pro $\gamma = 2$, per artificium infra in exemplis ostendendum determinare.

Hoc modo pergendo, manifesto tandem omnia $\frac{n-1}{\zeta}$ aggregata ζ terminorum habebuntur; evolvendo itaque per art. 348 [Corollary 12.2.13] aequationem ζ^{ti} gradus, cuius radices sint ζ radices ex Ω in (ζ, 1) contentae, huius coëfficientes omnes erunt quantitates cognitae; quodsi per resolutionem una eius radix quaecunque elicitur, hanc = [1] statuere licebit, omnesque reliquae radices Ω per huius potestates habebuntur. Si magis placet, etiam *omnes* radices illius aequationis per resolutionem erui, praetereaque per solutionem $\frac{n-1}{\zeta} - 1$ aliarum aequationum ζ^{ti} gradus, quae resp. omnes ζ radices in singulis reliquis periodis ζ terminorum contentas exhibent, omnes reliquae radices Ω inveniri poterunt.

Ceterum patet, simulac prima aequatio (A) soluta sit, sive simulac valores ominum α aggregatorum *a* terminorum habeantur, etiam resolutionem functionis **X** in α factores *a* dimensionum per art. 348 [Corollary 12.2.13] sponte haberi; porroque post solutionem aequ. (B), sive postquam valores omnium αβ aggregatorum *b* terminorum inventi sint, singulos illo factores iterum in β, sive **X** in αβ factores *b* dimensionum resolvi etc.

The previous theorems with their corollaries, contain the main basis of the whole theory and the way of finding the values of the roots Ω can now be presented in few words.

First, one needs to take a number g which is a primitive root [a generator] *for the module n and to compute the minimal residue of the powers of g up to g^{n-2} modulo n. Decompose n − 1 into factors and, if one wants to reduce the problem to equations of minimal degree, into prime factors; let them be (with an arbitrary order)* α, β, γ, ..., ζ, *and let us denote*

$$\frac{n-1}{\alpha} = \beta\gamma\ldots\zeta = a, \quad \frac{n-1}{\alpha\beta} = \gamma\ldots\zeta = b, \quad etc.$$

Distribute all the roots Ω in α periods of a terms; each of them again in β periods of b terms; each of them again in γ periods etc. Find by the previous article [Remark 12.2.16] *the equation (A) of degree α, whose roots are the α sums of a terms, of which the values will be obtained by solving this equation.*

But it there is a difficulty in that it is uncertain which root of equation (A) should be equated to which sum, that is, which root should be denoted by (a, 1), which by (a, g) etc.: we can remedy this problem in the following way. We can denote by (a, 1) any root of equation (A); in fact since any generic root of this equation is a sum of a roots in Ω and since it is completely arbitrary which root in Ω is denoted by [1], *obviously we can assume that* [1] *denotes any of the roots contained in any given root of equation (A), so that this root of equation (A) becomes (a, 1); but the root* [1] *has not yet been determined, and which root among those contained in (a, 1) is chosen to represent* [1] *is arbitrary and indeterminate. On the other hand, once (a, 1) is determined all the other periods of a terms can be rationally determined (article 346)* [Theorem 12.2.9]. *From this it follows that it is sufficient to find a unique root of this equation. One can also with this in mind apply the following less direct method. Take for* [1] *a determined root, i.e. put*

$$[1] = \cos\frac{k\pi}{n} + \sin\frac{k\pi}{n},$$

choosing an arbitrary integer k, provided that it is not divided by n. Then [2], [3], *etc. denote determined roots, so that the sums (a, 1)(a, g) etc. denote determined quantities.*

If from sine tables these quantities are sufficiently computed, that is with the precision to make it possible to decide which is the greater and which is the lesser, there can be no doubt in distinguishing the single roots of equation (A).

When in this way all α sums of a terms are found, compute by the previous article [Remark 12.2.16] the equation (B) of degree β, whose roots are the β sums of b terms, contained in $(a, 1)$; all the coefficients of this equation will be known quantities. Since we have not yet determined which of the $a = \beta b$ roots contained in $(a, 1)$ denotes [1], we can represent any root of equation (B) by $(b, 1)$, since obviously we can assume that some of the b roots of which it is composed are denoted by [1]. Find a root of equation (B) by solving it, denote it $(b, 1)$ and deduce, by article 346 [Theorem 12.2.9], all the other sums of b terms. In this way, we have a method for verifying the computation, since the sums of b terms which belong to the same period of a terms must give the known sum. In some cases the other $\alpha - 1$ equations of degree β, whose roots are resp. the different β sums of b terms contained in the other periods of a terms, (a, g), (a, g^2) etc., can be quickly constructed and all the roots of these equations and of equation B found by resolution: But then, in the same way as above, with the aid of the sine table one needs to decide which periods of b terms are equal to which roots so obtained. Other different tools, which are impossible to explain here, can also be applied; in the following examples it will be possible to discuss only one, for the case in which $\beta = 2$, which is the most important and which is better presented by example than by rules.

When in this way all αβ sums of b terms are found, in the same way, using equations of degree γ, all of the αβγ sums of c terms can be determined. Either by finding one equation of degree γ whose roots are the γ sums of c terms contained in $(b, 1)$ via article 350 [Remark 12.2.16]; by resolving this and computing a single root denoting it $(c, 1)$ and deducing, by article 346 [Theorem 12.2.9], all the other similar sums; or by developing all the αβ equations of degree γ, whose roots are respectively the γ sums of c terms contained in a single period of b terms, obtaining the values of all the roots of all these equations by resolution, and determining the order of these roots with the help of sine tables, or, for $\gamma = 2$, by the methods presented in the examples below.

Continuing in this way, we will obviously have $\frac{n-1}{\zeta}$ sums of ζ terms; finding by article 348 [Corollary 12.2.13] the equation of degree ζ, whose roots are the ζ roots from Ω contained in $(\zeta, 1)$, its coefficients are all known quantities; if by its resolution we compute one of its roots, we can denote it by [1] and its powers give all the other roots Ω. If one likes, all the other roots in Ω can be obtained by the above resolution all the roots of this equation and of the $\frac{n-1}{\zeta} - 1$ other equations of degree ζ, which respectively give all ζ roots contained in the other periods of ζ terms.

*Moreover, it is clear that when equation (A) is solved and one has all the values of the α sums of a terms, one has also factorized the function **X** in α factors of degree a by article 348 [Corollary 12.2.13]; and from the solution of equation (B) one gets the values of all αβ sums of b terms, and the factorization of each such factor into β factors; the factorization of **X** into αβ factors of degree b etc.*

Example 12.2.22 (Art. 353). Gauss then illustrates his algorithm using the example which has already been discussed, $n = 19$. Since $18 = 3 \cdot 3 \cdot 2$,

12.2 Cyclotomic Equations

computing Ω requires the resolution of two cubic and one quadratic equations.

Using 2 as the generator, we get Table 12.1 from which we produce

$$\Omega = (18,1) = \begin{cases} (6,1) \begin{cases} (2,1) & [1], [18] \\ (2,8) & [8], [11] \\ (2,7) & [7], [12] \end{cases} \\ (6,2) \begin{cases} (2,2) & [2], [17] \\ (2,16) & [3], [16] \\ (2,14) & [5], [14] \end{cases} \\ (6,4) \begin{cases} (2,4) & [4], [15] \\ (2,13) & [6], [13] \\ (2,9) & [9], [10]. \end{cases} \end{cases}$$

The equation (A) whose roots are $(6,1)$, $(6,2)$, $(6,4)$ is (cf. Example 12.2.18)

$$F_1(X) = X^3 + X^2 - 6X - 7$$

'solving' which, Gauss obtains

$$\mathsf{p}_6 = (6,1) = -1.2218761623\ldots$$

which allows us to compute (cf. Example 12.2.10)

$$\begin{aligned} (6,2) &= -(6,1)^2 + 4, \\ (6,4) &= (6,1)^2 - (6,1) - 5, \end{aligned}$$

substituting which in the polynomials $P_1(X), P_2(X), P_4(X)$ (cf. Example 12.2.14) we get a partial factorization of **X** in $\mathbb{C}[X]$[12].

The equation (B) whose roots are $(2,1)$, $(2,7)$, $(2,8)$ is (cf. Example 12.2.19)

$$G_1(X) = X^3 - \mathsf{p}_6 X^2 + (\mathsf{p}_6^2 - 5)X + \mathsf{p}_6^2 - 6,$$

solving which we get

$$\mathsf{p}_2 = \mathsf{q} := (2,1) = -1.3545631433\ldots$$

---

[12] And from the modern point of view, the computations of Example 12.2.14 give us the factorization

$$\mathbf{X} = P_1(X)P_2(X)P_4(X) \text{ in } \mathbb{K}[X] = \mathbb{Q}(\mathsf{p})[X].$$

With the algorithm presented in Theorem 12.2.9, using Gauss we obtain:

$$
\begin{aligned}
(2, 2) &= q^2 - 2, \\
(2, 3) &= q^3 - 3q, \\
(2, 4) &= q^4 - 4q^2 + 2, \\
(2, 5) &= q^5 - 5q^3 + 5q, \\
(2, 6) &= q^6 - 6q^4 + 9q^2 - 2, \\
(2, 7) &= q^7 - 7q^5 + 14q^3 - 7q, \\
(2, 8) &= q^8 - 8q^6 + 20q^4 - 16q^2 + 2, \\
(2, 9) &= q^9 - 9q^7 + 27q^5 - 30q^3 + 9q.
\end{aligned}
\qquad (12.2)
$$

Finally the roots [1] and [18] are obtained by solving

$$X^2 - (2, 1) + 1 = 0$$

and all the other roots are obtained by

- either repeatedly squaring [1]
- or solving all equations

$$X^2 - (2, i) + 1 = 0$$

and associating correctly the values to the roots [i] by comparing them with the rough evaluations given by the sine table[13].

Historical Remark 12.2.23. In his introduction Gauss quoted explicitly the value $n = 17$ for which, as he remarked, one has $n - 1 = 2^4$; computing roots with Gauss' Algorithm requires us to solve quadratic equations iteratively. Therefore *the 17th root of unity can be computed by solving quadratic equations*, which, in Gauss' time, meant that *the polygon with 17 sides can be constructed by ruler and compass,* an important result, to which Gauss proudly devoted the next example.

---

[13] In one sense the choice of these two approaches depends upon their complexity: in order to get a good approximation of all the roots is it better to repeatedly square [1] of which we need a better approximation, or to repeatedly extract roots of equations whose coefficients depend on values of which we again need a better approximation?
Correctly, Gauss did not want to limit his algorithm by an *a priori* choice.

12.2 Cyclotomic Equations

Table 12.2. *Logarithm table for* $\mathbb{Z}_{17}$

i	3^i	i	3^i
1	3	9	14
2	9	10	8
3	10	11	7
4	13	12	4
5	5	13	12
6	15	14	2
7	11	15	6
8	16	16	1

Example 12.2.24 (Art. 354). Setting $n = 17$ and taking 3 as the generator we obtain Table 12.2, from which we produce

$$\Omega = (16, 1) = \begin{cases} (8, 1) \begin{cases} (4, 1) \begin{cases} (2, 1) & [1], [16] \\ (2, 13) & [4], [13] \end{cases} \\ (4, 9) \begin{cases} (2, 9) & [8], [9] \\ (2, 15) & [2], [15] \end{cases} \end{cases} \\ (8, 3) \begin{cases} (4, 3) \begin{cases} (2, 3) & [3], [14] \\ (2, 5) & [5], [12] \end{cases} \\ (4, 10) \begin{cases} (2, 10) & [7], [10] \\ (2, 11) & [6], [11]. \end{cases} \end{cases} \end{cases}$$

Since $(8, 1)(8, 3) = -4$, the equation whose roots are $(8, 1)$ and $(8, 3)$ is

$$X^2 + X - 4,$$

whose roots are $-\frac{1}{2} \pm \frac{1}{2}\sqrt{17}$ which can then be approximated and one of them chosen to denote $\mathsf{p}_8 := (8, 1)$.

Since $(4, 1)(4, 9) = -1$, the equation whose roots are $(4, 1)$ and $(4, 9)$ is

$$X^2 - \mathsf{p}_8 X - 1.$$

With the algorithm presented in Theorem 12.2.9, Gauss expressed all the periods $(4, \lambda)$ in terms of $\mathsf{p}_4 := (4, 1)$ as:

$$\begin{align} (4, 3) &= -\tfrac{3}{2} + 3\mathsf{p}_4 - \tfrac{1}{2}\mathsf{p}_4^3, \\ (4, 10) &= \tfrac{3}{2} + 2\mathsf{p}_4 - \mathsf{p}_4^2 - \tfrac{1}{2}\mathsf{p}_4^3, \\ (4, 9) &= -1 - 6\mathsf{p}_4 + \mathsf{p}_4^2 + \mathsf{p}_4^3, \end{align}$$

so p_4 is approximately computed by root extraction and substituted into the above equations giving all the periods $(4, i)$.

Since $(2, 1)(2, 13) = (4, 3)$, the equation whose roots are $(2, 1)$ and $(2, 13)$ is

$$X^2 - \mathsf{p}_4 X + (4, 3).$$

Denoting $(2, 1) =: \mathsf{p}_2$ the formulas of Equation 12.2 represent $(2, i)$, $2 \leq i \leq 8$, in terms of p_2, so that all these values can be approximated by root(s) extraction.

Finally, solving the equation

$$X^2 - \mathsf{p}_2 X + 1$$

we obtain [1] from which all the roots can be obtained by squaring; or, alternatively, all the roots can be obtained by computing[14]

$$\frac{1}{2}(2, i) \pm \frac{1}{2}\sqrt{(2, i)^2 - 4} = \frac{1}{2}(2, i) \pm \frac{1}{2}\sqrt{(2, 2i) - 2}, \text{ for all } i,$$

and the values are correctly associated to the roots $[i]$ *per artificium*, whose discussion I omit here.

I want to summarize here, in Kronecker's language, the Gauss results presented in this section as:

Theorem 12.2.25. *If $n \in \mathbb{N}$ is an odd prime,*

$$n - 1 = p_1 p_2 \cdots p_\nu$$

is a factorization into prime factors, and

$$\mathbf{X}(X) := \sum_{i=0}^{n-1} X^i = \frac{X^n - 1}{X - 1},$$

the Gauss Algorithm produces a set of ν polynomials $\mathsf{f}_1, \ldots, \mathsf{f}_\nu \in \mathbb{C}[X]$ so that, denoting for each $i \leq \nu$

$\mathfrak{r}_i \in \mathbb{C}$ *a root of* f_i,
$\mathbb{K}_i := \mathbb{K}_{i-1}(\mathfrak{r}_i)$ *where* $\mathbb{K}_0 := \mathbb{Q}$,

it follows that

(1) $\deg(\mathsf{f}_i) = p_i$, *for all i;*
(2) $\mathbb{Q} \subset \mathbb{K}_1 \subset \cdots \subset \mathbb{K}_\nu \subset \mathbb{C}$;
(3) $\mathsf{f}_i(X) \in \mathbb{K}_{i-1}[X]$ *is irreducible, for all i;*

---

[14] Where we use

$$(2, i)^2 = (2, 2i) + (2, 0) = (2, 2i) + 2.$$

(4) $\mathbb{K}_i = \mathbb{K}_{i-1}[X]/\mathsf{f}_i(X)$;

(5) $\mathbb{K}_i$ is the splitting field of $\mathsf{f}_i(X)$;

(6) $\mathbb{K}_\nu$ is the splitting field of $\mathbf{X}(X)$;

(7) $\mathfrak{r}_\nu$ is a root of $\mathbf{X}(X)$.

Example 12.2.26. It is natural here, to apply Gauss' Algorithm to 'solve' the equation $\mathbf{X}(X)$ when $n = 17$ within Kronecker's Model.

The Gauss computations presented in Example 12.2.24, using only the formula of Theorem 12.2.9 and the linear algebra implied by Theorem 12.2.5, allow us to produce the equations

$$\begin{align}
\mathsf{f}_8 &:= X^2 + X - 4, \\
\mathsf{f}_4 &:= X^2 - \mathsf{p}_8 X - 1, \\
\mathsf{f}_2 &:= X^2 - \mathsf{p}_4 X - \frac{1}{2}\mathsf{p}_4^3 + 3\mathsf{p}_4 - \frac{3}{2}, \\
\mathsf{f}_1 &:= X^2 - \mathsf{p}_2 X + 1,
\end{align}$$

where p_i denotes the periods $(i, 1)$. Then we can build the tower

$$\mathbb{Q} \subset \mathbb{K}_8 \subset \mathbb{K}_4 \subset \mathbb{K}_2 \subset \mathbb{K}_1 \subset \mathbb{C}$$

where

$$\begin{align}
\mathbb{K}_8 &:= \mathbb{Q}/\mathsf{f}_8 = \mathbb{Q}(\mathsf{p}_8), \\
\mathbb{K}_4 &:= \mathbb{K}_8/\mathsf{f}_4 = \mathbb{Q}(\mathsf{p}_8, \mathsf{p}_4), \\
\mathbb{K}_2 &:= \mathbb{K}_4/\mathsf{f}_2 = \mathbb{Q}(\mathsf{p}_8, \mathsf{p}_4, \mathsf{p}_2), \\
\mathbb{K}_1 &:= \mathbb{K}_2/\mathsf{f}_1 = \mathbb{Q}(\mathsf{p}_8, \mathsf{p}_4, \mathsf{p}_2, \mathfrak{r}),
\end{align}$$

and we know that $\mathbb{Q}(\mathsf{p}_8, \mathsf{p}_4, \mathsf{p}_2, \mathfrak{r})$ is the splitting field of $\mathbf{X}(X)$ whose roots are

$$[i] := \mathfrak{r}^i, \quad 1 \leq i \leq 16,$$

and which we can easily compute by just using the relations

$$\begin{align}
\mathsf{p}_8^2 &= -\mathsf{p}_8 + 4, \\
\mathsf{p}_4^2 &= \mathsf{p}_8\mathsf{p}_4 + 1, \\
\mathsf{p}_2^2 &= \mathsf{p}_4\mathsf{p}_2 + \frac{1}{2}\mathsf{p}_4^3 - 3\mathsf{p}_4 + \frac{3}{2}, \\
\mathfrak{r}^2 &= \mathsf{p}_2\mathfrak{r} - 1.
\end{align}$$

An elementary *tour de force* then gives

$[1] = r,$

$[2] = r^2$

$ = p_2 r - 1,$

$[3] = p_2 r^2 - r$

$ = rp_2p_4 - \frac{1}{2}rp_4p_8 - \frac{1}{2}rp_4 + \frac{1}{2}rp_8 + \frac{1}{2}r - p_2,$

$[4] = \frac{1}{2}rp_2p_4p_8 - \frac{1}{2}rp_2p_4 + \frac{1}{2}rp_2p_8 + \frac{1}{2}rp_2 + \frac{1}{2}rp_4p_8 - \frac{1}{2}rp_4 - \frac{1}{2}rp_8$
$ -\frac{1}{2}r - p_2p_4 + \frac{1}{2}p_4p_8 + \frac{1}{2}p_4 - \frac{1}{2}p_8 - \frac{1}{2},$

$[5] = rp_2p_4 - rp_2 - \frac{1}{2}rp_4p_8 + \frac{1}{2}rp_4 + \frac{1}{2}rp_8 + \frac{1}{2}r - \frac{1}{2}p_2p_4p_8$
$ +\frac{1}{2}p_2p_4 - \frac{1}{2}p_2p_8 - \frac{1}{2}p_2 - \frac{1}{2}p_4p_8 + \frac{1}{2}p_4 + \frac{1}{2}p_8 + \frac{1}{2},$

$[6] = rp_2 + \frac{1}{2}rp_4p_8 + \frac{1}{2}rp_4 - \frac{1}{2}rp_8 - \frac{3}{2}r-$
$ -p_2p_4 + p_2 + \frac{1}{2}p_4p_8 - \frac{1}{2}p_4 - \frac{1}{2}p_8 - \frac{1}{2},$

$[7] = \frac{1}{2}rp_2p_4p_8 + \frac{1}{2}rp_2p_4 - \frac{1}{2}rp_2p_8 - \frac{1}{2}rp_2 - rp_4 + r$
$ -p_2 - \frac{1}{2}p_4p_8 - \frac{1}{2}p_4 + \frac{1}{2}p_8 + \frac{3}{2},$

$[8] = -\frac{1}{2}rp_2p_4p_8 + \frac{1}{2}rp_2p_4 + \frac{1}{2}rp_2p_8 + \frac{1}{2}rp_2 - \frac{1}{2}rp_4p_8 + \frac{3}{2}rp_4$
$ -\frac{1}{2}rp_8 - \frac{3}{2}r - \frac{1}{2}p_2p_4p_8 - \frac{1}{2}p_2p_4 + \frac{1}{2}p_2p_8 + \frac{1}{2}p_2 + p_4 - 1,$

$[9] = \frac{1}{2}rp_2p_4p_8 - \frac{1}{2}rp_2p_4 - \frac{1}{2}rp_2p_8 - \frac{1}{2}rp_2 + \frac{1}{2}rp_4p_8 - \frac{3}{2}rp_4 + \frac{1}{2}rp_8$
$ +\frac{3}{2}r + \frac{1}{2}p_2p_4p_8 - \frac{1}{2}p_2p_4 - \frac{1}{2}p_2p_8 - \frac{1}{2}p_2 + \frac{1}{2}p_4p_8 - \frac{3}{2}p_4$
$ +\frac{1}{2}p_8 + \frac{3}{2},$

$[10] = -\frac{1}{2}rp_2p_4p_8 - \frac{1}{2}rp_2p_4 + \frac{1}{2}rp_2p_8 + \frac{1}{2}rp_2 + rp_4 - r - \frac{1}{2}p_2p_4p_8$
$ +\frac{1}{2}p_2p_4 + \frac{1}{2}p_2p_8 + \frac{1}{2}p_2 - \frac{1}{2}p_4p_8 + \frac{3}{2}p_4 - \frac{1}{2}p_8 - \frac{3}{2},$

$[11] = -rp_2 - \frac{1}{2}rp_4p_8 - \frac{1}{2}rp_4 + \frac{1}{2}rp_8 + \frac{3}{2}r$
$ +\frac{1}{2}p_2p_4p_8 + \frac{1}{2}p_2p_4 - \frac{1}{2}p_2p_8 - \frac{1}{2}p_2 - p_4 + 1,$

$[12] = -rp_2p_4 + rp_2 + \frac{1}{2}rp_4p_8 - \frac{1}{2}rp_4 - \frac{1}{2}rp_8 - \frac{1}{2}r$
$ +p_2 + \frac{1}{2}p_4p_8 + \frac{1}{2}p_4 - \frac{1}{2}p_8 - \frac{3}{2},$

$[13] = -\frac{1}{2}rp_2p_4p_8 + \frac{1}{2}rp_2p_4 - \frac{1}{2}rp_2p_8 - \frac{1}{2}rp_2 - \frac{1}{2}rp_4p_8 + \frac{1}{2}rp_4$
$ +\frac{1}{2}rp_8 + \frac{1}{2}r + p_2p_4 - p_2 - \frac{1}{2}p_4p_8 + \frac{1}{2}p_4 + \frac{1}{2}p_8 + \frac{1}{2},$

$[14] = -rp_2p_4 + \frac{1}{2}rp_4p_8 + \frac{1}{2}rp_4 - \frac{1}{2}rp_8 - \frac{1}{2}r + \frac{1}{2}p_2p_4p_8$
$ -\frac{1}{2}p_2p_4 + \frac{1}{2}p_2p_8 + \frac{1}{2}p_2 + \frac{1}{2}p_4p_8 - \frac{1}{2}p_4 - \frac{1}{2}p_8 - \frac{1}{2},$

$[15] = -rp_2 + p_2p_4 - \frac{1}{2}p_4p_8 - \frac{1}{2}p_4 + \frac{1}{2}p_8 + \frac{1}{2},$

$[16] = -r + p_2,$

$[17] = 1.$

13
Sturm

Consider $f(X) := X^2 - 2 \in \mathbb{Q}[X]$ and the field

$$\mathbb{Q}[X]/f(X) = \mathbb{Q}[\alpha] \simeq \mathbb{Q}(\sqrt{2}) \subset \mathbb{R}.$$

As was discussed in Examples 8.1.1 and 8.1.2, the crucial point for the application of Kronecker's model to real numbers is, to crudely put, the ability to distingush whether α represents the positive root $\sqrt{2}$ or the negative root $-\sqrt{2}$.

In fact, there are two immersions $\mathbb{Q}[\alpha] \mapsto \mathbb{R}$ characterized by the image of α; these two immersions impose an ordering on $\mathbb{Q}[\alpha]$: in the one in which α represents $\sqrt{2}$ we have $\alpha > 0$; in the other, α represents $-\sqrt{2}$ and we have $\alpha < 0$.

Somehow, the difference between $\mathbb{R}$ and $\mathbb{C}$ is whether -1 is a sum of squares or not: in fact there is a strict relation among the possibility of imposing orderings on a field K and the possibility of representing -1 as a sum of squares in K; this relation will be discussed in Section 13.1 where I introduce the notion of *real closed fields* and of *real closure* of ordered fields.

Then, after introducing a set of notations (Section 13.2), I discuss the technique introduced by Sturm and generalized by Sylvester which allows us to count the real roots of a polynomial $f \in \mathbb{Q}[X]$ (Section 13.3) – and by generalization, the number of the roots $\alpha \in R$ of a polynomial $f \in K[X]$, where K is an ordered field and R its real closure.

The Sturm technique allows us to 'solve' a polynomial equation $f \in \mathbb{Q}[X]$ over $\mathbb{R}$ by producing, for each $\alpha \in \mathbb{R}$, $f(\alpha) = 0$, and each $\epsilon \in \mathbb{Q}$, $\epsilon > 0$, a rational number $c \in \mathbb{Q}$ such that $|\alpha - c| < \epsilon$. This suggests extending Kronecker's model to the reals by representing each algebraic real number $\alpha \in \mathbb{R}$ by introducing a squarefree polynomial $f(X) \in \mathbb{Q}[X]$ such that $f(\alpha) = 0$ and an interval $(a, b) \in \mathbb{Q}$ which contains only α among the roots of f. This *Sturm representation* allows us to perform arithmetical operations (including ordering ones) over algebraic real numbers *à la* Kronecker (Section 13.4).

An alternative technique for counting real roots of polynomial equations $f \in \mathbb{Q}[X]$ was proposed by Hermite using quadratic forms and has recently been proposed again[1] so as to solve the same problem but for a multivariate polynomial equation system which has finitely many roots (Section 13.5).

A different approach for representing algebraic real numbers is based on the remark that, denoting, for a polynomial $f(X) \in K[X]$ and an algebraic real number $\alpha \in R$ – where K is an ordered field and R its real closure – , the sequence

$$\mathsf{s}(f, \alpha) := (\mathrm{sgn}(f(\alpha)), \mathrm{sgn}(f'(\alpha)), \ldots, \mathrm{sgn}(f^{(i)}(\alpha)), \ldots)$$

of the values of the signs of the derivatives of f evaluated at α, each real root $\alpha \in R$ of f is identified uniquely by $\mathsf{s}(f, \alpha)$. This suggests introducing the *Thom codification* in which an algebraic real number $\alpha \in R$ is represented by introducing a squarefree polynomial $f(X) \in K[X]$, such that $f(\alpha) = 0$, and the sequence $\mathsf{s}(f, \alpha)$ (Section 13.6).

The Ben-Or, Kozen and Reif (BKR) Algorithm allows us, given $g_1, \ldots, g_q \in K[X_1, \ldots, X_n]$ and a multivariate polynomial equation system which has finitely many roots $\mathsf{x}_1, \ldots, \mathsf{x}_m \in R^n$, to compute for all j the sequences

$$(\mathrm{sgn}(g_1(\mathsf{x}_j)), \ldots, \mathrm{sgn}(g_i(\mathsf{x}_j)), \ldots, \mathrm{sgn}(g_q(\mathsf{x}_j)))$$

and, in particular, to compute the Thom codification of each root of a polynomial $f(X) \in K[X]$ (Section 13.7).

Finally I show how, using the Thom codification and the BKR Algorithm, to 'solve' polynomial equations $f \in K[X]$ and how to perform arithmetical operations with algebraic elements in a real closed field (Section 13.8).

13.1 Real Closed Fields

Definition 13.1.1. *By a* sum of squares *in a field K, we mean any expression*

$$\sum_{i=1}^{n} a_i^2, a_i \in K.$$

For this concept we have:

Lemma 13.1.2. *Let K be a field. Then:*

(1) If K is ordered, then -1 is not a sum of squares in K.
(2) The following conditions are equivalent:

  (a) -1 is not a sum of squares in K;
  (b) a sum of squares $\sum_{i=1}^{n} a_i^2$ in K is $0 \iff a_i = 0$, for all i.

---

[1] P. Pederson, M.-F. Roy, A. Szpirglas Complexity of computation with real algebraic numbers, J. Symb. Comp. **10** (1990) 1273–1278.

(3) *If -1 is a sum of squares in K and $\mathrm{char}(K) \neq 2$, then each $a \in K$ is a sum of squares.*
(4) *If -1 is a sum of squares in K, the set $\{a \in K : a$ is a sum of squares in $K\}$ is a field.*

Proof

(1) In fact if $\sum_{i=1}^{n} a_i^2$ is a sum of squares, since $a_i^2 > 0$, then $\sum_{i=1}^{n} a_i^2 > 0 > -1$.
(2) $(a) \implies (b)$ Assume $\sum_{i=1}^{n} a_i^2 = 0$ for some $a_i \in K$ and $a_n \neq 0$. Then, setting $b_i := a_i a_n^{-1}$ we have $\sum_{i=1}^{n-1} b_i^2 = -1$ contradicting the assumption.
$(b) \implies (a)$ Assume $-1 = \sum_{i=1}^{n} a_i^2$; then $1^2 + \sum_{i=1}^{n} a_i^2 = 0$; therefore 0 is a sum of squares of non-vanishing elements.
(3) Follows from $2^2 a = (1+a)^2 + (-1)(1-a)^2$.
(4) Let a and b be sums of squares; it is then obvious that $a + b$ and ab are also such sums. And so also is $-a = (-1)a$. To show that $\frac{a}{b}$ is one also, we have just to remark that

$$\frac{a}{b} = \frac{ab}{b^2} = ab(b^{-1})^2.$$

♃

In order to discuss the relationship between the possibility of imposing orderings on a field K and the possibility of representing -1 as a sum of squares in K, thus proving the converse result of Lemma 13.1.2(1), that is

if -1 is not a sum of squares in K then K is ordered,

let us introduce

Definition 13.1.3. *A field K is called*

- *formally real if -1 is not a sum of squares in it;*
- *real closed if it is formally real but no proper algebraic extension of K is formally real.*

To prove our claimed result, let us first prove the following

Lemma 13.1.4. *Let K be a formally real field.*
Let $\gamma \in K \setminus \{0\}$ be such that it is not the square of an element in K, so that $f(X) = X^2 - \gamma$ is irreducible over K and we can consider the field extension

$$K[\sqrt{\gamma}] := K[X]/f(X).$$

With the notation above:

(1) If $K[\sqrt{\gamma}]$ is not formally real, then

$$\text{there exists } a_i, b_i \in K : -1 = \gamma \sum_{i=1}^{n} a_i^2 + \sum_{i=1}^{n} b_i^2. \tag{13.1}$$

(2) If γ is a sum of squares in K, then $K[\sqrt{\gamma}]$ is formally real.
(3) Either $K[\sqrt{\gamma}]$ or $K[\sqrt{-\gamma}]$ is formally real.

Proof

(1) Since $K[\sqrt{\gamma}]$ is not formally real, there is a representation

$$-1 = \sum_{i=1}^{n}(a_i\sqrt{\gamma} + b_i)^2 = \gamma \sum_{i=1}^{n} a_i^2 + 2\sqrt{\gamma}\sum_{i=1}^{n} a_i b_i + \sum_{i=1}^{n} b_i^2, \; a_i, b_i \in K.$$

We have $2\sum_{i=1}^{n} a_i b_i = 0$ since, otherwise,

$$\sqrt{\gamma} = \frac{-1 - \gamma\sum_{i=1}^{n} a_i^2 - \sum_{i=1}^{n} b_i^2}{2\sum_{i=1}^{n}(a_i b_i)} \in K$$

contradicting the assumption that f is irreducible over K; therefore Equation 13.1 holds.

(2) Assume $K[\sqrt{\gamma}]$ is not formally real, so that Equation 13.1 holds. If γ is a sum of squares in K, then -1 could be represented as a sum of squares in K, contradicting the fact that K is a formally real field.

(3) Assuming $K[\sqrt{\gamma}]$ is not formally real, so that Equation 13.1 holds, we deduce

$$-\gamma = \frac{1^2 + \sum_{i=1}^{n} b_i^2}{\sum_{i=1}^{n} a_i^2} \tag{13.2}$$

and so, by Lemma 13.1.2, $-\gamma$ is a sum of squares, so that $K[\sqrt{-\gamma}]$ is formally real. ♃

Lemma 13.1.5. *Let K be a real closed field and let $\gamma \in K$, $\gamma \neq 0$. If γ is not the square of an element in K, then $K[\sqrt{\gamma}])$ is not formally real.*

Proof In fact, $K[\sqrt{\gamma}]$ is a proper algebraic extension of K; since K is real closed, $K[\sqrt{\gamma}]$ is not formally real. ♃

13.1 Real Closed Fields

Theorem 13.1.6. *Let K be a real closed field. Then:*

(1) If $\gamma \in K$, $\gamma \neq 0$, is a sum of squares in K, then there is $\beta \in K : \beta^2 = \gamma$.

(2) If $\gamma \in K$, either γ is a square in K or $-\gamma$ is a square in K, but the cases are mutually exclusive.

(3) K can be ordered in one and only one way.

Proof

(1) Assume γ is not the square of an element in K; then $K[\sqrt{\gamma}]$ is not formally real and so Equation 13.1 holds. Substituting for γ in Equation 13.1 its representation as a sum of squares in K, we obtain a representation of -1 as a sum of squares in K, and so a contradiction.

(2) If γ is not a square, $K[\sqrt{\gamma}]$ is not formally real and Equations 13.1 and 13.2 hold, from which we deduce that $-\gamma$ is a square.
On the other hand, if both γ and $-\gamma$ are squares, i.e. there exists $b, c \in K$ such that $\gamma = b^2$, $-\gamma = c^2$, we deduce $-1 = (\frac{c}{b})^2$, contradicting the assumption that K is real closed.

(3) To order K it is sufficient to decide for each $a \in K \setminus \{0\}$ which of a or $-a$ is in the positive cone; the choice is completely determined since only one of them is a square and so is in the positive cone.

♄

Example 13.1.7. It is worthwhile discussing further the case considered in Examples 8.1.1 and 8.1.2 of the field

$$K_1 := \mathbb{Q}[X]/f_1(X) = \mathbb{Q}[\alpha] \simeq \mathbb{Q}(\sqrt{2}) \subset \mathbb{R}$$

where $f_1(X) := X^2 - 2 \in \mathbb{Q}[X]$, which has two orderings according to whether $\alpha > 0$ or $\alpha < 0$.

Let us now extend K_1 to the field

$$K_2 := K_1[X]/f_2(X) = K_1[\beta] = \mathbb{Q}[\alpha, \beta] = \mathbb{Q}[\beta] = \mathbb{Q}[X]/(X^4 - 2),$$

where

$$f_2(X) := X^2 - \alpha \in K_1[X].$$

The polynomial $X^4 - 2$ has four roots in $\mathbb{C}$, the positive real root $\sqrt[4]{2}$, the negative real root $-\sqrt[4]{2}$, and the two conjugate complex roots $\pm i\sqrt[4]{2}$, so that there are two copies of K_2 embedded in $\mathbb{C}$, both having, of course, two $\mathbb{Q}$-automorphisms:

$$\mathbb{Q}[\sqrt[4]{2}] \simeq \mathbb{Q}[-\sqrt[4]{2}] \subset \mathbb{R},$$
$$\mathbb{Q}[i\sqrt[4]{2}] \simeq \mathbb{Q}[-i\sqrt[4]{2}] \not\subset \mathbb{R}.$$

In both, of course, -1 is not a sum of squares: in fact, -1 is obviously not this in the real copy; therefore if it were such in the other copy

$$-1 = \sum_i (a_i + b_i\beta + c_i\alpha + d_i\alpha\beta)^2$$

we only have to substitute α with $-\alpha$ to find a representation

$$-1 = \sum_i (a_i + b_i\beta - c_i\alpha - d_i\alpha\beta)^2$$

in the real copy and a contradiction.

Apparently, up to now we have not been able to distinguish whether, in the representation $K_1 = \mathbb{Q}[\alpha]$, α represents the positive real number $\sqrt{2}$ or the negative one $-\sqrt{2}$.

However, paradoxically, our choice of extending K_1 by a square root of α has fixed α to represent $\sqrt{2}$, the single positive root of f, since the negative root cannot be a square!

In fact, if we now try to extend K_2 by adding a root of $f_3(X) := X^2 + \alpha \in K_2[X]$ giving

$$K_3 = K_2[\gamma] = \mathbb{Q}[\beta, \gamma]$$

we deduce

$$\left(\frac{\beta}{\gamma}\right)^2 = -1$$

so that K_3 is neither formally real nor ordered, since, with this extension, we pretend that both α and $-\alpha$ are squares; obviously we have $\frac{\beta}{\gamma} = i$.

In a real closed field the following holds, as a consequence of Theorem 13.1.6,

Corollary 13.1.8. *Let K be a real closed field; then for each positive $a \in K$, there is $b \in K : b^2 = a$.*

and also

Lemma 13.1.9. *Let K be a real closed field; then each polynomial $f(X) \in K[X]$ of odd degree has a root in K.*

13.1 Real Closed Fields

Proof The proof will be by induction on $\deg(f) = n$, the case $\deg(f) = 1$ being trivial.

Let then $f(X) \in K[X]$ be an irreducible polynomial of odd degree $\deg(f) = n > 1$.

Let us consider the field extension $K[\alpha] := K[X]/f(X)$; then $K[\alpha]$ is not formally real so that we have

$$-1 = \sum_{i=1}^{v} g_i(\alpha)^2, \, g_i(X) \in K[X], \quad \deg(g_i) < n;$$

therefore there is $g(X) \in K[X]$ such that

$$-1 = g(X)f(X) + \sum_{i=1}^{v} g_i(X)^2; \tag{13.3}$$

now $\sum_{i=1}^{v} g_i(X)^2$ is of even degree, since the leading coefficients of each $g_i(X)^2$ are squares and therefore cannot cancel in the addition; the degree is also positive, otherwise the representation

$$-1 = \sum_{i=1}^{v} g_i(\alpha)^2 = \sum_{i=1}^{v} g_i(0)$$

of -1 is a sum of squares in the real closed field K, giving a contradiction.

Therefore $\deg(g) \leq n-2$ is odd, and by the assumptions it has a root $\beta \in K$; evaluating Equation 13.3 in β we get

$$-1 = \sum_{i=1}^{v} g_i(\beta)^2$$

which again gives a contradiction. ♄

From Corollary 13.1.8, Lemma. 13.1.9, Theorem 12.1.6 we deduce

Theorem 13.1.10. *Let K be a real closed field and let $K[i] := K[X]/(X^2 + 1)$. Then:*

$K[i]$ is algebraically closed;
each polynomial $f \in K[X]$ factorizes over K in linear and quadratic factors,

♄

of which we have a sort of converse:

Theorem 13.1.11. *Let K be a formal real field. If $K[i]$ is algebraically closed, then K is a real closed field.*

Proof There exists no field K' such that $K \subset K' \subset K[i]$, therefore the only algebraic extensions of K are K itself and $K[i]$. Since $K[i]$ cannot be formally real, K is real closed. ♄

In a real closed field, we can generalize the Weierstrass Theorem of continous functions as:

Lemma 13.1.12. *Let K be a real closed field and let $f(X) \in K[X]$. Let $a, b \in K$ be such that*

$$f(a) < 0 < f(b).$$

Then there is $c \in K$ between a and b such that $f(c) = 0$.

Proof An irreducible factor

$$g(X) = X^2 + pX + q = \left(X + \frac{p}{2}\right)^2 + (q - p^2/4) \in K(X)$$

is everywhere positive in K since the first term is a square and the second is positive because $g(X)$ is irreducible in K and so its determinant $p^2 - 4q < 0$. Since f factors into linear and quadratics factors and the quadratic factors are everywhere positive, a change of sign of f between a and b depends on a change of sign of a linear factor $(X - c)$, and so implies the existence of a root of f between a and b. ♄

We note here, omitting the proof, the following fact:

Fact 13.1.13. *Let K be an ordered field; then there is a unique algebraic extension field $\mathbf{K} \supset K$ which is really closed and whose ordering extends that of K. Such a field is called the* real closure *of K.*

If $\mathbf{K}$ is the algebraic closure of K then $\mathbf{K} = \mathsf{K}[i]$. ♄

Note that the algebraic closure of $\mathbb{Q}$ is

$$\mathbb{A} := \{\alpha \in \mathbb{C} : \exists g(X) \in \mathbb{Q}[X], g(\alpha) = 0\}$$

and its real closure is $\mathbb{A} \cap \mathbb{R}$.

I want to introduce here some ordered and real closed fields which have great applications in real algebraic geometry.

Let K be an ordered field and let us consider the simple transcendental extension $K(\epsilon)$.

13.1 Real Closed Fields

First let us remark that in order to decide whether a rational function

$$\frac{a(\epsilon)}{b(\epsilon)} \in K(\epsilon), a(\epsilon), b(\epsilon) \in K[\epsilon] \setminus \{0\}$$

is positive or negative we can solve the same problem for the polynomial function $a(\epsilon)b(\epsilon) \in K[\epsilon]$ since

$$\frac{a(\epsilon)}{b(\epsilon)} = \frac{a(\epsilon)b(\epsilon)}{b^2(\epsilon)} > 0 \iff a(\epsilon)b(\epsilon) > 0;$$

therefore, in order to impose an ordering over $K(\epsilon)$ it is sufficient to impose one on the polynomial ring $K[\epsilon]$.

On $K[\epsilon]$ we can impose different orderings by saying that the polynomial $f(\epsilon) := \sum_{i=0}^{n} a_i \epsilon^i$ is positive iff

$a_n > 0$; in this case we have

$$a < \epsilon < \epsilon^2 < \cdots < \epsilon^n < \cdots$$

for all $a \in K$;

$a_n < 0$; in this case we have

$$a > \epsilon > \epsilon^2 > \cdots > \epsilon^n > \cdots$$

for all $a \in K$; in both cases ϵ is said to be an *infinity* over K;

$a_0 > 0$; in this case we have

$$a > \epsilon > \epsilon^2 > \cdots > \epsilon^n > \cdots > 0$$

for all $a > 0$ in K;

$a_0 < 0$; in this case we have

$$a < \epsilon < \epsilon^2 < \cdots < \epsilon^n < \cdots < 0$$

for all $a < 0$ in K; in both cases ϵ is said to be an *infinitesimal* over K.

A Pouiseux series in X with coefficients in K is a series with rational exponents

$$P(X) := \sum_{\substack{i \geq i_0 \\ i \in \mathbb{Z}}} a_i X^{\frac{i}{q}}, i_0 \in \mathbb{Z}, a_i \in K, q \in \mathbb{N}.$$

Denote by $K\langle\epsilon\rangle$ the ring of all the Pouiseux series in ϵ with coefficients in K and impose an ordering on it by $P(\epsilon) > 0 \iff a_{i_0} > 0$; let $K\langle\epsilon\rangle_{\text{alg}}$ be the set of all Pouiseux series $P(\epsilon)$ for which there is $f(X, Y) \in K[X, Y]$ such that $f(\epsilon, P(\epsilon)) = 0$.

With this notation we have:

Fact 13.1.14. *The ring $K\langle\epsilon\rangle$ is real closed, and $K\langle\epsilon\rangle_{\mathrm{alg}}$ is the real closure of $K(\epsilon)$ endowed with the ordering such that ϵ is a positive infinitesimal, i.e. $0 < \epsilon < a$, for all $a > 0$ in K.* ♄

13.2 Definitions

In order to introduce Sturm theory, we need to introduce a series of definitions and notations, where D is a domain, its fraction field K is an ordered field, R is the real closure and $C = R[i]$ the algebraic closure of K.

Definition 13.2.1.

- *The* sign *of the elements of R can be interpreted as a function* $\mathrm{sgn} : R \mapsto \mathbb{Z}_3$ *defined by*

$$\mathrm{sgn}(x) := \begin{cases} -1 & \text{if } x < 0 \\ 0 & \text{if } x = 0 \\ 1 & \text{if } x > 0. \end{cases}$$

- *A* sign-vector *is an element of $\mathbb{Z}_3^n$.*
- *If $\mathcal{S} := \{S_1, \ldots, S_n\} \subset D[X_1, \ldots, X_k]$ is a sequence of polynomials, the* sign pattern *of $\mathcal{S}$ at $\mathbf{x} := (x_1, \ldots, x_k) \in R^k$ is the vector*

$$\mathrm{sgn}(\mathbf{x}, \mathcal{S}) := (\mathrm{sgn}(S_1(\mathbf{x})), \ldots, \mathrm{sgn}(S_n(\mathbf{x}))).$$

- *For a sign-vector $\mathbf{s} := \{s_1, \ldots, s_n\} \neq \{0, 0, \ldots, 0\}$ the* number of sign changes *of $\mathbf{s}$, $V(\mathbf{s})$, is defined as follows*

    if $s_i \neq 0$ for all i, the definition is by induction on n via

    - *if $n = 1$, then $V(\mathbf{s}) := 0$;*
    - *otherwise, let $\mathbf{s}' := \{s_1, \ldots, s_{n-1}\}$; then*

    $$V(\mathbf{s}) := \begin{cases} V(\mathbf{s}') + 1 & \text{if } s_{n-1}s_n = -1 \\ V(\mathbf{s}') & \text{otherwise.} \end{cases}$$

    Otherwise let $\mathbf{s}'$ be the sequence which is obtained by $\mathbf{s}$ dropping all the occurrences of 0; then $V(\mathbf{s}) := V(\mathbf{s}')$.

- *For a sequence of polynomials $\mathcal{S} \subset D[X_1, \ldots, X_k]$, for each $\mathbf{x} \in R^k$, we will put*

$$W_{\mathcal{S}}(\mathbf{x}) := V(\mathrm{sgn}(\mathbf{x}, \mathcal{S})).$$

13.2 Definitions

Remark 13.2.2. Let $\mathcal{S} := \{S_1(X), \ldots, S_n(X)\} \in D[X]$ be a sequence of polynomials. Let $M \in R$ be a positive element such that for each root α of each polynomial S_i we have $-M < \alpha < M$. Let us denote for all i, a_i to be the leading coefficient and d_i the degree of S_i.

Then it is obvious that for all i, for all $\alpha > M$ we have

$$\text{sgn}(S_i(\alpha)) = \text{sgn}(a_i), \text{sgn}(S_i(-\alpha)) = \text{sgn}((-1)^{d_i} a_i);$$

therefore with an abuse of notation, we will denote

$$\begin{aligned}
\text{sgn}(\infty, \mathcal{S}) &:= (\text{sgn}(a_1), \ldots, \text{sgn}(a_n)), \\
\text{sgn}(-\infty, \mathcal{S}) &:= (\text{sgn}((-1)^{d_1} a_1), \ldots, \text{sgn}((-1)^{d_n} a_n)), \\
W_{\mathcal{S}}(\infty) &:= V(\text{sgn}(\infty, \mathcal{S})), \\
W_{\mathcal{S}}(-\infty) &:= V(\text{sgn}(-\infty, \mathcal{S})).
\end{aligned}$$

Example 13.2.3. To present the above notation let us consider the polynomials

$$\begin{aligned}
P(X) &:= X^4 - \frac{17}{4}X^2 + 1, \\
S_1(X) &:= X^4 - \frac{17}{4}X^2 + 1, \\
S_2(X) &:= 4X^3 - \frac{17}{2}X, \\
S_3(X) &:= \frac{17}{8}X^2 - 1, \\
S_4(X) &:= \frac{225}{34}X, \\
S_5(X) &:= 1
\end{aligned}$$

in $\mathbb{R}[X]$ and the sequence $\mathcal{S} := \{S_1, S_2, S_3, S_4, S_5\}$.

Then the sign patterns of $\mathcal{S}$ at $x = \pm\infty, \pm\frac{1}{2}, \pm 2$ are

$$\begin{aligned}
\text{sgn}(-\infty, \mathcal{S}) &= (1, -1, 1, -1, 1), \\
\text{sgn}(-2, \mathcal{S}) &= (0, -1, 1, -1, 1), \\
\text{sgn}\left(-\frac{1}{2}, \mathcal{S}\right) &= (0, 1, -1, -1, 1), \\
\text{sgn}\left(\frac{1}{2}, \mathcal{S}\right) &= (0, -1, -1, 1, 1), \\
\text{sgn}(2, \mathcal{S}) &= (0, 1, 1, 1, 1), \\
\text{sgn}(\infty, \mathcal{S}) &= (1, 1, 1, 1, 1),
\end{aligned}$$

and the number of sign changes of $\mathrm{sgn}(x, \mathcal{S})$ for $x = \pm\infty, \pm\frac{1}{2}, \pm 2$ are

$$W_{\mathcal{S}}(-\infty) = 4,$$
$$W_{\mathcal{S}}(-2) = 3,$$
$$W_{\mathcal{S}}\left(-\frac{1}{2}\right) = 2,$$
$$W_{\mathcal{S}}\left(\frac{1}{2}\right) = 1,$$
$$W_{\mathcal{S}}(2) = 0,$$
$$W_{\mathcal{S}}(\infty) = 0.$$

Definition 13.2.4. *Let*

$$\mathcal{F} := \{F_1, \ldots, F_m\} \subset D[X_1, \ldots, X_m] \text{ be a system of monic polynomials}$$
$$F_i(X_i) \in D[X_i], \quad \text{for all } i,$$

so that there also are finitely many solutions

$$\mathbf{x} := (x_1, \ldots, x_m) \in C^m : F_1(x_1) = \cdots = F_m(x_m) = 0;$$

$\mathcal{Z}(\mathcal{F}) := \{\mathbf{x} \in C^m : F(\mathbf{x}) = 0, \text{ for all } F \in \mathcal{F}\}$,
$Z(\mathcal{F}) := \mathcal{Z}(\mathcal{F}) \cap R^m$;
$\mathcal{Q} := \{Q_1, \ldots, Q_n\} \subset D[X_1, \ldots, X_m]$ *be a sequence of polynomials;*
$\mathbf{s} \in \mathbb{Z}_3^n$ *be a sign-vector;*
$\mathbf{A} \subset R^m$,
$\mathcal{C} := \{\mathbf{x} \in \mathcal{Z}(\mathcal{F}) \cap \mathbf{A} : \mathrm{sgn}(\mathbf{x}, \mathcal{Q}) = \mathbf{s}\};$

with this notation, let us denote

$$c(\mathcal{F}, \mathcal{Q}, \mathbf{s}, \mathbf{A}) := \mathrm{card}\,(\mathcal{C}),$$

which counts the number of the roots $\mathbf{x}$ *of the system* $\mathcal{F}$ *in the region* $\mathbf{A}$ *which satisfy the sign conditions* $\mathrm{sgn}(\mathbf{x}, \mathcal{Q}) = \mathbf{s}$.

Example 13.2.5. With the notation of Example 13.2.3, setting

$Q(X) := X^2 - 3X$,
$\mathcal{F} := \{Q\}$,
$\mathcal{Z}(\mathcal{F}) = \{0, 3\}$,
$\mathcal{Q} := \{S_1, S_2, S_3, S_4, S_5\}$,
$\mathbf{s} := (1, 0, -1, 0, 1)$,
$\mathbf{A} := \mathbb{R}$,

we have
$$\text{sgn}(0, \mathcal{Q}) = (1, 0, -1, 0, 1),$$
$$\text{sgn}(3, \mathcal{Q}) = (1, 1, 1, 1, 1),$$

so that
$$C = \{0\}, c(\mathcal{F}, \mathcal{Q}, \mathbf{s}, \mathsf{A}) = 1.$$

13.3 Sturm

Let K be an ordered field and R its real closure. Let $P(X), Q(X) \in K[X]$ and let us define the sequence
$$S_0 := P, \quad S_1 := P'Q, \quad S_2 := -\mathbf{Rem}(S_0, S_1),$$
and, inductively,
$$S_{i+1} := -\mathbf{Rem}(S_{i-1}, S_i)$$
while $S_i \neq 0$; let r be the last index such that $S_r \neq 0$ and also denote $Q_i := \mathbf{Quot}(S_{i-1}, S_i)$. Finally let us denote
$$U_i := S_i/S_r, \quad \text{for all } i.$$

Definition 13.3.1. *The sequence*
$$\mathcal{S}(P, Q) := \{S_0, \ldots, S_r\}$$
is called the Sylvester sequence *of P and Q.*

When $Q = 1$, *i.e.* $P_1 := P'$ *the sequence is known as the* Sturm sequence *of P.*

Remark 13.3.2. The definition of the Sylvester sequence $\mathcal{S}(P, Q)$ of P and Q is similar to that of the polynomial remainder sequences of $S_0 = P$ and $S_1 = P'Q$.

The only difference is that, at each step, the remainder of a division *has its sign changed.*

As a consequence we deduce that:
$$\begin{aligned} S_r &= \gcd(P, P'Q), \\ S_{k-2} &= S_{k-1}Q_{k-1} - S_k, \\ S_{r-1} &= S_r Q_r, \\ U_{k-2} &= U_{k-1}Q_{k-1} - U_k, \\ U_{r-1} &= Q_r, \\ U_r &= 1. \end{aligned}$$

It is also easy to verify that the relations between the Sylvester sequence $\mathcal{S}(P, Q)$ and the polynomial remainder sequence $P_0 := S_0$, $P_1 := S_1$, $P_2, \ldots, P_r$ are given by the formula $S_i = \begin{cases} P_i & i \equiv 0, 1 \\ -P_i & i \equiv 2, 3 \end{cases} \pmod{4}$.

Finally let us denote $\mathcal{S} := \mathcal{S}(P, Q)$ and

$$\mathcal{U} := \{U_0, \ldots, U_r\}$$

and, for any interval $\mathsf{I} := (a, b) \subset \mathbb{R}$, let us use the following shorthand for the notion introduced in Definition 13.2.4

$$
\begin{aligned}
c_+(P, Q, \mathsf{I}) &:= c(\{P\}, \{Q\}, \{1\}, \mathsf{I}) \\
&= \operatorname{card}(\{x : P(x) = 0, Q(x) > 0, a < x < b\}), \\
c_0(P, Q, \mathsf{I}) &:= c(\{P\}, \{Q\}, \{0\}, \mathsf{I}) \\
&= \operatorname{card}(\{x : P(x) = 0, Q(x) = 0, a < x < b\}), \\
c_-(P, Q, \mathsf{I}) &:= c(\{P\}, \{Q\}, \{-1\}, \mathsf{I}) \\
&= \operatorname{card}(\{x : P(x) = 0, Q(x) < 0, a < x < b\}), \\
c(P, \mathsf{I}) &:= \operatorname{card}(\{x : P(x) = 0, a < x < b\}), \\
c(P) &:= \operatorname{card}(\{x : P(x) = 0, x \in \mathbb{R}\}).
\end{aligned}
$$

Lemma 13.3.3. *With the above notation, we have*

(1) If $x \in \mathbb{R} \cup \{\infty, -\infty\}$ is such that either $P(x) \neq 0$ or $P'(X)Q(X) \neq 0$, then $W_{\mathcal{S}}(x) = W_{\mathcal{U}}(x)$;

(2) No two successive elements U_i, U_{i+1}, for all $a \in \mathbb{R}$, are such that $U_i(a) = U_{i+1}(a) = 0$;

(3) Let $a_1, a_2 \in \mathbb{R}$ and denote $\mathcal{U}' := \{U_0, U_1\}$, $U(X) := \frac{U_0(X)}{U_1(X)}$; then

$$W_{\mathcal{U}'}(a_1) - W_{\mathcal{U}'}(a_2) = \frac{1}{2}(\operatorname{sgn}(U(a_2)) - \operatorname{sgn}(U(a_1))).$$

Proof

(1) The assumption implces $S_r(x) \neq 0$, so the result is obvious.

(2) Otherwise we deduce $U_{i+2}(a) = 0$ and, by induction, $1 = U_r(a) = 0$.

(3) It is sufficient to check the case

$$\operatorname{sgn}(a_1, \mathcal{U}') = \{1, 1\}, \operatorname{sgn}(a_2, \mathcal{U}') = \{1, -1\}, U(a_1) = 1, U(a_2) = -1.$$

♄

13.3 Sturm

Lemma 13.3.4 (Sturm). *With the above notation:*

(1) *if* $U_j(r) = 0$, $j > 0$, *then there exists* $\delta \in R$, $\delta > 0$, *such that*
$$U_{j-1}(x)U_{j+1}(x) < 0, \quad \text{for all } x \in (r - \delta, r + \delta);$$

(2) *if* $U_j(r) = 0$, $j > 0$, *then there exists* $\delta \in R, \delta > 0$, *such that*
$$W_{\mathcal{U}'}(x) = 1, \quad \text{for all } x \in (r - \delta, r + \delta),$$

where $\mathcal{U}' := \{U_{j-1}, U_j, U_{j+1}\}$;

(3) *if* $P(r) = Q(r) = 0$, *then* $U_0(r) \neq 0$;

(4) *if* $P(r) = 0$, $Q(r) \neq 0$, *then, denoting* $\mathcal{U}' := \{U_0, U_1\}$, *there exists* $\delta \in R, \delta > 0$, *such that*
$$W_{\mathcal{U}'}(r - \delta) - W_{\mathcal{U}'}(r + \delta) = \text{sgn}(Q(r)).$$

Proof

(1) Since $U_j(x) = 0$ we deduce that $U_{j-1}(x) \neq 0$ and $U_{j+1} \neq 0$ by Lemma 13.3.3.2. Therefore

there exists $\delta \in R, \delta > 0 : U_{j-1}(x) \neq 0, U_{j+1}(x) \neq 0$,

for all $x \in (r - \delta, r + \delta)$.

Since
$$U_{j-1}(r) = U_j(r)Q_j(r) - U_{j+1}(r) = -U_{j+1}(r)$$

and the signs of U_{j-1} and U_{j+1} are constant in $(r - \delta, r + \delta)$, the claim follows obviously.

(2) To fix the argument, let us say that $\text{sgn}(U_{j-1}(r)) = 1$ so that

for all $x \in (r - \delta, r + \delta) : \text{sgn}(U_{j-1}(x)) = 1, \quad \text{sgn}(U_{j+1}(x)) = -1$;

clearly for all $x \in (r - \delta, r + \delta)$ $\text{sgn}(x, \mathcal{U}')$ will be one of $(1, 1, -1)$, $(1, 0, -1)$ and $(1, -1, -1)$.

(3) Let $l \in \mathbb{N}$, $l > 0$ be the multiplicity of r in P, and let $F(X) \in R[X]$ such that
$$P(X) = (X - r)^l F(X), \quad F(r) \neq 0.$$

If $Q(r) = 0$, there are $G(X), H(X) \in R[X]$ so that
$$Q(X) = (X - r)G(X), H(r) \neq 0$$

and
$$\begin{aligned} S_0 &= (X - r)^l F(X), \\ S_1 &= (X - r)^l (F'(X)Q(X) + lF(X)G(X)), \\ S_r(X) &= (X - r)^l H(X); \end{aligned}$$

therefore
$$U_0(r) = \frac{F(r)}{H(r)} \neq 0.$$

(4) As above, let $l \in \mathbb{N}$, $l > 0$ and $F(X) \in R[X]$ be such that
$$P(X) = (X - r)^l F(X), \quad F(r) \neq 0.$$
Since $Q(r) \neq 0$, then there is $H(X) \in R[X]$, $H(r) \neq 0$, so that
$$\begin{aligned} S_0 &= (X - r)^l F(X), \\ S_1 &= (X - r)^{l-1}\big((X - r)F'(X) + lF(X)\big)Q(X), \\ S_r(X) &= (X - r)^{l-1} H(X); \end{aligned}$$

therefore, denoting $G(X) := \frac{F(X)}{H(X)}$, we have
$$\begin{aligned} U_0(X) &= (X - r)G(X), \\ U_1(X) &= \frac{(X - r)F'(X) + lF(X)}{H(X)} Q(X) \\ &= (X - r)\frac{F'(X)}{H(X)} Q(X) + lG(X)Q(X), \end{aligned}$$

so that $U_1(r) = lG(r)Q(r) \neq 0$.

Let $\delta \in R$, $\delta > 0$ be such that $\operatorname{sgn}(U_1(x))$ and $\operatorname{sgn}(Q(x))$, and so also $\operatorname{sgn}(G(x))$, are constant in $[r - \delta, r + \delta]$.

Let us consider
$$U(X) := \frac{U_0(X)}{U_1(X)};$$
then for $x \in [r - \delta, r + \delta]$ we have
$$\operatorname{sgn}(U(x)) = \frac{\operatorname{sgn}(U_0(x))}{\operatorname{sgn}(U_1(x))} = \operatorname{sgn}\left(\frac{(x - r)}{lQ(r)}\right).$$

Therefore we have
$$\operatorname{sgn}(U(r - \delta)) = -\operatorname{sgn}(Q(r)), \quad \operatorname{sgn}(U(r + \delta)) = +\operatorname{sgn}(Q(r))$$
so that, by Lemma 13.3.3.3 we have
$$\begin{aligned} \operatorname{sgn}(Q(r)) &= \frac{1}{2}\big(\operatorname{sgn}(U(r + \delta)) - \operatorname{sgn}(U(r - \delta))\big) \\ &= W_{U'}(r - \delta) - W_{U'}(r + \delta). \end{aligned}$$

♄

13.3 Sturm

Theorem 13.3.5 (Sylvester's Theorem). *Let K be an ordered field, R its real closure; let $P(X), Q(X) \in K[X]$, and S be the Sylvester sequence of P and Q.*
Then, for each interval $I := (a, b) \subset R \cup \{\infty, -\infty\}$, $P(a) \neq 0$, $P(b) \neq 0$,

$$W_S(a) - W_S(b) = c_+(P, Q, I) - c_-(P, Q, I).$$

Proof The roots of all the polynomials in S divide the interval (a, b) into subintervals. If $r_1, \ldots, r_{s-1}$ denote these roots and we set $r_0 := a, r_s := b$ we have s intervals (r_{i-1}, r_i), $1 \leq i \leq s$ in each of which the sign of each S_i is constant.

Clearly the roots $x \in R$ such that $S_0(x) = S_1(x) = \cdots = S_i(x) = \cdots S_r(x) = 0$ cannot influence our analysis; except for them, the roots of all the polynomials in S coincide with the roots of all the polynomials in $\mathcal{U}$ and $W_S(x) = W_{\mathcal{U}}(x)$, for all $x \in R$. Therefore it is sufficient to discuss the variations of the function $W_{\mathcal{U}}(x)$ at each point r_i:

if $U_j(r_i) = 0$, $j > 0$, we know that the number of sign changes of the sign-vector $(\text{sgn}(U_{j-1}(x)), \text{sgn}(U_j(x)), \text{sgn}(U_{j+1}(x)))$ is constant in the interval (r_{i-1}, r_{i+1});
if $U_0(r_i) = 0$ then $P(r_i) = 0$, $Q(r_i) \neq 0$, and, denoting $\mathcal{U}' := \{U_0, U_1\}$, for any elements $r_i' \in (r_{i-1}, r_i)$ and $r_i'' \in (r_i, r_{i+1})$

$$W_{\mathcal{U}'}(r_i') - W_{\mathcal{U}'}(r_i'') = \text{sgn}(Q(r)).$$

From these results, the claim follows. ♄

Corollary 13.3.6 (Sturm's Theorem). *Let K be an ordered field and R its real closure and let $P(X) \in K[X]$, and let S be the Sturm sequence of P.*
Then, for each interval $I := (a, b) \subset R \cup \{\infty, -\infty\}$, $P(a) \neq 0$, $P(b) \neq 0$,

$$W_S(a) - W_S(b) = c(P, I)$$

Proof When $Q = 1$, we have $c(P, I) = c_+(P, Q, I)$ and $c_-(P, Q, I) = 0$. ♄

Historical Remark 13.3.7. The results of this chapter were developed by Sturm in 1835 for the case $Q = 1$, i.e. for the case for which the Sturm sequence was computed by adapting the polynomial remainder sequence of P and P'; Sylvester showed in 1853 how to generalize it by substituting $P'Q$ for P'. Obviously, today it is easier to present the result in the reverse order.

Example 13.3.8. Using the notation and the results of Example 13.2.3, it is sufficient to check that the roots of

$$P(X) = X^4 - \frac{17}{4}X^2 + 1$$

are $\pm\frac{1}{2}, \pm 2$ and to glance at the sequence

$$\begin{aligned} W_S(-\infty) &= 4 \\ W_S(-2) &= 3 \\ W_S\left(-\frac{1}{2}\right) &= 2 \\ W_S\left(\frac{1}{2}\right) &= 1 \\ W_S(2) &= 0 \\ W_S(\infty) &= 0, \end{aligned}$$

in order to obtain a confirmation of Sturm's results.

13.4 Sturm Representation of Algebraic Reals

Sturm's Theorem is the central tool for solving numerically real polynomial equations; to discuss its application let us consider a squarefree polynomial $P(X) \in \mathbb{R}[X]$, $d := \deg(P)$, and let

$$Z := \{\alpha_1, \ldots, \alpha_d\} = \{\alpha \in \mathbb{R} : P(\alpha) = 0\},$$

$M, m \in \mathbb{Q}$ be such that[2] $M \geq |\alpha|$, for all $\alpha \in Z$,

$$m \leq \min\{|\alpha - \beta| : \alpha, \beta \in Z, \alpha \neq \beta\},$$

S be the Sturm sequence of P; finally let $\mathsf{I} := (a, b) \subseteq (-M, M)$, $c := \frac{b+a}{2}$, $\mathsf{I}_l := (a, c), \mathsf{I}_r := (c, b)$.

Algorithm 13.4.1. With the above notation, let us assume that

$$W_S(a) - W_S(b) = c(P, \mathsf{I}) > 1$$

and let us compute $W_S(c)$; then one of the following holds

$\text{sgn}(P(c)) = 0$ and $c \in Z$;

$0 < c(P, \mathsf{I}_l) < c(P, \mathsf{I})$, so that $0 < c(P, \mathsf{I}_r) < c(P, \mathsf{I})$ and the roots of P in I are partitioned into the two non-empty subsets $\mathsf{I}_l \cap Z, \mathsf{I}_r \cap Z$;

---

[2] M can be estimated via the results of Section 18.4; similar estimations are also available for m; they are not reported here because the algorithms we are discussing do not need m's evaluation.

13.4 Sturm Representation of Algebraic Reals

Fig. 13.1. Solving real polynomial equations I

$[I_1, \ldots, I_d] := \mathbf{Sturm1}(P)$
where
 $P \in \mathbb{R}[X]$, is a squarefree polynomial,
 $d := \deg(P)$,
 $\mathsf{Z} := \{\alpha_1, \ldots, \alpha_d\} = \{\alpha \in \mathbb{R} : P(\alpha) = 0\}$,
 for all i, $1 \leq i \leq d$ either
  $I_i = [c, c]$ and $P(c) = 0$, or
  $I_i = (a_i, b_i)$ and $\exists! \alpha \in I_i \cap \mathsf{Z}$.
$L := \emptyset$
Let S be the Sturm sequence of P
Let $M : M \geq |\alpha|$, for all $\alpha \in \mathsf{Z}$
Compute $W_S(M), W_S(-M)$
If $W_S(-M) - W_S(M) = 1$ **then** $L := L \cup [(-M, M)]$
If $W_S(-M) - W_S(M) > 1$ **then** $L_0 := [(-M, M)]$
While $L_0 \neq \emptyset$ **do**
 $(a, b) := \mathbf{First}(L_0)$, $L_0 := \mathbf{Right}(L_0)$
 $c := \frac{b+a}{2}$
 Compute $W_S(c)$
 If $P(c) = 0$ **then** $L := L \cup [[c, c]]$
 If $W_S(c) - W_S(b) = 1$ **then** $L := L \cup [(c, b)]$
 If $W_S(c) - W_S(b) > 1$ **then** $L_0 := L_0 \cup [(c, b)]$
 If $W_S(a) - W_S(c) = 1$ **then** $L := L \cup [(a, c)]$
 If $W_S(a) - W_S(c) > 1$ **then** $L_0 := L_0 \cup [(a, c)]$
L

$0 = c(P, \mathsf{I}_r)$ (respectively $c(P, \mathsf{I}_l)$) and $c(P, \mathsf{I}_l)$ (respectively $c(P, \mathsf{I}_r)$) $= c(P, \mathsf{I})$; in this case we have $\mathsf{I}_l \cap \mathsf{Z} = \mathsf{I} \cap \mathsf{Z}$ but the length of the interval containing this set is halved, since $c - a = \frac{b-a}{2}$.

Since, for each interval $\mathsf{I} := (a, b)$, by definition

$$b - a < m \implies c(P, \mathsf{I}) \leq 1,$$

then, by repeated bisection, we will obtain d intervals $\mathsf{I}_i = (a_i, b_i)$ such that $c(P, \mathsf{I}_i) = 1$, for all i, in a finite number of steps, so that, up to a renumeration,

$$a_i < \alpha_i < b_i, \quad \text{for all } i.$$

This algorithm, which allows us to approximate all the roots of P, is described in Figure 13.1.

Algorithm 13.4.2. If we are given an interval $\mathsf{I} = (a, b)$ such that $c(P, \mathsf{I}) = 1$ and a value $\epsilon \in \mathbb{R}$, $\epsilon > 0$, let us denote by α the single root of P in the

Fig. 13.2. Solving real polynomial equations II

$[c_1, \ldots, c_d] := \mathbf{Sturm2}(P, \epsilon)$
where
  $P \in \mathbb{R}[X]$, is a squarefree polynomial,
  $\epsilon \in \mathbb{R}, \epsilon > 0$,
  $d := \deg(P)$,
  $Z := \{\alpha_1, \ldots, \alpha_d\} = \{\alpha \in \mathbb{R} : P(\alpha) = 0\}$,
  for all i, $c_i \in \mathbb{Q}$ is such that $|c_i - \alpha_i| < \epsilon$
$L_0 := \mathbf{Sturm1}(P)$
$L := \emptyset$
While $L_0 \neq \emptyset$ **do**
  $\mathsf{I} := \mathbf{First}(L_0), L_0 := \mathbf{Right}(L_0)$
  If $\mathsf{I} = [c, c]$ **then** $L := L \cup [c]$
  If $\mathsf{I} = [a, b]$ **then**
    While $b - a > 2\epsilon$ **do**
      $c := \frac{b+a}{2}$
      Compute $W_\mathcal{S}(c)$
      If $P(c) = 0$ **then** $L := L \cup [c]$
      If $W_\mathcal{S}(c) - W_\mathcal{S}(b) = 1$ **then**
        $a := c$
      else
        $b := c$
    $c := \frac{b+a}{2}, L := L \cup [c]$
L

---

interval I. It is clear that a further dichotomy, by repeatedly computing $c := \frac{b+a}{2}$ and $W_\mathcal{S}(c)$, allows us to produce an interval $\mathsf{I}' = (a', b') \subseteq \mathsf{I}$ such that

$$|b' - a'| < 2\epsilon$$
$$c(P, \mathsf{I}') = 1$$

and so an approximation $c := \frac{b+a}{2}$ of α such that $|c - \alpha| < \epsilon$.

This algorithm, which allows us to approximate each root of P, is described in Figure 13.2.

Example 13.4.3. With the notation of Example 13.3.8, let us see how we can compute the real roots of

$$P(X) := X^4 - \frac{17}{4}X^2 + 1;$$

we get

- $M := 5, W_\mathcal{S}(5) = 0, W_\mathcal{S}(-5) = 4$,
- $L_0 := [(-5, 5)], L := \emptyset$,
- $\mathsf{I} := (-5, 5), c := 0, W_\mathcal{S}(0) = 2$,

13.4 Sturm Representation of Algebraic Reals

- $L_0 := [(-5, 0), (0, 5)], L := \emptyset$,
- $I := (-5, 0), c := \frac{-5}{2}, W_S(\frac{-5}{2}) = 4$,
- $L_0 := [(0, 5), (\frac{-5}{2}, 0)], L := \emptyset$,
- $I := (0, 5), c := \frac{5}{2}, W_S(\frac{5}{2}) = 0$,
- $L_0 := [(\frac{-5}{2}, 0), (0, \frac{5}{2})], L := \emptyset$,
- $I := (\frac{-5}{2}, 0), c := \frac{-5}{4}, W_S(\frac{-5}{4}) = 3$,
- $L_0 := [(0, \frac{5}{2})], L := [(\frac{-5}{2}, \frac{-5}{4}), (\frac{-5}{4}, 0)]$,
- $I := (0, \frac{5}{2}), c := \frac{5}{4}, W_S(\frac{5}{4}) = 1$,
- $L_0 := \emptyset, L := [(\frac{-5}{2}, \frac{-5}{4}), (\frac{-5}{4}, 0), (0, \frac{5}{4}), (\frac{5}{4}, \frac{5}{2})]$,

and, in fact, the four roots of P are

$$-2 \in \left(\frac{-5}{2}, \frac{-5}{4}\right), \quad -\frac{1}{2} \in \left(\frac{-5}{4}, 0\right), \quad \frac{1}{2} \in \left(0, \frac{5}{4}\right), \quad 2 \in \left(\frac{5}{4}, \frac{5}{2}\right).$$

Sturm's algorithm directly suggests a possible representation of real algebraic numbers:

Definition 13.4.4. *A Sturm representation (P, a, b) of a real algebraic number $\alpha \in \mathbb{A} \cap \mathbb{R}$ is the assignment of*

a squarefree polynomial $P(X) \in \mathbb{Q}[X]$,
two numbers $a, b \in \mathbb{Q}$

such that

$P(\alpha) = 0$,
$a < \alpha < b$,
$c(P, [a, b]) = 1$.

We now need to show that this definition allows us to effectively perform the real operations; so let us assume we are given two real algebraic numbers α_1, α_2 by means of the Sturm representations (P_1, a_1, b_1) and (P_2, a_2, b_2).

- In order to get the Sturm representation (P_3, a_3, b_3) of $\alpha_3 := \alpha_1 \pm \alpha_2$ we compute (cf. Corollary 6.7.5)

    $$P_3(Y) := \text{Res}(P_1(Y \mp X), P_2(X)), \quad a_3 := a_1 \pm a_2, \quad b_3 := b_1 \pm b_2,$$

    and we verify whether $c(P_3, [a_3, b_3]) = 1$; If this is not the case, by repeated dichotomy, we compute better approximations $[a_i, b_i], = 1, 2$ of α_i until $c(P_3, [a_3, b_3]) = 1$.
- The Sturm representation of $\alpha_1 \alpha_2$ and $\frac{\alpha_1}{\alpha_2}$ can be similarly computed using the opportune resultants presented in Corollary 6.7.5.

- If $\alpha_1 \neq 0$, then $\alpha_3 := \frac{1}{\alpha_1}$ has the Sturm representation
$$\left(X^{\deg(P_1)} P_1(1/X), \frac{1}{b_1}, \frac{1}{a_1} \right)$$
- The equality $\alpha_1 = \alpha_2$ holds iff $P_1 = P_2$ and $[a_1, b_1] \cap [a_2, b_2] \neq \emptyset$.
- To decide $\text{sgn}(\alpha_1)$ we need only to compute, by repeated dichotomy, a Sturm representation (P_1, a_1, b_1), such that $0 \notin [a_1, b_1]$; then $\text{sgn}(\alpha_1) = \text{sgn}(a_1)$.

Following Kronecker's Philosophy, we must be able to solve (i.e. apply Sturm's algorithm to) a polynomial $P(X) = \sum_i \alpha_i X^i \in \mathbb{R}[X]$ where each α_i is given via a Sturm representation (P_i, a_i, b_i). In order to do so, we need at least to compute the Sturm sequence of P – which can be done by the arithmetical tools discussed above – and to evaluate in a rational number $q \in \mathbb{Q}$ a polynomial $Q(X) = \sum_j \beta_j X^j \in \mathbb{R}[X]$ where each β_j is given via a Sturm representation. This problem can be avoided in two ways:

by refining the Sturm representations of the β_is so that it is possible to decide the sign of $Q(q)$;

by substituting $P(X) \in \mathbb{R}[X]$ with an opportune multiple $P^*(X) \in \mathbb{Q}(X)$; this can be done by considering the splitting field $\mathbb{F} \supset \mathbb{Q}$ of the polynomial $\prod_i P_i$ and the Galois group $G(\mathbb{F}/\mathbb{Q})$ (cf. the next chapter), and computing

$$P^*(X) := \prod_{\phi \in \mathfrak{G}} \phi(P)(X).$$

13.5 Hermite's Method

Let D be a domain whose fraction field K is an ordered field, R be the real closure and $C = R[i]$ the algebraic closure of K.

Given two polynomials $P(X), Q(X) \in K[X]$ Sylvester's Theorem allows us to compute

$$c_+(P, Q, R) - c_-(P, Q, R) :=$$
$$\#(\{x \in R : P(x) = 0, Q(x) > 0\}) - \#(\{x \in R : P(x) = 0, Q(x) < 0\})$$

by computing $c_+(P, Q, R) - c_-(P, Q, R) = W_\mathcal{S}(-\infty) - W_\mathcal{S}(+\infty)$ where $\mathcal{S}$ is the Sylvester sequence of P and Q.

We now need to solve the generalization of this problem for multivariate polynomials. Therefore let us use the notation introduced in Definition 13.2.4, where we will assume that

$\mathcal{Q} = \{Q\}$, with $Q(X_1, \ldots, X_m) \in D[X_1, \ldots, X_m]$,
$\mathsf{A} = R^m$

and we will use the shorthand notation:

$$\begin{aligned}
c_+(\mathcal{F}, Q) &:= c(\mathcal{F}, \{Q\}, \{1\}, R^m) \\
&= \mathrm{card}(\{\mathbf{x} \in \mathcal{Z} : Q(\mathbf{x}) > 0\}) \\
c_0(\mathcal{F}, Q) &:= c(\mathcal{F}, \{Q\}, \{0\}, R^m) \\
&= \mathrm{card}(\{\mathbf{x} \in \mathcal{Z} : Q(\mathbf{x}) = 0\}) \\
c_-(\mathcal{F}, Q) &:= c(\mathcal{F}, \{Q\}, \{-1\}, R^m) \\
&= \mathrm{card}(\{\mathbf{x} \in \mathcal{Z} : Q(\mathbf{x}) < 0\}).
\end{aligned}$$

Our aim is to compute

$$c_+(\mathcal{F}, Q) - c_-(\mathcal{F}, Q);$$

in order to do that, It has been proposed that one of Hermite's univariate solutions is generalized based on quadratic forms.

Fact 13.5.1. *Let M be a symmetric $n \times n$ matrix with entries in R. Then*

(1) There is an invertible square matrix Δ with entries in R such that

$$\Delta M \Delta^t = \begin{pmatrix} I_p & 0 & 0 \\ 0 & -I_m & 0 \\ 0 & 0 & 0 \end{pmatrix}.$$

(2) The value $p - m$ is called the signature *of M.*
(3) The eigenvalues of M are real.

♮

Let $d := \prod_i \deg(F_i)$,

$\mathbf{B} := \{X_1^{a_1} \cdots X_n^{a_n} : a_i < \deg(F_i)\} \subset R[X_1, \ldots, X_n] = \{b_1, b_2, \ldots b_d\}$

and let us consider the quadratic form

$$\mathfrak{B}(\mathcal{F}, Q) := \sum_{\mathbf{x} \in \mathcal{Z}(\mathcal{F})} Q(\mathbf{x}) \left(\sum_{i=1}^d b_i(\mathbf{x}) Y_i \right)^2 = \sum_{i,j=1}^d a_{ij} Y_i Y_j,$$

for which:

Theorem 13.5.2 (Petersen, Roy, Szpirglas). *With the above notation:*

(1) The matrix $M := (a_{ij})$ is symmetric with entries in D.
(2) The rank of $\mathfrak{B}(\mathcal{F}, Q)$ is the number of the roots of $\mathcal{F}$ which are not roots of Q.

(3) *The signature $v(\mathcal{F}, Q)$ of $\mathfrak{B}(\mathcal{F}, Q)$ satisfies*

$$v(\mathcal{F}, Q) := c_+(\mathcal{F}, Q) - c_-(\mathcal{F}, Q).$$

Proof

(1) We have

$$\mathfrak{B}(\mathcal{F}, Q) = \sum_{i,j=1}^{d} \sum_{\mathbf{x} \in \mathcal{Z}(\mathcal{F})} Q(\mathbf{x}) b_i(\mathbf{x}) b_j(\mathbf{x}) Y_i Y_j$$

so that

$$a_{ij} = \sum_{\mathbf{x} \in \mathcal{Z}(\mathcal{F})} Q(\mathbf{x}) b_i(\mathbf{x}) b_j(\mathbf{x}), \forall i, j.$$

Since each a_{ij} is symmetric in each root of each monic polynomial $F_k(X_k)$, it can be expressed in terms of the coefficients of the F_ks.

(2) Let us enumerate the elements of $\mathcal{Z}(\mathcal{F})$ as

$$\mathbf{y}_1, \ldots, \mathbf{y}_r, \mathbf{z}_1, \bar{\mathbf{z}}_1, \ldots, \mathbf{z}_s, \bar{\mathbf{z}}_s,$$

where each $\mathbf{y}_k \in R^m$ with multiplicity m_k, while the conjugate roots $\mathbf{z}_l, \bar{\mathbf{z}}_l \in C^m \setminus R^m$ have multiplicity n_l.

For each $\mathbf{x} \in C^m$ let $\ell(\mathbf{x})$ be the linear form

$$\ell(\mathbf{x}) := \sum_{i=1}^{d} b_i(\mathbf{x}) Y_i;$$

then

$$\ell(\mathbf{y}_1), \ldots, \ell(\mathbf{y}_r), \ell(\mathbf{z}_1), \ell(\bar{\mathbf{z}}_1), \ldots, \ell(\mathbf{z}_s), \ell(\bar{\mathbf{z}}_s)$$

are linearly independent[3].

Now let us express the quadratic form $\mathfrak{B}(\mathcal{F}, Q)$ in terms of them obtaining

$$\mathfrak{B}(\mathcal{F}, Q) = \sum_{k=1}^{r} m_k Q(\mathbf{y}_k) \ell(\mathbf{y}_k)^2 + \sum_{l=1}^{s} n_l \left(Q(\bar{\mathbf{z}}_l) \ell(\bar{\mathbf{z}}_l)^2 + Q(\mathbf{z}_l) \ell(\mathbf{z}_l)^2 \right),$$

from which the thesis follows.

---

[3] Let $\Delta = \Delta(\mathcal{F})$ be the matrix whose rows are indexed by the elements $b_i \in \mathbf{B}$ and whose columns are indexed by the roots $\mathbf{x}_j \in \mathcal{Z}(\mathcal{F})$ and whose (i, j)th entry is $b_i(\mathbf{x}_j)$; we must prove that $\det(\Delta) \neq 0$. The argument is a repeated application of the Vandermonde determinant via induction on the number m of variables.

If $m = 1$ we have $b_j = X_1^j$ and the result follows from the Vandermonde determinant.

If $m > 1$, let us consider the system of equations

$$\mathcal{F}_* := \{F_1(X_1), \ldots, F_{m-1}(X_{m-1})\},$$

whose corresponding matrix $\Delta_* := \Delta(\mathcal{F}_*)$ is such that $\det(\Delta_*) \neq 0$, and the equation

13.5 Hermite's Method

(3) Let $\gamma_l \in C$ be such that $\gamma_l^2 = Q(\bar{z}_l)$ and let $\ell'_l, \ell''_l \in R[Y_1, \ldots, Y_d]$ be the linear forms such that
$$\gamma_l \ell(\bar{z}_l) = \ell'_l + i\ell''_l;$$

then we have
$$\begin{aligned} Q(\bar{z}_l)\ell(\bar{z}_l)^2 + Q(z_l)\ell(z_l)^2 &= (\ell'_l + i\ell''_l)^2 + (\ell'_l - i\ell''_l)^2 \\ &= 2\ell'^2_l - 2\ell''^2_l, \end{aligned}$$

so that
$$\mathfrak{B}(\mathcal{F}, Q) = \sum_{k=1}^{r} m_k Q(y_k)\ell(y_k)^2 + \sum_{l=1}^{s} 2n_l \ell'^2_l + \sum_{l=1}^{s} -2n_l \ell''^2_l.$$

Therefore we have $p = s + c_+(\mathcal{F}, Q)$, $m = s + c_-(\mathcal{F}, Q)$.

♄

To complete this approach we need to discuss

how to compute the entries a_{ij} of $\mathfrak{B}(\mathcal{F}, Q)$ and
how to compute the signature of $\mathfrak{B}(\mathcal{F}, Q)$.

To solve the problem of computing
$$a_{ij} = \sum_{\mathbf{x} \in \mathcal{Z}(\mathcal{F})} Q(\mathbf{x}) b_i(\mathbf{x}) b_j(\mathbf{x}) \in D$$

we consider the polynomial
$$A_{ij}(X_1, \ldots, X_n) := Q b_i b_j \in D[X_1, \ldots, X_n]$$

$F_m(X_m)$. If $\alpha_j \in C$, $1 \le j \le \delta := \deg(F_m)$ are the roots of $F_m(X_m) = 0$ then we have

$$\Delta := \begin{pmatrix} \alpha_1 \Delta_* & \cdots & \alpha_j \Delta_* & \cdots & \alpha_\delta \Delta_* \\ \vdots & \ddots & \vdots & \ddots & \vdots \\ \alpha_1^i \Delta_* & \cdots & \alpha_j^i \Delta_* & \cdots & \alpha_\delta^i \Delta_* \\ \vdots & \ddots & \vdots & \ddots & \vdots \\ \alpha_1^\delta \Delta_* & \cdots & \alpha_j^\delta \Delta_* & \cdots & \alpha_\delta^\delta \Delta_* \end{pmatrix}$$

so that
$$\det(\Delta) = \det(\Delta_*)^\delta \begin{vmatrix} \alpha_1 & \cdots & \alpha_j & \cdots & \alpha_\delta \\ \vdots & \ddots & \vdots & \ddots & \vdots \\ \alpha_1^i & \cdots & \alpha_j^i & \cdots & \alpha_\delta^i \\ \vdots & \ddots & \vdots & \ddots & \vdots \\ \alpha_1^\delta & \cdots & \alpha_j^\delta & \cdots & \alpha_\delta^\delta \end{vmatrix} \ne 0.$$

and we compute $A'_{ij} := \mathbf{Reduction}(A_{ij}; \{F_1, \ldots, F_m\})$ (cf. Figure 8.1) obtaining a polynomial $A'_{ij} := \sum_{k=1}^{d} c_k b_k \in D[X_1, \ldots, X_n]$ so that

$$a_{ij} = \sum_{k=1}^{d} \sum_{\mathbf{x} \in \mathcal{Z}(\mathcal{F})} c_k b_k(\mathbf{x});$$

our task is now reduced to expressing the generic symmetric products

$$\sum_{(x_1,\ldots,x_n) \in \mathcal{Z}(\mathcal{F})} x_1^{a_1} \cdots x_n^{a_n} = \prod_{i=1}^{n} \sum_{(x_1,\ldots,x_n) \in \mathcal{Z}(\mathcal{F})} x_i^{a_i},$$

in terms of the coefficients of the F_is, which can be done using the Waring functions.

To solve the second problem, we just need to compute the characteristic polynomials of $\mathfrak{B}(\mathcal{F}, Q)$, which only have real roots, and apply Descartes's rule of signs[4] in order to compute the number of the positive and negative real roots with their multiplicity.

13.6 Thom Codification of Algebraic Reals (1)

Let D be a domain whose fraction field K is an ordered field, R be the real closure and $C = R[i]$ the algebraic closure of K.

An alternative proposal to the Sturm representation for algebraic real numbers $\alpha \in R$ was suggested by M. Coste and M.-F. Roy in 1988[5] based on

Lemma 13.6.1 (Thom's Lemma). *Let $P(X) \in R[X]$, $n = \deg(P)$. Let $\mathcal{P} := \{P, P', P'', \ldots, P^{(i)}, \ldots, P^{(n-1)}\}$ and let*

$$\mathbf{s} := (s_0, s_1, \ldots, s_{n-1}) \in \mathbb{Z}_3^n$$

be a sign-vector.
Then

$$\mathcal{Z}(\mathbf{s}) := \{x \in R : \mathrm{sgn}(x, \mathcal{P}) = \mathbf{s}\}$$

is void, or a point or an interval.

---

[4] Which states that
Let $P(X) := \sum_{i=0}^{n} a_i X^i \in R[X]$ be a polynomial whose roots are all in R and consider the sign-vector $\mathbf{s} := \{\mathrm{sgn}(a_0), \ldots, \mathrm{sgn}(a_i), \ldots, \mathrm{sgn}(a_n)\}$. Then $V(\mathbf{s})$ is the number of positive roots of P, counted with their multiplicity.

[5] M. Coste, M.-F. Roy *Thom's Lemma, the coding of real algebraic numbers and the topology of semi-algebraic sets* J. Synb. Comp. **5** (1988) 121–129.

13.6 Thom Codification of Algebraic Reals (1)

Proof Is by induction on n, the case $n = 1$ being trivial.
If $n > 1$ let
$$\mathcal{Z} := \{x \in R : \text{sgn}(P^{(i)}(x)) = s_i, \text{ for all } i, 1 \leq i \leq n-1\}.$$

If $\mathcal{Z}$ is void, then so is $\mathcal{Z}(\mathbf{s})$; if $\mathcal{Z}$ is a point, then $\mathcal{Z}(\mathbf{s})$ is either void or a point. We are therefore left to consider the case in which $\mathcal{Z}$ is an interval; in this case, since P' has a fixed constant sign in $\mathcal{Z}$, P is monotonic in $\mathcal{Z}$, whence the claim. ♄

Corollary 13.6.2 (Coste, Roy).
With the same notation as above, let $\alpha, \alpha' \in R$ and let for all i, $0 \leq i \leq n-1$, $s_i := \text{sgn}(P^{(i)}(\alpha))$, $s_i' := \text{sgn}(P^{(i)}(\alpha'))$. Then:

1. *if $P(\alpha) = P(\alpha') = 0$, then $\alpha = \alpha' \iff s_i = s_i'$, for all i.*
2. *if there exists $i : s_i \neq s_i'$, let k be such that*
$$s_k \neq s_k', \quad s_i = s_i', \quad \text{for all } i > k;$$

then

(a) $s_{k+1} = s_{k+1}' \neq 0$;
(b) *if $s_{k+1} = s_{k+1}' = 1$ then*
$$\alpha > \alpha' \iff P^{(k)}(\alpha) > P^{(k)}(\alpha');$$
(c) *if $s_{k+1} = s_{k+1}' = -1$ then*
$$\alpha > \alpha' \iff P^{(k)}(\alpha) < P^{(k)}(\alpha').$$

Proof Let
$$\mathbf{s} := (s_0, \ldots, s_{n-1}), \mathbf{s}' := (s_0', \ldots, s_{n-1}').$$

If $\mathbf{s} = \mathbf{s}'$ and $s_0 = s_0' = 0$, then $\mathcal{Z}(\mathbf{s}) = \mathcal{Z}(\mathbf{s}')$ is a finite set, i.e. a point, and (1) holds.
Assume
$$s_{k+1} = P^{(k+1)}(\alpha) = P^{(k+1)}(\alpha') = s_{k+1}' = 0;$$

then, applying (1) to $P^{(k+1)}$, we have $\alpha = \alpha'$, and the contradiction implies (2a).
Let
$$\mathcal{Z} = \{x \in R : \text{sgn}(P^{(i)}(x)) = s_i, k+1 \leq i \leq n-1\};$$

then $\mathcal{Z}$ is an interval containing both α and α' and in which $P^{(k+1)}$ is positive (respectively negative), and so $P^{(k)}$ is increasing (respectively decreasing), which proves (2b) (respectively (2c)). ♮

Definition 13.6.3 (Coste, Roy). *A Thom codification (P, s) of a real algebraic number $\alpha \in R$ is the assignment of*

a squarefree polynomial $P(X) \in D[X]$,
the sign-vector $\mathsf{s} := (s_1, \ldots, s_{n-1})$

such that

(1) $P(\alpha) = 0$,
(2) $n = \deg(P)$,
(3) $\operatorname{sgn}(P^{(i)}(\alpha)) = s_i$, *for all i, $1 \leq i \leq n-1$.*

13.7 Ben-Or, Kozen and Reif Algorithm

Let D be a domain whose fraction field K is an ordered field, R be the real closure and $C = R[i]$ the algebraic closure of K.

On the basis of the above definition it is clear that the polynomial $P(X) \in D[X]$ is 'solved' if it is possible to compute the Thom codification of each real root $\alpha \in R$ of P.

The Ben-Or, Kozen and Reif (BKR) Algorithm solves a more general problem: given two sets $\mathcal{F}$ and $\mathcal{Q} := \{Q_1, \ldots, Q_m\}$ satisfying the conditions of Definition 13.2.4 – and using the notation of that definition – it allows us to compute

$$\mathsf{S} := \{\operatorname{sgn}(\mathsf{x}, \mathcal{Q}) : \mathsf{x} \in \mathsf{Z}(\mathcal{F})\},$$

and

$$c_\mathsf{s} := c(\mathcal{F}, \mathcal{Q}, \mathsf{s}, R^m) = \operatorname{card}\{\mathsf{x} \in \mathsf{Z}(\mathcal{F}) : \operatorname{sgn}(\mathsf{x}, \mathcal{Q}) = \mathsf{s}\}, \quad \text{for all } \mathsf{s} \in \mathsf{S}.$$

Let us begin by discussing, using the same shorthand of the previous chapter, the easiest case, in which $m = 1$ and we need to count the number of real roots in $\mathsf{Z}(\mathcal{F})$, at which the single polynomial $Q := Q_1$ is (respectively) positive, zero or negative.

Lemma 13.7.1. *With the notation above, and of Theorem 13.5.2(3)*

$$\begin{aligned} v(\mathcal{F}, 1) &= +c_+ + c_0 + c_-, \\ v(\mathcal{F}, Q) &= +c_+ - c_-, \\ v(\mathcal{F}, Q^2) &= +c_+ + c_-, \end{aligned}$$

13.7 Ben-Or, Kozen and Reif Algorithm

and, therefore, by solving the above system we obtain

$$
\begin{aligned}
c_0 &= v(\mathcal{F}, 1) - v(\mathcal{F}, Q^2), \\
c_+ &= \tfrac{1}{2}v(\mathcal{F}, Q^2) + \tfrac{1}{2}v(\mathcal{F}, Q), \\
c_- &= \tfrac{1}{2}v(\mathcal{F}, Q^2) - \tfrac{1}{2}v(\mathcal{F}, Q).
\end{aligned}
$$

Proof The formula for $v(\mathcal{F}, Q^2)$ follows from the fact that $v(\mathcal{F}, Q^2)$ is the difference between the number of the real roots of $\mathcal{F}$ at which Q^2 is positive (i.e. Q is non-zero) and the number of the real roots of $\mathcal{F}$ at which Q^2 is negative (which is obviously zero). ♄

In principle we can approach the general problem in the same way, deriving a linear system

$$A_k \bar{c}^t = \bar{b}$$

where

- the unknowns are the c_ss where s runs in $\mathbb{Z}_3^k$, i.e. the number of real roots x of $\mathcal{F}$ satisfying the sign condition $\mathrm{sgn}(x, \mathcal{Q}) = s$;
- $\bar{b}$ is a vector of suitable $v(\mathcal{F}, Q)$s, where Q runs over suitable products of the Q_is and/or their squares; and
- the matrix A_k is inductively defined by means of easy theoretical considerations.

For instance for $k = 2$, denoting

$$c_{++} = \#\{x \in Z(\mathcal{F}) : Q_1(x) > 0, Q_2(x) > 0\}$$

and, analogously, $c_{+0}, c_{+-}, c_{0+}, c_{00}, c_{0-}, c_{-+}, c_{-0}, c_{--}$, and

$$v_{ij} := v(\mathcal{F}, Q_1^i Q_2^j), \quad 0 \le i, j \le 2,$$

so that

$$v_{00} = v(\mathcal{F}, 1), v_{10} = v(\mathcal{F}P, Q_1), v_{20} = v(\mathcal{F}, Q_1^2), \ldots$$

we have the linear system

$+c_{++}$	$+c_{0+}$	$+c_{-+}$	$+c_{+0}$	$+c_{00}$	$+c_{-0}$	$+c_{+-}$	$+c_{0-}$	$+c_{--}$	$= v_{00}$
$+c_{++}$		$-c_{-+}$	$+c_{+0}$		$-c_{-0}$	$+c_{+-}$		$-c_{--}$	$= v_{10}$
$+c_{++}$		$+c_{-+}$	$+c_{+0}$		$+c_{-0}$	$+c_{+-}$		$+c_{--}$	$= v_{20}$
$+c_{++}$	$+c_{0+}$	$+c_{-+}$				$-c_{+-}$	$-c_{0-}$	$-c_{--}$	$= v_{01}$
$+c_{++}$		$-c_{-+}$				$-c_{+-}$		$+c_{--}$	$= v_{11}$
$+c_{++}$		$+c_{-+}$				$-c_{+-}$		$-c_{--}$	$= v_{21}$
$+c_{++}$	$+c_{0+}$	$+c_{-+}$				$+c_{+-}$	$-c_{0-}$	$+c_{--}$	$= v_{02}$
$+c_{++}$		$-c_{-+}$				$+c_{+-}$		$-c_{--}$	$= v_{12}$
$+c_{++}$		$+c_{-+}$				$+c_{+-}$		$+c_{--}$	$= v_{22}.$

The problem with this approach is that in order to decide the signs of k polynomials at the roots of P, we have to solve a linear system of dimension 3^k, so that the resulting algorithm is of exponential complexity in k. On the other hand, at most $d := \prod_{i=1}^{n} \deg(F_i)$ different sign conditions can be actually satisfied.

It is clear that if $(s_1, \ldots, s_{k-1}) \in \mathbb{Z}_3^{k-1}$ is not satisfied by $\{Q_1, \ldots, Q_{k-1}\}$ for any real root of $\mathcal{F}$, and $s_k \in \mathbb{Z}_3$, then $(s_1, \ldots, s_{k-1}, s_k) \in \mathbb{Z}_3^k$ is not satisfied by $\{Q_1, \ldots, Q_k\}$ for any real root of $\mathcal{F}$.

The BKR Algorithm is an iterative version of the above algorithm, which makes full use of these remarks and so has complexity polynomial in $\min(d, 3^k) = d(k)$.

Algorithm 13.7.2 (Ben-Or, Kozen, Reif). The algorithm is by induction on k. So, setting

$$\mathcal{Q}_k := \{Q_1, \ldots, Q_k\}$$

and assuming we explicitly know

$\mathsf{S}(\mathcal{F}, \mathcal{Q}_{k-1}) := \{\mathsf{s}_1, \ldots, \mathsf{s}_s\}$, where $s := \leq d(k-1)$,

for all i, $1 \leq i \leq s$, $c_{k-1,i} := c_{\mathsf{s}_i} > 0$,

an invertible $s \times s$ matrix A_{k-1},

polynomials $R_{k-1,1}, \ldots, R_{k-1,s}$,

$v_{k-1,i} := v(\mathcal{F}, R_{k-1,i})$, for all $i, 1 \leq i \leq s$,

so that, denoting

$\underline{v}_{k-1}$ the vector $(v_{k-1,1}, \ldots, v_{k-1,s})$ and
$\underline{c}_{k-1}$ the vector $(c_{k-1,1}, \ldots, c_{k-1,s})$,

we have

$$A_{k-1}\underline{c}_{k-1}^t = \underline{v}_{k-1}^t.$$

we can therefore describe how to compute

$\mathsf{S}(\mathcal{F}, \mathcal{Q}_k) := \{\mathsf{s}_1, \ldots, \mathsf{s}_\sigma\}$, where $\sigma \leq d(k)$,

for all i, $1 \leq i \leq \sigma$, $c_{k,i} := c_{\mathsf{s}_i} > 0$,

an invertible $\sigma \times \sigma$ matrix A_k,

polynomials $R_{k,1}, \ldots, R_{k,\sigma}$,

$v_{k,i} := v(\mathcal{F}, R_{k-,i})$, for all $i, 1 \leq i \leq \sigma$,

so that denoting

$\underline{v}_k$ the vector $(v_{k,1}, \ldots, v_{k,\sigma})$ and
$\underline{c}_k$ the vector $(c_{k,1}, \ldots, c_{k,\sigma})$,

13.7 Ben-Or, Kozen and Reif Algorithm

we have
$$A_k \underline{c}_k^t = \underline{v}_k^t.$$

For $k = 1$, according to Lemma 13.7.1 we compute

$$\begin{aligned} c_0 &= v(\mathcal{F}, 1) - v(\mathcal{F}, Q_1^2) \\ c_+ &= \tfrac{1}{2} v(\mathcal{F}, Q_1^2) + \tfrac{1}{2} v(\mathcal{F}, Q_1) \\ c_- &= \tfrac{1}{2} v(\mathcal{F}, Q_1^2) - \tfrac{1}{2} v(\mathcal{F}, Q_1) \end{aligned}$$

If some among c_0, c_+, c_- are zero, we choose from the matrix

$$\begin{pmatrix} 1 & 1 & 1 \\ 1 & 0 & -1 \\ 1 & 0 & 1 \end{pmatrix}$$

the columns corresponding to the non-zero values of c_+, c_0, c_- and rows in order to obtain a non-zero minor A_1.

Inductively, in the case k, let us define

B to be the matrix

$$B := \begin{pmatrix} A_{k-1} & A_{k-1} & A_{k-1} \\ A_{k-1} & 0 & -A_{k-1} \\ A_{k-1} & 0 & A_{k-1} \end{pmatrix},$$

for each $s_i \in S(\mathcal{F}, Q_{k-1})$

$$\begin{aligned} s_{i+} &:= \#\{x \in Z(\mathcal{F}) : \text{sgn}(x, Q_{k-1}) = s_i, Q_k(x) > 0\}, \\ s_{i0} &:= \#\{x \in Z(\mathcal{F}) : \text{sgn}(x, Q_{k-1}) = s_i, Q_k(x) = 0\}, \\ s_{i-} &:= \#\{x \in Z(\mathcal{F}) : \text{sgn}(x, Q_{k-1}) = s_i, Q_k(x) < 0\}, \end{aligned}$$

$\underline{c}_+ := (c_{1+}, \ldots, c_{s+})$,
$\underline{c}_0 := (c_{10}, \ldots, c_{s0})$,
$\underline{c}_- := (c_{1-}, \ldots, c_{s-})$,
$\underline{c} := (c_{1+}, \ldots, c_{s+}, c_{10}, \ldots, c_{s0}, c_{1-}, \ldots, c_{s-})$,
$T_{i0} := R_i, T_{i1} := R_i Q_k, T_{i2} := R_i Q_k^2, \quad \text{for all } i, 1 \leq i \leq s$,
$v_{ij} := v(\mathcal{F}, T_{ij})$ for $0 \leq j \leq 2$ and for all i,
$\underline{v}_0 := (v_{10}, \ldots, v_{s0})$,
$\underline{v}_1 := (v_{11}, \ldots, v_{s1})$,
$\underline{v}_2 := (v_{12}, \ldots, v_{s2})$,
$\underline{v} := (v_{10}, \ldots, v_{s0}, v_{11}, \ldots, v_{s1}, v_{12}, \ldots, v_{s2})$.

Proposition 13.7.3. *Under this notation*
$$B\underline{c}^t = \underline{v}^t.$$
Moreover B is invertible.

Proof The matrix B is invertible since it can be transformed by row operations into
$$\begin{pmatrix} A_{k-1} & A_{k-1} & A_{k-1} \\ 0 & A_{k-1} & 0 \\ 0 & 0 & 2A_{k-1} \end{pmatrix}$$
whose determinant is $2\det(A_{k-1})^3$.

Denote $A_{k-1} = (a_{ij})_{ij}$, $B = (b_{ij})_{ij}$ and let $\underline{b}_i$ be the ith row of B.
We have for all $i, j, 1 \leq i \leq s, \quad 1 \leq i \leq s$

$$\underline{b}_i \, \underline{c}^t = \sum_{j=1}^{s} a_{ij}(c_{j0} + c_{j1} + c_{j2}) = v(\mathcal{F}, R_i) = v(\mathcal{F}, T_{i0}) = v_{i0},$$

$$\underline{b}_{s+i} \, \underline{c}^t = \sum_{j=1}^{s} a_{ij}(c_{j+} - c_{j-}) = v(\mathcal{F}, R_i Q) = v(\mathcal{F}, T_{i1}) = v_{i1}$$

$$\underline{b}_{2s+i} \, \underline{c}^t = \sum_{j=1}^{s} a_{ij}(c_{j+} + c_{j-}) = v(\mathcal{F}, R_i Q^2) = v(\mathcal{F}, T_{i2}) = v_{i2},$$

so that $B\underline{c}^t = \underline{v}^t$. ♄

Since each $v_{ij} := v(\mathcal{F}, T_{ij})$ can be computed via Hermite's method and B is available and invertible, it is therefore possible to compute $\underline{c}$.

Removing the zero entries we obtain $\mathsf{S}(\mathcal{F}, \mathcal{Q}_k)$; the matrix A is obtained from B by choosing the columns corresponding to the non-zero entries in $\underline{c}$ and rows in order to obtain a non-zero minor A_1; again, the elements $R_{k,i}$ are those corresponding to the non-zero entries in $\underline{c}$ and their corresponding values $v_{k,i}$ are available.

13.8 Thom Codification of Algebraic Reals (2)

As in Section 13.4 where we have shown that the Sturm representation allows us to solve polynomial equations and to effectively perform the real operations, we now show the same for the Thom codification.

Let us use the same notation and setting as in the previous section and let us denote the BKR Algorithm by $\mathsf{S} := \mathbf{BKR}(\mathcal{F}, \mathcal{Q})$.

13.8 Thom Codification of Algebraic Reals (2)

Definition 13.8.1. *The* Thom sequence $\mathcal{T}(P(X))$ *of a polynomial* $P(X) \in K[X]$, $\deg(P) = n$, *is the sequence*

$$\mathcal{T}(P(X)) := \{P(X), P'(X), \ldots, P^{(i)}(X), \ldots, P^{(n-1)}(X)\} \in K[X].$$

Algorithm 13.8.2. With this definition it is clear that in order to 'solve' in R a squarefree polynomial $P(X) \in D[X]$, it is sufficient to compute $\mathsf{S} := \mathbf{BKR}(\{P\}, \mathcal{T}(P(X)))$, whose solution satisfies

$$\mathsf{S} = \{\mathrm{sgn}(\mathsf{x}, \mathcal{T}(P)) : x \in R, P(x) = 0\},$$
$$c_\mathsf{s} := \mathrm{card}\{x \in R : P(x) = 0, \mathrm{sgn}(x, \mathcal{T}(P)) = \mathsf{s}\} = 1, \quad \text{for all } \mathsf{s} \in \mathsf{S},$$

so that the set of the real solutions of $P(X)$ in their Thom codification is $\{(P, \mathsf{s}) : \mathsf{s} \in \mathsf{S}\}$.

Now let us briefly discuss how to effectively perform the real operations over real algebraic numbers represented by Thom codification. So let us assume that the real algebraic numbers $\alpha_1, \alpha_2 \in R$ are represented by the Thom codifications $(P_1(X_1), \mathsf{s}_1)$ and $(P_2(X_2), \mathsf{s}_2)$.

- In order to obtain the Thom codification $(P_3(X_3), \mathsf{s}_3)$ of $\alpha_3 := \alpha_1 \pm \alpha_2$ we compute (cf. Corollary 6.7.5) $P_3(X_1) := \mathrm{Res}(P_1(X_1 \mp X_2), P_2(X_2))$ and

$$\mathsf{S} := \mathbf{BKR}\left(\{P_1, P_2\}, [\mathcal{T}(P_1(X_1)), \mathcal{T}(P_2(X_2)), \mathcal{T}(P_3(X_1 \pm X_2))]\right).$$

Denoting, for all $(x_1, x_2) \in R^2$,

$$\mathsf{s}(x_1, x_2) := \Big(\mathrm{sgn}(x_1, \mathcal{T}(P_1)), \mathrm{sgn}(x_2, \mathcal{T}(P_2)), \mathrm{sgn}(x_1 \pm x_2, \mathcal{T}(P_3))\Big)$$

the result will be

$$\mathsf{S} = \{\mathsf{s}(\beta_1, \beta_2) : (\beta_1, \beta_2) \in \mathsf{Z}(\{P_1, P_2\})\}.$$

Therefore there is a single sign-vector $\mathsf{s}(\beta_1, \beta_2) \in \mathsf{S}$ such that

$$\mathrm{sgn}(\beta_1, \mathcal{T}(P_1)) = \mathsf{s}_1 \text{ and } \mathrm{sgn}(\beta_2, \mathcal{T}(P_2)) = \mathsf{s}_2;$$

the Thom codification of α_3 is then

$$(P_3, \mathrm{sgn}(\beta_1 \pm \beta_2, \mathcal{T}(P_3))).$$

- The Thom codification of $\alpha_1 \alpha_2$ and $\frac{\alpha_1}{\alpha_2}$ can be computed similarly using the opportune resultants presented in Corollary 6.7.5.
- To decide the ordering relation between α_1 and α_2, we compute

$$\mathsf{S} := \mathbf{BKR}\left(\{P_1, P_2\}, [\mathcal{T}(P_1(X_1)), \mathcal{T}(P_2(X_2)), X_1 - X_2]\right);$$

there is a single $\mathsf{s} \in \mathsf{S}$ such that

$$\mathsf{s} = (\mathsf{s}_1, \mathsf{s}_2, s), s \in \mathbf{Z}_3;$$

then $\text{sgn}(\alpha_1 - \alpha_2) = s$.

- If $\alpha_1 \neq 0$ and $\mathcal{T}(P_1) = \{Q_1(X_1), \ldots Q_d(X_1)\}$, the Thom codification of $\alpha_3 := \frac{1}{\alpha_1}$ is obtained by computing

  $P_3(X_1) := X_1^{\deg(P_1)} P_1(1/X_1);$

  for each $i \leq d$, $Q_i^*(X_1) := Q_i((1/X_1)X_1^{\delta_i}$ where $\delta_i \geq \deg(Q_i)$ is even so that

  - $Q_i^*(X_1) \in K[X_1]$ and
  - $\text{sgn}(Q_i^*(x)) = \text{sgn}(Q_i(x)), \forall x \in R, x \neq 0;$

  $\mathcal{T}^* := \{Q_1^*(X_1), \ldots, Q_d^*(X_1)\};$

  $\mathsf{S} := \mathbf{BKR}\left(\{P_3\}, \left[\mathcal{T}^*, \mathcal{T}(P_3)\right]\right);$

  $\mathsf{s}_3 \in \mathbb{Z}_3^d$ to be the unique element such that $(\mathsf{s}_1, \mathsf{s}_3) \in \mathsf{S}$.

  Then the Thom codification of α_3 is (P_3, s_3).

14
Galois II

> Je me suis souvent hasardé dans ma vie à avancer des propositions dont je n'étais pas sûr; mais tous ce que j'ai écrit là est depuis bientôt un an dans ma tête, et il est trop de mon intérêt de ne pas me tromper pour qu'on me soupçonne d'énouncer des théorèmes dont je n'aurais pas la démonstration complète.
>
> Tu prieras publiquement Jacobi et Gauss de donner leur avis, non sur la vérité, mais sur l'importance des théorèmes.
>
> Après cela, il y aura, j'espère, des gens qui trouveront leur profit à déchiffrer tout ce gâchis.
>
> E. Galois

This chapter is devoted to the Galois approach to solving polynomial equations.

After introducing the settings of this research, i.e. normal separable extensions $K \supset k$ and the group $G(K/k)$ of the k-automorphisms of K (Section 14.1), I discuss the correspondence between the intermediate fields F, $K \supseteq F \supseteq k$, and the subgroups of $G(K/k)$; in this biunivocal correspondence, a field F corresponds to the subgroup of the k-automorphisms which leave F invariant and a group G corresponds to the subfield of the elements which are kept invariant by all the elements of G (Section 14.2), and we characterize the subgroups which are equivalent to the normal extensions $F \supset k$.

This Galois correspondence allows us to characterize the polynomial equations $f(X) \in k[X]$ which are 'solvable' (in the pre-Abel–Ruffini meaning) in terms of the structure of the Galois group $G(K/k)$ where $K \supset k$ is the splitting field of f; moreover it provides an algorithm which allows us to 'solve' any solvable polynomial equation f provided its Galois group $G(K/k)$ is available

(Section 14.3). In particular, this theory permits a proof of the Abel–Ruffini Theorem (Section 14.4) and gives a characterization of those geometric objects which are constructable by ruler and compass (Section 14.5).

14.1 Galois Extension

Let $K \supset k$ be a field extension. Then the set of all the k-automorphisms of K is clearly a group.

Theorem 14.1.1. *Let $K \supset k$ be a finite field extension and let $G(K/k)$ be the group of the k-automorphisms of K. Then the following conditions are equivalent:*

(1) $k = \{\alpha \in K : \alpha = \sigma(\alpha), \text{ for all } \sigma \in G(K/k)\}$;
(2) K is a normal separable extension;
(3) for each $\xi \in K$, its minimal polynomial $f(X) \in k[X]$ over k has simple roots, all contained in K;
(4) $\#G(K/k) = [K : k]$.

Proof

(2) $\iff$ (3) This follows obviously from the definitions.

(2) $\implies$ (4) This follows from Corollary 10.3.3.

(4) $\implies$ (2) By assumption, K possesses $[K : k]$ k-automorphisms. Therefore, for any extension field $\mathbf{K} \supset K$, every k-isomorphism of K into $\mathbf{K}$ is an automorphism, so that K is normal by Corollary 10.3.7. Then K is the splitting field of a polynomial $f(X) \in k[X]$ and, if we express $\deg(f) = n_0 p^e$, where $p = \text{char}(k)$, $\gcd(p, n_0) = 1$, we know by Corollary 10.3.7 that $[K : k] = \#G(K/k) = n_0$, i.e. that K is separable.

(1) $\implies$ (3) Let $G(K/k) := \{\sigma_i : 1 \le i \le n_0\}$. Let

$$C := \{\sigma_i(\xi) : 1 \le i \le r \le n_0\}$$

be the set of all the $r \le n_0$ distinct conjugates[1] of ξ in K, and let $g(X) := \prod_{\zeta \in C}(X - \zeta)$.

Since, for all $\sigma \in G(K/k)$, $\sigma(f) = f$, and $f(\sigma(\xi)) = \sigma(f(\xi)) = 0$, then ζ is a root of f for each $\zeta \in C$. Therefore $g(X)$ divides f. Moreover, since

for all $\sigma \in G(K/k)$, for all $\zeta_1, \zeta_2 \in C$, $\sigma(\zeta_1) = \sigma(\zeta_2) \implies \zeta_1 = \zeta_2$,

---

[1] Note that in this setting it is possible that $r < n_0$: just consider $k := \mathbb{Q}$, $K = \mathbb{Q}[i, \sqrt{2}]$ where $\#G(K/k) = 4$ and $\sqrt{2}$ has just 2 conjugates.

14.1 Galois Extension

i.e. each $\sigma \in G(K/k)$ is a permutation of C, then $\sigma(g) = g$, for all $\sigma \in G(K/k)$ and $g \in k[X]$. Since f is irreducible in $k[X]$, we conclude that $g = f$.

Therefore f has simple roots, all contained in K.

(3) $\implies$ (1) Let $\xi \in K \setminus k$ and let f be its minimal polynomial; since $\xi \notin k$, then $\deg(f) \geq 2$, and, since all its roots are simple and contained in K, there is $\zeta \in K$, $\zeta \neq \xi$, which is conjugate to ξ. Therefore, by Corollary 10.3.6, there is $\Xi \in G(K/k)$ such that $\Xi(\xi) = \zeta \neq \xi$.

♄

Remark 14.1.2. In view of the above result we consider a finite normal separable extension $K \supset k$ and recall that there is a separable element $\xi \in K$ and a separable polynomial $f(X) \in k[X]$ such that

$$K = k[\xi] = k[X]/f(X);$$

then there are $n = n_0 = [K : k] = \deg(f)$ conjugate elements $\xi =: \xi_1, \ldots, \xi_n$ of ξ over k in K; the $n_0 = G(K/k)$ k-automorphisms of K are those defined by

$$\phi_i \left(\sum_{j=0}^{n-1} a_j \xi^j \right) := \sum_{j=0}^{n-1} a_j \xi_i^j$$

for each element $\sum_{j=0}^{n-1} a_j \xi^j \in k[\xi] = K$.

The above result leads us to introduce the following

Definition 14.1.3. *Let $K \supset k$ be a finite field extension; the group $G(K/k)$ of the k-automorphisms of K is called the* Galois group *of K over k.*

If $K \supset k$ satisfies the condition of Theorem 14.1.1 it is called a Galois extension.

If $f(X) \in k[X]$ is squarefree and $K \supset k$ is the splitting field of f, the Galois group *of f over k is $G(K/k)$.*

We explicitly remark that the proof of the (1) $\implies$ (3) in Theorem 14.1.1 implies

Corollary 14.1.4. *Let $K \supset k$ be a Galois extension and let $f(X)$ be a separable polynomial of which K is the splitting field.*

Then each element of $G(K/k)$ is a permutation over the set of all the $n := \deg(f)$ roots of f, i.e. $G(K/k) \subset \mathsf{S}_n$, where S_n denotes the symmetric group of all the permutations over a set of n elements.

Proof Let ξ in K be a separable root of f and let $\xi =: \xi_1, \ldots, \xi_n$ be all the n distinct conjugates of ξ in K.

Since for all $\sigma \in G(K/k)$ and for each pair of conjugates ξ_i, ξ_j we have

$$\sigma(\xi_i) = \sigma(\xi_j) \implies \xi_i = \xi_j,$$

then each $\sigma \in G(K/k)$ is a permutation of C.

Since $K = k[\xi_1, \ldots, \xi_n]$, if $\sigma \in G(K/k)$ is such that $\sigma(\xi_i) = \xi_i$, for all i, then

$$\sigma(\alpha) = \alpha, \text{ for all } \alpha \in K,$$

i.e. σ is the identity and $G(K/k)$ is a subgroup of S_n. ♄

Moreover Corollary 10.3.7 implies

Corollary 14.1.5. *Let $f(X) \in k[X]$ be squarefree and let G be its Galois group over k.*

Then f is irreducible iff G is transitive[2].

Proof If f is irreducible, using the notation above, for each ξ_i conjugate with ξ we know by Corollary 10.3.7 that there is $\Xi \in G$ such that $\Xi(\xi) = \xi_i$.

Conversely, let α be the root of f such that for each other root β of f there is $\phi \in G : \beta = \phi(\alpha)$.

Let h be the minimal polynomial of α over k, which is irreducible; then for each root β of f

$$f(\beta) = f(\phi(\alpha)) = \phi(f(\alpha)) = 0;$$

as a consequence f and h are associate and f is irreducible. ♄

14.2 Galois Correspondence

Let $K \supset k$ be a Galois extension and let G be its Galois group. Let $\mathfrak{G}$ be the set of all the subgroups $H \subseteq G$ and let $\mathfrak{F}$ be the set of all the fields F such that $K \supseteq F \supseteq k$.

For each $H \in \mathfrak{G}$ let

$$\mathsf{I}(H) := \{\alpha \in K : \sigma(\alpha) = \alpha, \text{ for all } \sigma \in H\}$$

which is an element of $\mathfrak{F}$.

---

[2] A group G of permutations over a set C is called *transitive* iff there is an element $\alpha \in C$ such that

$$\text{for all } \beta \in C, \text{ there exists } \phi \in G : \beta = \phi(\alpha).$$

14.2 Galois Correspondence

For each $F \in \mathfrak{F}$ let

$$\mathsf{A}(F) := \{\sigma \in G : \sigma(\alpha) = \alpha, \text{ for all } \alpha \in F\}$$

which is an element of $\mathfrak{G}$.

Lemma 14.2.1. *Under the above notation*

(1) for all $H_1, H_2 \in \mathfrak{G}$, $H_1 \subset H_2 \implies \mathsf{I}(H_1) \supset \mathsf{I}(H_2)$;
(2) for all $F_1, F_2 \in \mathfrak{F}$, $F_1 \subset F_2 \implies \mathsf{A}(F_1) \supset \mathsf{A}(F_2)$;
(3) for all $H \in \mathfrak{G}$, $\mathsf{A}(\mathsf{I}(H)) \supset H$;
(4) for all $F \in \mathfrak{F}$, $\mathsf{I}(\mathsf{A}(F)) \supset F$.

♮

Corollary 14.2.2. *Under the above notation*

(1) for all $H \in \mathfrak{G}$, $\mathsf{I}(\mathsf{A}(\mathsf{I}(H))) = \mathsf{I}(H)$;
(2) for all $F \in \mathfrak{F}$, $\mathsf{A}(\mathsf{I}(\mathsf{A}(F))) = \mathsf{A}(F)$.

Proof

(1) If $H \in \mathfrak{G}$, by Lemma 14.2.1.(4) applied to $\mathsf{I}(H)$ we deduce

$$\mathsf{I}(\mathsf{A}(\mathsf{I}(H))) \supset \mathsf{I}(H).$$

By Lemma 14.2.1.(3), we deduce $\mathsf{A}(\mathsf{I}(H)) \supset H$ and, via Lemma 14.2.1.(1),

$$\mathsf{I}(\mathsf{A}(\mathsf{I}(H))) \subset \mathsf{I}(H).$$

(2) By the dual proof of (1).

♮

Lemma 14.2.3. *Under the above notation, $\mathsf{I}(\mathsf{A}(F)) = F$, for each $F \in \mathfrak{F}$.*

Proof In fact, since $K \supset k$ is a separable normal extension, then $K \supset F$ is a separable normal extension too. As a consequence by Theorem 14.1.1

$$F = \{\alpha \in K : \alpha = \sigma(\alpha), \text{ for all } \sigma \in \mathsf{A}(F)\} = \mathsf{I}(\mathsf{A}(F)).$$

♮

Remark 14.2.4. In the proof we use the obvious fact that if $K \supset k$ is a Galois extension and F is such that $K \supset F \supset k$, then $K \supset F$ is also a Galois extension.

The dual result of Lemma 14.2.3 for subgroups $H \in \mathfrak{G}$ requires a more difficult argument:

Lemma 14.2.5. *Under the above notation, let $H \in \mathfrak{G}$ and let $F := \mathsf{I}(H)$. Then $[K : F] \leq \#H =: n$.*

Proof Let $H = \{\sigma_1, \ldots, \sigma_n\}$ and let $\alpha_1, \ldots, \alpha_{n+1} \in K$. We need to show that they are linearly dependent over F; we can, of course, assume that none of them is zero.

Let us consider the system of linear equations

$$\sum_{i=1}^{n+1} \sigma_j(\alpha_i) x_i = 0, \quad j = 1, \ldots, n, \tag{14.1}$$

which has non-null solutions $(x_1, \ldots, x_{n+1})$ in K; among them let us choose one which has a minimal number r of non-zero elements; since we can re-order the elements of H, the equations are homogeneous and the identity is one of the elements of H, we can assume wlog that:

$x_i \neq 0$, $i \leq r$;
$x_i = 0$, $r < i \leq n+1$;
$x_1 = 1$;
$\sum_{i=1}^{n+1} \alpha_i x_i = \sum_{i=1}^{r} \alpha_i x_i = 0$;
$r > 1$ – since otherwise $\alpha_1 = 0$, contrary to our assumption.

Moreover, for each $\sigma \in H$, $(\sigma(x_1), \ldots, \sigma(x_{n+1}))$ is another solution of Equation 14.1, since

$$\sum_{i=1}^{n+1} \sigma \sigma_j(\alpha_i) \sigma(x_i) = \sigma \left(\sum_{i=1}^{n+1} \sigma_j(\alpha_i) x_i \right) = 0, \quad j = 1, \ldots, n,$$

and $H = \{\sigma \sigma_1, \ldots, \sigma \sigma_n\}$.
Then we have

$$0 = \sum_{i=1}^{n+1} \sigma_j(\alpha_i) x_i - \sum_{i=1}^{n+1} \sigma_j(\alpha_i) \sigma(x_i) = \sum_{i=1}^{n+1} \sigma_j(\alpha_i)(x_i - \sigma(x_i)), \quad j = 1, \ldots, n;$$

therefore, since $x_1 = 1 = \sigma(1)$, we have that

$x_1 - \sigma(x_1) = 0$,
$(x_1 - \sigma(x_1), \ldots, x_{n+1} - \sigma(x_{n+1}))$ is a solution of Equation 14.1, which has less than r non-zero elements since $x_i - \sigma(x_i) = 0$ if $i > r$ and $i = 1$.

14.2 Galois Correspondence

As a consequence we can deduce

$$x_i - \sigma(x_i) = 0, \text{ for all } i, \text{ for all } \sigma \in H,$$

i.e. $x_i \in F$, for all i, and the relation $\sum_{i=1}^{n+1} \alpha_i x_i = 0$, is a linear dependence of $\alpha_1, \ldots, \alpha_{n+1}$ over F. ♮

Corollary 14.2.6. *Under the notation above, for all $H \in \mathfrak{G}$,*

$\mathsf{A}(\mathsf{I}(H)) = H$;
$[K : \mathsf{I}(H)] = \#H$.

Proof Let us denote $F := \mathsf{I}(H)$. Since $K \supset F$ is a Galois extension, we have

$$\#\mathsf{A}(\mathsf{I}(H)) = \#\mathsf{A}(F) = [K : F] \leq \#H,$$

the last inequality following from Lemma 14.2.5.
Since, by Lemma 14.2.1.3, $\mathsf{A}(\mathsf{I}(H)) \supset H$, the claim follows. ♮

Theorem 14.2.7 (Fundamental Theorem of Galois Theory).
Let $K \supset k$ be a Galois extension and let G be its Galois group. Let $\mathfrak{G}$ be the set of all the subgroups $H \subseteq G$ and let $\mathfrak{F}$ be the set of all the fields F such that $K \supseteq F \supseteq k$.
For each $H \in \mathfrak{G}$ let

$$\mathsf{I}(H) := \{\alpha \in K : \sigma(\alpha) = \alpha, \text{ for all } \sigma \in H\}.$$

For each $F \in \mathfrak{F}$ let

$$\mathsf{A}(F) := \{\sigma \in G : \sigma(\alpha) = \alpha, \text{ for all } \alpha \in F\}.$$

Then:

(1) the maps $\mathsf{I} : \mathfrak{G} \mapsto \mathfrak{F}, \mathsf{A} : \mathfrak{F} \mapsto \mathfrak{G}$ are inverses and are bijections between $\mathfrak{G}$ and $\mathfrak{F}$;
(2) and they are order inverting with respect to inclusion.
(3) Let $F \in \mathfrak{F}, H \in \mathfrak{G}$ be such that $\mathsf{A}(F) = H, \mathsf{I}(H) = F$. Then $[K : F] = \#H$ and $[F : k]$ is the index of H in G, i.e. the number of its cosets in G.

Proof

(1) Follows from Lemma 14.2.3 and Corollary 14.2.6.
(2) Follows from Lemma 14.2.1.1 and 2.

(3) $[K : F] = \#H$ follows from Corollary 14.2.6, and the second statement follows from the facts that the index of H in G is $\frac{\#G}{\#H}$ and that

$$[K : k] = [K : F][F : k].$$

♮

We recall

Definition 14.2.8. *Let G be a group and $H \subset G$ be a subgroup. Then H is called* normal *or* invariant *iff*

$$\text{for all } g \in G : gH := \{gh : h \in H\} = \{hg : h \in H\} =: Hg$$

or, equivalently,

$$\text{for all } g \in G, \text{ for all } h \in H : ghg^{-1} \in H,$$

in which case

$$G/H = \{gH : g \in G\}$$

is a group, the factor group *of G with respect to H.*

Lemma 14.2.9. *Let $K \supset k$ be a Galois extension and let G be its Galois group. Let $F \in \mathfrak{F}$ and let $H := \mathsf{A}(F) \in \mathfrak{G}$. Let $\sigma \in G$. Then*

$$\mathsf{A}(\sigma(F)) = \sigma H \sigma^{-1} = \{\sigma \tau \sigma^{-1} : \tau \in H\}.$$

Proof Let $F' := \sigma(F)$, $H' := \mathsf{A}(F')$, and let $\tau \in H$; then $\sigma \tau \sigma^{-1} \in H'$, since, for all $\alpha' \in F'$, denoting $\alpha := \sigma^{-1}(\alpha') \in F$, we have

$$\sigma \tau \sigma^{-1}(\alpha') = \sigma \tau(\alpha) = \sigma(\alpha) = \alpha';$$

therefore $\sigma H \sigma^{-1} \subset H'$ and, by symmetry, $\sigma H \sigma^{-1} = H'$.

♮

Theorem 14.2.10. *Let $K \supset k$ be a Galois extension and let G be its Galois group. Let $F \in \mathfrak{F}$ and let $H := \mathsf{A}(F) \in \mathfrak{G}$.*

Then F is a Galois extension of k iff H is a normal subgroup of G. In such a case $G(F/k) = G/H$.

Proof Assume H is normal; to prove that F is normal, we need to prove, for each $x \in F$ and for each $y \in K$ which is conjugate of x over k, that $y \in F$: let $\sigma \in G$ be such that $y = \sigma(x)$; since $H = \sigma^{-1} H \sigma$, for each $\tau \in H$ we have $x = \sigma^{-1} \tau \sigma(x) = \sigma^{-1} \tau(y)$, $y = \sigma(x) = \sigma \sigma^{-1} \tau(y) = \tau(y)$; therefore $y \in F$.

14.3 Solvability by Radicals

Moreover each k-automorphism of K induces, by restriction, a k-automorphism of F. Such a restriction is then a map $\pi : G \mapsto G(F/k)$ whose kernel is the set of the elements $\sigma \in G$ whose restriction to F is the identity, i.e. H; moreover, since K is Galois over F, each k-isomorphism of F extends to a k-isomorphism[3] of K, so that π is surjective; therefore $G/H = G(F/k)$. Conversely, if F is Galois, then, for all $\sigma \in G, \sigma(F) = F$ by Corollary 10.3.7, so that

$$H = \mathsf{A}(F) = \mathsf{A}(\sigma(F)) = \sigma H \sigma^{-1}.$$

♄

Example 14.2.11. As an example let us continue the computation developed in Examples 5.2.5, 5.2.6, 5.5.1, 10.3.8, 10.3.11 and 10.4.2, using the same notation as Example 10.4.2.

We see that $G(K/k) = \mathsf{S}_3$ which has 4 non-trivial subgroups:

$$\begin{aligned}
H_1 &:= \{\Phi_{123}, \Phi_{132}\} &= \mathsf{A}(\mathbb{Q}(\alpha_1)); \\
H_2 &:= \{\Phi_{123}, \Phi_{321}\} &= \mathsf{A}(\mathbb{Q}(\alpha_2)); \\
H_3 &:= \{\Phi_{123}, \Phi_{213}\} &= \mathsf{A}(\mathbb{Q}(\alpha_3)); \\
\mathsf{A}_3 &:= \{\Phi_{123}, \Phi_{231}, \Phi_{312}\} &= \mathsf{A}(\mathbb{Q}(\delta)).
\end{aligned}$$

It is easy to verify that H_i are non-normal, that $H_i = \mathsf{A}(\mathbb{Q}(\alpha_i))$ and the validity of Lemma 14.2.9; in order to exhibit the field corresponding to the subgroup A_3, the alternating group over 3 elements, thereby proving that $\mathsf{A}_3 = \mathsf{A}(\mathbb{Q}(\delta))$, let us note that $[\mathsf{I}(\mathsf{A}_3) : \mathbb{Q}] = 2$ and that

$$\Delta = 6\delta = (\alpha_2 - \alpha_1)(\alpha_3 - \alpha_1)(\alpha_3 - \alpha_2)$$

is such that $\sigma(\Delta) = \begin{cases} \Delta & \text{if } \sigma \in \mathsf{A}_3 \\ -\Delta & \text{if } \sigma \notin \mathsf{A}_3 \end{cases}$ while

$$\Delta^2 = 36\delta^2 = -108 = \mathrm{Disc}(f) \in \mathbb{Q},$$

whence the claim.

14.3 Solvability by Radicals

Throughout this section, I will restrict myself to fields of characteristic 0 to avoid the complications of inseparability and some difficulties with roots of unity.

---

[3] In fact K is a splitting field over F of a polynomial $f(X) \in k[X]$; therefore for any k-isomorphism Φ of F, K is a splitting field of $\Phi(f)$ over F and, by Proposition 5.5.4, Φ extends to a k-isomorphism of K.

Definition 14.3.1. *A field extension* $K \supset k$ *is called a* radical extension *or* root tower *if there is a tower*

$$k = K_0 \subseteq K_1 \subseteq \cdots \subseteq K_{r-1} \subseteq K_r = K$$

such that

for all i, there exists $x_i \in K_i, n_i \in \mathbb{N}, n_i \neq 0, : K_i = K_{i-1}[x_i],$
$x_i^{n_i} := y_i \in K_{i-1}.$

An algebraic number α over k is solvable by radicals *iff there is a root tower $K \supset k$ such that $\alpha \in K$.*

A polynomial $f(X) \in k[X]$ is solvable by radicals *iff its splitting field can be embedded into a radical extension.*

Remark 14.3.2.

(1) The notion of solvability by radicals is nothing more than a formulation in Kronecker's language of the concept of 'solvability' of the pre-Abel–Ruffini culture, when solving meant computing the roots by applying the five operations to the coefficients of the equations.
(2) In the notion of root tower we can (and we need to!) assume wlog that k contains all the n_ith primitive roots of unity for all i: we can in fact perform a preliminary extension of k adding an nth primitive root of unity, where $n = \mathrm{lcm}_i(n_i)$.
(3) In the notion of root towers we can wlog assume that each n_i is prime, since an n_ith root of y_i can be obtained by a sequence of extractions of prime roots.
(4) On the basis of the definition of root tower, it is natural to ask how to solve à la Kronecker the polynomial $X_i^{n_i} - y_i \in K_{i-1}[X_i]$; the answer is in Lemma 14.3.5.
(5) The notions of solvability by radicals for an algebraic number α and for its minimal polynomial $f(X)$ coincide as we will show via the next lemma.

Lemma 14.3.3. *Any radical extension $K \supset k$ is contained in a finite normal radical extension $L \supset k$.*

Proof The proof is by induction on the length r of the root tower. If $r = 1$ and $\xi \in k$ denotes an n_1th primitive root of unity, the conjugate roots of $X^{n_1} - y_1$ are $x_1 \xi^j$, $1 \leq j \leq n_1$, all of them being in $k[x_1, \xi] = K_1[\xi]$ which is therefore a radical extension.

Then let us assume, by induction, that, given a root tower

$$k = K_0 \subseteq K_1 \subseteq \cdots \subseteq K_{r-1} \subseteq K_r = K,$$

14.3 Solvability by Radicals

we have a finite normal extension $L^{(r-1)}$ such that

$K_{r-1} \subseteq L^{(r-1)}$,
$L^{(r-1)}$ has a root tower

$$k = L_0 \subseteq L_1 \subseteq \cdots \subseteq L_{s-1} \subseteq L_s = L^{(r-1)},$$

and let us show that we can produce a finite normal extension $L^{(r)}$ which satisfies $K_r \subseteq L^{(r)}$, and is a rational extension $L^{(r)} \supseteq L^{(r-1)}$, so that there is a root tower

$$L^{(r-1)} = M_0 \subseteq M_1 \subseteq \cdots \subseteq M_{t-1} \subseteq M_t = L^{(r)};$$

as a consequence we would then obtain a root tower

$$k = L_0 \subseteq L_1 \subseteq \cdots \subseteq L_{s-1} \subseteq L^{(r-1)} \subseteq M_1 \subseteq \cdots \subseteq M_{t-1} \subseteq M_t = L^{(r)}.$$

Since $L^{(r-1)}$ is normal, we can consider all the conjugates of $z := y_r \in K_{r-1}$ over k, which we will denote by $z =: z_1, \ldots z_m$ and which are in $L^{(r-1)}$; for each of them let us consider the polynomial $g_j(X) := X^{n_r} - z_j$. Let

$$G(X) := \prod_{j=1}^{m} g_j(X) \in L^{(r-1)}[X]$$

and let $L^{(r)}$ be the splitting field of $G(X)$ over $L^{(r-1)}$. Let us denote the roots of G_j by $w_i \in L^{(r)}$, $1 \le i \le mn_r := t$, so that

for all i, there exists j : $w_i^{n_r} = z_j$.

Let us note that:

there exists j : $w_j = x_r$, so that $K_r \subseteq L^{(r)}$;
by definition, G is invariant under each k-automorphism of $L^{(r-1)}$ so that $G(X) \in k[X]$;
since $L^{(r-1)} \supset k$ is a normal extension, there is $f(X) \in k[X]$ of which $L^{(r-1)}$ is the splitting fields; therefore $L^{(r)}$ is the splitting field of $fG \in k[X]$ so that $L^{(r)} \supset k$ is a normal extension;
if we recursively define $M_i := M_{i-1}[w_i]$ since, there exists j such that

$$w_i^{n_r} = z_j \in L^{(r-1)} \subseteq M_{i-1},$$

then

$$L^{(r-1)} = M_0 \subseteq M_1 \subseteq \cdots \subseteq M_{t-1} \subseteq M_t = L^{(r)}$$

is a root tower.

This proves our claim. ♃

Corollary 14.3.4. *Let α be an algebraic number over k and let $f(X) \in k[X]$ be its minimal polynomial over k; then f is solvable by radicals iff α is also.*

Proof The root tower $K \supset k$ such that $\alpha \in K$ is embedded into a finite normal radical extension $L \supset k$; therefore all the conjugates of $\alpha \in K$ are in L, so that the splitting field of f is contained in it. ♄

Lemma 14.3.5. *Let n be a prime, K be a field which contains an nth primitive root of unity ξ, and let $f(X) := X^n - y \in K[X]$; then either*

f is irreducible over K, or
there is $x \in K : f(x) = 0$.

Proof In its splitting field $L \supset K$, f has a factorization $f = \prod_{i=0}^{n-1}(X - z\xi^i)$ for a suitable $z \in L$ such that $z^n = y$.
Therefore if f is reducible as $f = gh$ over K, then g must be a product of certain factors $X - z\xi^i$ and the constant term $c \in K$ of it must have the form $c = (-1)^d \xi^\delta z^d$, where $d = \deg(g) < n$ and δ is a suitable integer; therefore, we have $(-1)^{dn} c^n = z^{dn} = y^d$.
Since n is prime and $d < n$, we have $\gcd(d, n) = 1$ and there are s, t such that $ds + tn = 1$ so that

$$y = y^{ds} y^{tn} = (-1)^{dns} c^{ns} y^{tn} = \left((-1)^{ds} c^s y^t\right)^n,$$

i.e. y is a power of an element $x = (-1)^{ds} c^s y^t \in K$ and f has the factorization $f(X) = (X - x) \sum_{i=1}^{n} X^{n-i} x^i$ in $k[X]$. ♄

In our analysis of root towers, for any field extensions $K_i = K_{i-1}[x_i]$ – where $x_i^{n_i} := y_i \in K_{i-1}$ and n_i is prime –, we need to evaluate the Galois group $G(K_i / K_{i-1})$ via the following

Lemma 14.3.6. *Let $K \supset k$ be a field extension $K = k[\alpha] = k[X]/f(X)$ where $f(X) = X^n - y \in k[X]$ and n is prime. If k contains an nth root ξ of unity, then*

K is the splitting field of f,
$K \supset k$ is a Galois extension,
$G(K/k)$ is the cyclic group of order n, i.e. the additivity group $(\mathbb{Z}_n, +)$.

14.3 Solvability by Radicals

Proof In fact, the roots of f are $\alpha, \xi\alpha, \ldots, \xi^j\alpha, \ldots, \xi^{n-1}\alpha$ so that K is the splitting field of f.

Denoting $\phi_j : K \mapsto K$, $0 \le j \le n-1$, to be the k-automorphisms defined by $\phi_j(\alpha) = \xi^j \alpha$, the Galois group $G(K/k)$ is the set $\{\phi_j : 0 \le j \le n-1\}$, which is isomorphic to the group $(\mathbb{Z}_n, +)$ via the correspondence $j \leftrightarrow \phi_j$. ♄

The inverse also holds:

Lemma 14.3.7. *Let $K \supset k$ be a Galois field extension such that $G(K/k) = (\mathbb{Z}_n, +)$, where n is prime. If k contains an nth primitive root ξ of unity, then there is $x \in K$ such that $y := x^n \in k$ and $K = k[x] = k[X]/f(X)$ where $f(X) = X^n - y \in k[X]$.*

Proof Let σ be the generator of $G(K/k)$ and, for each $\alpha \in K$, let us consider the *Lagrange resolvent*

$$x := (\xi, \alpha) := \alpha + \xi\sigma(\alpha) + \xi^2\sigma^2(\alpha) + \cdots + \xi^{n-1}\sigma^{n-1}(\alpha).$$

By Theorem 10.3.16, we know that the elements of $G(K/k)$ are linearly independent, so that there is α such that $(\xi, \alpha) \ne 0$. For such α

$$\begin{aligned}\sigma(x) &= \sigma(\alpha) + \xi\sigma^2(\alpha) + \xi^2\sigma^3(\alpha) + \cdots + \xi^{n-1}\sigma^n(\alpha) \\ &= \xi^{-1}\left(\xi\sigma(\alpha) + \xi^2\sigma^2(\alpha) + \xi^3\sigma^3(\alpha) + \cdots + \alpha\right) \\ &= \xi^{-1}x,\end{aligned}$$

from which we deduce $\sigma^j(x) = \xi^{-j}x$, for all j, so that x is invariant only under the identity.

Setting $y := x^n$ we have

$$\sigma^j(y) = \left(\sigma^j(x)\right)^n = (\xi^{-j}x)^n = x^n = y, \text{ for all } j$$

so that $y \in k$. ♄

An extension $K \supset k$ satisfying the condition above is called a *Kummer extension*

Remark 14.3.8. If $K \subset k$ is a Kummer extension, we can prove the existence of $\alpha \in K : (\xi, \alpha) \ne 0$, using the discriminant: we know that there is a primitive element $\chi \in K$ such that $K = k[\chi]$, whose minimal polynomial is $f(X) \in k[X]$ and whose conjugates are $\chi_j := \sigma^j(\chi)$, $1 \le i \le n$, where σ is the generator of $G(K/k)$. If we consider the Lagrange resolvants

$$(\xi, \chi^i) = \sum_{j=1}^{n} \xi^j \chi_j^i, \quad 0 \le i \le n-1,$$

by Theorem 10.6.5 we deduce that not all the resolvants (ξ, χ^i) are zero since the discriminant $\operatorname{Disc}(f) \ne 0$. Therefore $(\xi, \chi^i) = 0$, for all i, would imply the contradiction $\xi^j = 0$, for all j.

Remark 14.3.9. Let us also note that, with the above assumptions and notations, if $\alpha \in K$ is such that $(\xi, \alpha) \ne 0$, then the following formula holds:

$$\sum_{i=0}^{n-1} (\xi^i, \alpha) = n\alpha;$$

in fact, since each ξ^j, $1 \le j \le n-1$, is a root of $\sum_{i=0}^{n-1} X^i$,

$$\begin{aligned}
\sum_{i=0}^{n-1} (\xi^i, \alpha) &= \sum_{i=0}^{n-1} \sum_{j=1}^{n} \xi^{ij} \sigma^j(\alpha) \\
&= n\alpha + \sum_{j=1}^{n-1} \sigma^j(\alpha) \sum_{i=0}^{n-1} (\xi^j)^i \\
&= n\alpha.
\end{aligned}$$

Since our result requires that k contains roots of unity, and in our root towers we have field extensions $K_i = K_{i-1}[x_i]$, where x_i is a primitive root of unity, let us evaluate the Galois group $G(K_{i-1}/K_i)$ for this case.

Let us fix a prime $n \ne 2$ and let us recall that

the nth primitive root ξ of unity is a root of the nth cyclotomic polynomial Φ_n, which satisfies
$\Phi_n = \sum_{i=0}^{n-1} X^i$, since n is prime,
and is irreducible if $K = \mathbb{Q}$ (Proposition 7.6.10).

On the basis of this, let K be a field such that $\operatorname{char}(K) = 0$, and which does not contain an nth primitive root ξ of unity. Let $\Phi_n(X) \in K[X]$ be the nth cyclotomic polynomial, F be the splitting field of $\Phi_n(X)$, $\xi \in F$ any nth primitive root ξ of unity, $K^{(n)} := K[\xi]$.

Note that, when $K = \mathbb{Q}$, we have $\mathbb{Q}^{(n)} := \mathbb{Q}[\xi] = \mathbb{Q}[X]/\Phi_n(X)$.

14.3 Solvability by Radicals

Lemma 14.3.10. *With the above notation:*

(1) $K^{(n)}$ is a Galois extension.
(2) $G(K^{(n)}/K)$ is a cyclic group.
(3) If $K = \mathbb{Q}$, then $G(\mathbb{Q}^{(n)}/\mathbb{Q})$ is the cyclic group of order $n-1$, i.e. the multiplicative group $(\mathbb{Z}_n \setminus \{0\}, \cdot)$.

Proof The roots of $\Phi_n(X)$ are a subset R of $\{\xi^j, 1 \leq j < n\} \subset K^{(n)}$ which is therefore a Galois extension. Moreover, when $K = \mathbb{Q}$, that set consists of all the roots of $\Phi_n(X)$.
For each element $\xi^j \in$ R, let $\psi_j : K^{(n)} \mapsto K^{(n)}$ be the K-automorphism defined by $\psi_j(\xi) = \xi^j$.
Then the Galois group $G(K^{(n)}/K)$ is the set $\{\psi_j : \xi^j \in$ R$\}$, which is isomorphic to a subgroup of the group $(\mathbb{Z}_n \setminus \{0\}, \cdot)$ via the immersion $\psi_j \mapsto j$, since

$$\psi_j(\psi_k(\xi)) = \psi_j(\xi^k) = \xi^{kj} = \psi_{kj}(\xi),$$
ψ_1 is the identity.

Since $(\mathbb{Z}_n \setminus \{0\}, \cdot)$ is cyclic, so is any subgroup of it and in particular $G(K^{(n)}/K)$.
When $K = \mathbb{Q}$, we have $\#G(\mathbb{Q}^{(n)}/\mathbb{Q}) = [\mathbb{Q}^{(n)} : \mathbb{Q}] = n - 1$ so that $G(\mathbb{Q}^{(n)}/\mathbb{Q}) \cong (\mathbb{Z}_n \setminus \{0\}, \cdot)$. ♃

Remark 14.3.11. If $K \supsetneq \mathbb{Q}$, we have $K^{(n)} = \mathbb{Q}^{(n)}$ and $\mathfrak{g} := G(\mathbb{Q}^{(n)}/K)$ is a cyclic subgroup of $\mathfrak{G} := G(\mathbb{Q}^{(n)}/\mathbb{Q})$; we have

$$\#\mathfrak{G} = [\mathbb{Q}^{(n)} : \mathbb{Q}] = n - 1, \quad \#\mathfrak{g} = [\mathbb{Q}^{(n)} : K] := f,$$

so that setting $e := \frac{n-1}{f}$ and $\mathfrak{h} := G(K/\mathbb{Q})$, we have

$\#\mathfrak{h} = [K : \mathbb{Q}] = e$;
$\mathfrak{g} = \{\psi_{je} : 1 \leq j \leq f\}$;
if g is a generator of $\mathbb{Z}_n \setminus \{0\}$, ψ_g is a generator of $\mathfrak{G}$ and ψ_{g^e} generates $\mathfrak{g}$;
the periods

$$\eta_i := (f, g^i) := \xi^{g^i} + \xi^{g^{i+e}} + \cdots + \xi^{g^{i+je}} + \cdots + \xi^{g^{i+(f-1)e}}, \quad 0 \leq i < e,$$

are elements in $\mathsf{I}(\mathfrak{g}) = K$ so that $K = \mathbb{Q}[\eta_1]$.

Proposition 14.3.12. *Let $K \supset k$ be a radical extension and let*

$$k = K_0 \subseteq K_1 \subseteq \cdots \subseteq K_{r-1} \subseteq K_r = K$$

be its root tower, where

for all i, there exists $x_i \in K_i, n_i \in \mathbb{N}, n_i \neq 0$:

$$K_i = K_{i-1}[x_i], x_i^{n_i} := y_i \in K_{i-1},$$

for all i, n_i is prime and either

*$y_i \neq 1$ and K_{i-1} contains an n_ith primitive root of unity, or
$y_i = 1$.*

Let $\mathfrak{G}_i := G(K_r/K_i)$, for all i, and $\mathfrak{H}_i = \mathfrak{G}_i/\mathfrak{G}_{i+1}, 0 \leq i < r$. Then there is a chain of subgroups

$$G(K/k) = \mathfrak{G}_0 \supseteq \mathfrak{G}_1 \supseteq \cdots \supseteq \mathfrak{G}_{r-1} \supseteq \mathfrak{G}_r = \{1\}, \tag{14.2}$$

which satisfies the following conditions:

(1) each $\mathfrak{G}_{i+1}$ is invariant in $\mathfrak{G}_i$;
(2) $\mathfrak{H}_i$ is cyclic of prime order.

Proof Equation 14.2 holds from Theorem 14.2.7(2). For each i, K_{i+1} is a Galois extension of K_i and $G(K_{i+1}/K_i)$ is cyclic of order prime, either

by Lemma 14.3.6, if $y_{i+1} \neq 1$ and K_i contains an $n\text{th}_{i+1}$ primitive root of unity, or
by Lemma 14.3.10 if $y_{i+1} = 1$.

This implies that, for each i, $\mathfrak{G}_{i+1}$ is invariant in $\mathfrak{G}_i$ and $\mathfrak{H}_i = G(K_{i+1}/K_i)$ (Theorem 14.2.10), being cyclic. ♄

On the basis of the above result, let us introduce

Definition 14.3.13. *A group $\mathfrak{G}_0$ which has a chain (Equation 14.2) of subgroups satisfying conditions (1) and (2) of the above proposition is called solvable,*

which satisfies

Fact 14.3.14. *Let $\mathfrak{G}$ be solvable and $\mathfrak{g}$ be a subgroup of $\mathfrak{G}$. Then $\mathfrak{g}$ is solvable and, if $\mathfrak{g}$ is invariant, $\mathfrak{G}/\mathfrak{g}$ is also solvable.*

Lemma 14.3.15. *Let $K \supset k$ be the splitting field of $f(X) \in k[X]$ over k and let $K[\xi] \supset K$ be an algebraic extension; then $K[\xi] \supset k[\xi]$ is the splitting field of f over $k[\xi]$ and $G(K[\xi]/k[\xi])$ is isomorphic to a subgroup of $G(K/k)$.*

Proof To each $\sigma \in G(K[\xi]/k[\xi])$ its restriction σ' to K is an element of $G(K/k)$; therefore there is a homomorphism $\phi : G(K[\xi]/k[\xi]) \mapsto G(K/k)$,

$\phi(\sigma) = \sigma'$, which is isomorphic since, if $\phi(\sigma) = \sigma'$ is the identity, then $\sigma(\beta) = \sigma'(\beta) = \beta$ for each root β of f, i.e. σ is the identity. $\boxed{\text{♄}}$

Theorem 14.3.16 (Galois). *A squarefree polynomial $f(X) \in k[X]$ is solvable by radicals over a field k of characteristic 0 iff its Galois group is solvable.*

Proof

$\Rightarrow$ Let $f(X) \in k[X]$ be solvable by radicals and let N be its splitting field which can be embedded into a radical extension $L \supset k$ which, by Lemma 14.3.3, we can assume to be normal. Then L has a root tower

$$k = L_0 \subseteq L_1 \subseteq \cdots \subseteq L_{r-1} \subseteq L_r = L$$

where for all i, there exists $x_i \in L_i$, n_i a prime: such that

$$L_i = L_{i-1}[x_i], x_i^{n_i} := y_i \in L_{i-1}.$$

Let $n := \operatorname{lcm} n_i$ and let $\xi \in k^{(n)}$ be an nth primitive root of unity; we can obtain a root tower

$$k = K_0 \subseteq K_1 \subseteq \cdots \subseteq K_{s-1} \subseteq K_s = k[\xi].$$

This root tower can be extended as

$$\begin{aligned}
k = K_0 &\subseteq K_1 \subseteq \cdots \subseteq K_{s-1} \subseteq K_s = k[\xi] = L_0[\xi] \\
&\subseteq K_{s+1} = K_s[x_1] = L_1[\xi] \subseteq K_{s+2} = K_{s+1}[x_2] = L_2[\xi] \\
&\subseteq \cdots \subseteq K_{s+r-1} = K_{s+r-2}[x_{r-1}] = L_{r-1}[\xi] \\
&\subseteq K_{s+r} = K_{s+r-1}[x_r] = L_r[\xi] = L[\xi].
\end{aligned}$$

By Proposition 14.3.12 we can deduce that $G(L[\xi]/k)$ is solvable. Let us now remark that, by Theorem 14.2.10, $\mathfrak{g} = (L[\xi]/N)$ is a normal subgroup of $\mathfrak{G} = (L[\xi]/k)$ and that $G(N/k) = \mathfrak{G}/\mathfrak{g}$, so that the claim follows from Fact 14.3.14.

$\Leftarrow$ Conversely, let us assume that the Galois group $\mathfrak{G} := G(K/k)$ of $f(X) \in k[X]$ is solvable, where K is the splitting field of f. Let $n = \#\mathfrak{G}$ and ξ be a primitive nth root of unity. Then (Lemma 14.3.15) $K[\xi]$ is the splitting field of f over $k[\xi]$ and $G(K[\xi]/k[\xi])$ is isomorphic to a subgroup of $\mathfrak{G}$ and is solvable by Fact 14.3.14.

Therefore there is a chain of subgroups

$$G(K[\xi]/k[\xi]) = \mathfrak{G}_0 \supseteq \mathfrak{G}_1 \supseteq \cdots \supseteq \mathfrak{G}_{r-1} \supseteq \mathfrak{G}_r = \{1\},$$

where each $\mathfrak{G}_{i+1}$ is invariant in $\mathfrak{G}_i$ and $\mathfrak{G}_i/\mathfrak{G}_{i+1}$ is cyclic of prime order n_i.
Setting $K_i := \mathsf{I}(\mathfrak{G}_i)$ we obtain the chain

$$k[\xi] = K_0 \subseteq K_1 \subseteq \cdots \subseteq K_{r-1} \subseteq K_r = K[\xi]; \qquad (14.3)$$

since each K_i contains the primitive nth root ξ of unity, it contains also the primitive n_ith root ξ^{n/n_i} of unity; since each $\mathfrak{G}_{i+1}$ is normal, K_{i+1} is normal over K_i and $G(K_{i+1}/K_i) = \mathfrak{G}_i/\mathfrak{G}_{i+1}$ which is cyclic of prime order n_i, i.e. K_{i+1} is a Kummer extension and Equation 14.3 is a root tower. ♄

Remark 14.3.17. When $\text{char}(k) = p \neq 0$ and f is separable, the Galois Theorem holds, provided the characteristic does not appear among

the values $n_i := [K_i : K_{i-1}]$ in the root tower of radical extension $L \supset k$ in which the splitting field of f is embedded,
the indexes of $\mathfrak{G}_{i+1}$ in $\mathfrak{G}_i$.

14.4 Abel–Ruffini Theorem

Let us fix a field k of characteristic 0 and let us consider the generic equation of degree n

$$f(X) = X^n + \sum_{i=1}^{n}(-1)^i a_i X^{n-i} = \prod_{i=1}^{n}(X - \alpha_i)$$

in the field

$$\mathsf{k} := k(a_1, \ldots, a_n) \subset k(\alpha_1, \ldots, \alpha_n) =: \mathsf{K}$$

of which K is the splitting field.

Clearly the Galois group $G(\mathsf{K}/\mathsf{k})$ is the symmetric group S_n by Corollary 6.2.6.

Let us recall the notion of the *alternating group* $\mathsf{A}_n \subset \mathsf{S}_n$, which can be given in terms of the discriminant:

Definition 14.4.1. *Let*

$$\Delta := \prod_{i>j}(\alpha_i - \alpha_j)$$

14.4 Abel–Ruffini Theorem

so that $\Delta^2 = \text{Disc}(f)$. Then

$$A_n := \{\sigma \in S_n : \sigma(\Delta) = \Delta\},$$

from which follow

$$\text{for all } \sigma \in S_n : \sigma(\Delta) = \begin{cases} \Delta & \text{if } \sigma \in A_n \\ -\Delta & \text{if } \sigma \notin A_n \end{cases},$$

and its principal property:

Fact 14.4.2. *A_n is a normal subgroup of S_n, whose index is 2 and which contains, if $n > 4$, no normal subgroup except the identity and itself.*

As a consequence of this we directly obtain

Corollary 14.4.3 (Abel–Ruffini). *The generic equation of degree $n \geq 5$ is not solvable by radicals.*

Proof If $n \geq 5$, A_n is not solvable and so, by Fact 14.3.14, neither is S_n. ♩

The generic equation of degree $n \leq 4$ is indeed solvable by radicals, since we have the chains

$$S_3 \supset A_3 \supset \{1\},$$
$$S_4 \supset A_4 \supset V_4 \supset V_2 \supset \{1\},$$

where V_4 is the *Viergruppe*,

$$V_4 = \{(1), (12)(34), (13)(24), (14)(23)\},$$

and V_2 is any one of its subgroups of order 2.

Now let us discuss the solution of the generic equation of degree $n = 3$,

$$f(X) = X^3 - a_1 X^2 + a_2 X - a_3 = \prod_{i=1}^{3}(X - \alpha_i);$$

via the linear change of coordinates $X \mapsto X + \frac{a_1}{3}$, we can reduce it to the form

$$f(X) = X^3 + pX + q = \prod_{i=1}^{3}(X - \beta_i),$$

where $p = a_2 - \frac{1}{3}a_1^2$, $q = a_3 - \frac{1}{3}a_1 a_2 + \frac{2}{27}a_1^3$ and $\beta_i = \alpha_i - \frac{a_1}{3}$.

The polynomial f is solvable since S_3 is, having the chain

$$S_3 \supset A_3 \supset \{1\}$$

from which we obtain the root tower

$$k \subset k[\Delta] \subset K$$

and we can use the Galois Theorem and its proof to 'solve' it.

Denoting $\xi := \frac{-1+\sqrt{-3}}{2}$ the 3rd primitive root of unity, so that $\xi^2 := \frac{-1-\sqrt{-3}}{2}$, let us recall that A_3 is generated by the permutation (123); therefore the Lagrange resultants are

$$\begin{aligned}
(\xi, \beta_1) &= \beta_1 + \xi\beta_2 + \xi^2\beta_3, \\
(\xi^2, \beta_1) &= \beta_1 + \xi^2\beta_2 + \xi\beta_3, \\
(1, \beta_1) &= \beta_1 + \beta_2 + \beta_3 = 0;
\end{aligned}$$

adding them together given

$$3\beta_1 = (\xi, \beta_1) + (\xi^2, \beta_1).$$

First, using Newton's results, we can compute[4]

$$\begin{aligned}
\Delta^2 &= -4p^3 - 27q^2, \\
\sum_i \beta_i^3 &= -3q, \\
\sum_{ij} \beta_i^2 \beta_j &= 3q, \\
(\xi, \beta_1)^3 &= \sum_i \beta_i^3 - \frac{3}{2}\sum_{ij}\beta_i^2\beta_j + 6\beta_1\beta_2\beta_3 + \frac{3\sqrt{-3}}{2}\Delta \\
&= \frac{-27}{2}q + \frac{3\sqrt{-3}}{2}\Delta, \\
(\xi^2, \beta_1)^3 &= \frac{-27}{2}q - \frac{3\sqrt{-3}}{2}\Delta,
\end{aligned}$$

obtaining

$$K = k[\Delta, y] = k[\Delta][X]/(X^3 - x),$$

where $y := (\xi, \beta_1)$ and $x := -\frac{27}{2}q + \frac{3\sqrt{-3}}{2}\Delta$; therefore

$$(\xi, \beta_1) = \sqrt[3]{\frac{-27}{2}q + \frac{3\sqrt{-3}}{2}\Delta}$$

---

[4] Using standard shorthand notation $\sum_{i_1 i_2 \ldots i_r} t_{i_1}^{d_1} \cdots t_{i_r}^{d_r}$ denotes the expression

$$\sum_{i_1=1}^{n} \sum_{i_2=1}^{n} \cdots \sum_{i_r=1}^{n} t_{i_1}^{d_1} \cdots t_{i_r}^{d_r}.$$

$$(\xi^2, \beta_1) = \sqrt[3]{\frac{-27}{2}q - \frac{3\sqrt{-3}}{2}\Delta}$$

$$\beta_1 = \sqrt[3]{\frac{-q}{2} + \sqrt{\left(\frac{q}{2}\right)^2 + \left(\frac{p}{3}\right)^3}} + \sqrt[3]{\frac{-q}{2} - \sqrt{\left(\frac{q}{2}\right)^2 + \left(\frac{p}{3}\right)^3}},$$

i.e. Tartaglia's formula.

Note that the other two roots can be similarly obtained using the equalities:

$$3\beta_2 = \xi^2(\xi, \beta_1) + \xi(\xi^2, \beta_1),$$
$$3\beta_3 = \xi(\xi, \beta_1) + \xi^2(\xi^2, \beta_1).$$

Ferrari's formula for the biquadratic generic equation could be described with a similar approach.

On the basis of the Abel–Ruffini Theorem, it is natural to ask if there is a polynomial $f(X) \in \mathbb{Z}[X]$ of degree 5 whose Galois group is S_5 and so is not 'solvable'. To present such a polynomial we need to recall

Fact 14.4.4. *If $H \subseteq S_n$ is a transitive group which contains a transposition, then $H = S_n$.*

and to prove that

Corollary 14.4.5. *Let $k \subset \mathbb{R}$ be a field and let $f(X) \in k[X]$ be a polynomial such that*

f is irreducible,
$\deg(f) := n$ is prime,
f has exactly two roots in $\mathbb{C} \setminus \mathbb{R}$;

then the Galois group of f is S_n so that f is not solvable by radicals.

Proof Let $H \subset S_n$ be the Galois group of f and let K be its splitting field. Since f is irreducible, H is transitive.
Let $\alpha_1, \ldots, \alpha_n \in \mathbb{C}$ be the roots of f and let us assume that wlog $\alpha_1, \alpha_2 \notin \mathbb{R}$; these two roots are conjugate so that $\overline{\alpha_1} = \alpha_2$ while $\overline{\alpha_i} = \alpha_i$, for all $i \geq 3$; therefore the restriction of the complex conjugation to K is a k-automorphism of it which corresponds to the transposition (12). The claim follows from the above fact. $\boxed{\text{ħ}}$

Remark 14.4.6. On the basis of the above result, we can explicitly present a series of polynomials $f(X) \in \mathbb{Z}[X]$ which are not 'solvable'; for any prime p

let us consider the polynomial $f_p(X) := X^5 + pX + p$, which is not 'solvable' since it satisfies the above condition, because

- it is irreducible via the Eisenstein criterion[5],
- it has three real roots, since the Sturm sequence (cf. Section 13) is $X^5 + pX + p, 5X^4 + p, 4pX + 5p, 1$.

14.5 Constructions with Ruler and Compass

Throughout this section, by 'geometric object' we mean any point, line and circle in the plane.

Given a set of geometric objects let us consider the following operations:

(1) construct the line passing through two given points,
(2) construct the circle whose centre is a given point and which passes through another given point,
(3) determine the intersection of two given lines,
(4) determine the intersections of a given line with a given circle,
(5) determine the intersections of two given circles.

Definition 14.5.1. *Given a set $\mathcal{O}$ of geometric objects, a geometric object $\mathbf{O}$ is said to be* constructable by ruler and compass *in terms of $\mathcal{O}$, if there is a finite sequence $\mathbf{O}_1, \mathbf{O}_2, \ldots, \mathbf{O}_n = \mathbf{O}$ such that for all i either*

$\mathbf{O}_i \in \mathcal{O}$, *or*
there are $i_1, i_2 < i$ such that $\mathbf{O}_i$ is obtained from $\mathbf{O}_{i_1}$ and $\mathbf{O}_{i_2}$ by the five constructions listed above.

---

[5] Which asserts that a polynomial $f(X) := \sum_{i=0}^{n} a_i X^i \in D[X]$, D a domain, is irreducible in $D[X]$, if there is a prime $p \in D$ such that

$a_n \not\equiv 0 \pmod{p}$,
$a_i \equiv 0 \pmod{p}$, for all $i < n$,
$a_0 \not\equiv 0 \pmod{p^2}$.

In fact, assuming f has the non-trivial factorization $f = gh$ where $g(X) =: \sum_{i=0}^{r} b_i X^i$, $h(X) =: \sum_{i=0}^{s} c_i X^i$, from $b_0 c_0 = a_0 \equiv 0 \pmod{p}$, and $b_0 c_0 = a_0 \not\equiv 0 \pmod{p^2}$ we deduce that wlog $b_0 \equiv 0 \pmod{p}$ and $c_0 \not\equiv 0 \pmod{p}$; since not all the coefficients of g are multiples of p, since $a_n \not\equiv 0 \pmod{p}$, let i be the least integer such that $b_i \not\equiv 0 \pmod{p}$, so that

$$0 \equiv a_i = \sum_{j=0}^{i} b_j c_{j-i} \equiv b_i c_0 \not\equiv 0 \pmod{p},$$

giving a contradiction.

14.5 Constructions with Ruler and Compass

Example 14.5.2. Given two points O_1 and O_2, the middle point O of the segment joining O_1 and O_2 is constructable by ruler and compass, as follows: let

O_3 be the line passing through O_1 and O_2,
O_4 be the circle whose centre is O_1 and which passes through O_2,
O_5 be the circle whose centre is O_2 and which passes through O_1,
O_6 and O_7 the intersections of O_4 and O_5,
O_8 be the line passing through O_6 and O_7,
O be the intersection of O_3 and O_8.

Remark 14.5.3. Among the constructions by ruler and compass, we can list the following:

(1) given a line O_1 and a point O_2, construct the line O which is perpendicular to O_1 and passing through O_2;
(2) given a line O_1 and a point O_2, construct the line O which is parallel to O_1 and passing through O_2;
(3) given two points O_1 and O_2, construct the points which divide into n parts the segment joining O_1 and O_2;
(4) given two points O_1 and O_2, construct the circle O whose diameter is the segment based on O_1 and O_2;
(5) given three points O_1, O_2 and O_3, construct the circle O which passes through them;
(6) given two points O_1 and O_2, a line O_3 and a point O_4 contained in it, construct a point O over O_3 such that the two segments joining O_1 and O_2 (respectively O_4 and O) have the same length;
(7) given an angle[6] O_1, construct an angle O which is half of it;
(8) given two angles O_1 and O_2, construct an angle O which is the sum of them;
(9) given two lines O_1 and O_2 intersecting in O_3, construct the line O passing through O_3 which is symmetric to O_2 with respect to O_1.

Remark 14.5.4. Given four points O_1, O_2, O_3 and O_4, it is evident that it is possible to construct by ruler and compass

(1) two points O_5, O_6 such that

L_1 denotes the line through O_1 and O_2;
O_5 is in L_1 and the two segments joining O_5 and O_2 (respectively O_3 and O_4) have the same length;

---

[6] We consider an angle O to be given (respectively constructed) if two lines O_1 and O_2 are given (respectively constructed) which intersect at a point O_3.

L_2 denotes the line passing through O_2 which is perpendicular to L_1;
C is the circle whose diameter is the segment based on O_1 and O_5;
O_6 is an intersection between L_2 and C;

(2) two points O_5, O_6 such that

L_1 denotes the line through O_1 and O_2;
L_2 denotes the line passing through O_2 which is perpendicular to L_1;
O_6 is in L_2 and the two segments joining O_6 and O_2 (respectively O_3 and O_4) have the same length;
L_3 is the line perpendicular to L_2 and passing through O_6;
O_5 is the intersection of L_1 and L_3.

Denoting

a the length of the segment joining O_1 and O_2,
b the length of the segment joining O_2 and O_5,
c the length of the segment joining O_2 and O_6,

in both cases we have the relation $a : c = c : b$, so that when

(1) we are given a and b, we obtain $c = \sqrt{ab}$;
(2) we are given a and c, we obtain $b = \frac{c^2}{a}$.

Remark 14.5.5. It is sufficient to choose two points O_1 and O_2 in order to impose a cartesian coordinate system: in fact we only have to construct

the line O_3 through O_1 and O_2,
the line O_4 perpendicular to O_3 and passing through O_1,

and choose the length of the segment joining O_1 and O_2 as unity.

From now on, we will assume that a cartesian coordinate system is fixed in the plain and we say that

Definition 14.5.6. *A complex number $x + iy \in \mathbb{C}$ is called* constructable *if the point $O = (x, y)$ is constructable by ruler and compass starting from O_1 and O_2.*

Proposition 14.5.7. *Let*

$$\mathcal{C} := \{c \in \mathbb{C} : c \text{ is constructable }\} \subset \mathbb{C}.$$

Then

(1) $\mathbb{Z} \subset \mathcal{C}$;
(2) $\mathbb{Q} \subset \mathcal{C}$;
(3) for all $x, y \in \mathbb{R}$, $x + iy \in \mathcal{C}$ is constructable iff both x and y are such;

14.5 Constructions with Ruler and Compass

(4) $\mathcal{C}$ is an additive group;
(5) $\mathcal{C} \cap \mathbb{R}$ is a field;
(6) if α is a positive real constructable number, then $\sqrt{\alpha} \in \mathcal{C}$;
(7) $\mathcal{C}$ is a field;
(8) if $\alpha \in \mathcal{C}$, then $\sqrt{\alpha} \in \mathcal{C}$.

Proof

(1) The integer number 2 is obtained by constructing, via construction (4) of Remark 14.5.3, the circle whose centre is O_2 and which passes through O_1 determining the other intersection. All the integers are obtained by iterating this construction.
(2) In order to represent the rational $\frac{n}{m}$ we only have to construct the point O which divides into n parts the segment joining O_1 and $(m, 0)$, via construction (3) of Remark 14.5.3.
(3) The intersection of the lines parallel to O_3 (respectively O_4) and passing through $(0, y)$ (respectively $(x, 0)$) is (x, y). Conversely $(x, 0)$ (respectively $(0, y)$) is the intersection between O_3 (respectively O_4) and a line perpendicular to it and passing through (x, y).
(4) If $\mathsf{P}_i = (x_i, y_i)$, $i = 1, 2$, are constructable, to obtain the point

$$O := (x_1 + x_2, y_1 + y_2)$$

we only have to perform vector addition using construction (6) of Remark 14.5.3.
(5) Given the constructed real positive numbers α, β, via the construction of Remark 14.5.4 we obtain:

$b := \alpha^2$ setting $a := 1, c := \alpha$;
$b := \beta^2$ setting $a := 1, c := \beta$;
$c := \alpha\beta$ setting $a := \alpha^2, b := \beta^2$;
$b := \alpha^{-1}$ setting $a := \alpha, c := 1$.

(6) Via the construction of Remark 14.5.4 we obtain

$$c := \sqrt{\alpha} \text{ setting } a := \alpha, b := 1.$$

(7) Let $c_i := r_i\big(\cos(\phi_i) + i\sin(\phi_i)\big)$, $i = 1, 2$, be two constructable numbers, so that

$$\begin{aligned} c_1 c_2 &= r_1 r_2 \big(\cos(\phi_1 + \phi_2) + i\sin(\phi_1 + \phi_2)\big), \\ c_1^{-1} &= r_1^{-1} \big(\cos(-\phi_1) + i\sin(-\phi_1)\big) \end{aligned}$$

are constructable since

r_1^{-1} and $r_1 r_2$ are and
the angles $\phi_1 + \phi_2$ and $-\phi_1$ can be constructed via constructions (8) and (9) of Remark 14.5.3.

(8) If $\alpha = r\bigl(\cos(\phi) + i\sin(\phi)\bigr)$ then $\sqrt{\alpha} = \sqrt{r}\bigl(\cos(\frac{\phi}{2}) + i\sin(\frac{\phi}{2})\bigr)$, where $\sqrt{r}$ is constructable and $\frac{\phi}{2}$ can be constructed via construction (7) of Remark 14.5.3.

$\boxed{\text{♄}}$

Remark 14.5.8. If we are given a set of points $\mathsf{P}_i := (x_i, y_i)$, $1 \le i \le n$, clearly the lines and circles which can be constructed using the basic constructions by ruler and compass are defined by polynomials in

$$\mathbb{Q}(x_1, y_1, \ldots, x_n, y_n)[X, Y].$$

In fact

the line passing through (x_i, y_i) and (x_j, y_j) has equation

$$(x_i - x_j)(Y - y_j) - (y_i - y_j)(X - x_j);$$

the circle whose centre is (x_i, y_i) and passing through (x_j, y_j) has equation

$$(X - x_j)^2 + (Y - y_j)^2 - (x_i - x_j)^2 - (y_i - y_j)^2.$$

As a consequence we have:

Theorem 14.5.9. *Given a set $\mathcal{O}$ of points $\mathsf{P}_i := (x_i, y_i), 1 \le i \le n$, and a point $\mathsf{P} := (x, y)$, then the following conditions are equivalent:*

(1) P *is constructable by ruler and compass in terms of $\mathcal{O}$,*
(2) there is a root tower

$$\mathbb{Q}(x_1, y_1, \ldots, x_n, y_n) =: K_0 \subset K_1 \subset \cdots \subset K_{r-1} \subset K_r$$

satisfying

$x + iy \in K_r$,
$[K_i : K_{i-1}] = 2$, *for all i.*

14.5 Constructions with Ruler and Compass

Proof

(1) $\implies$ (2) By definition there is a finite sequence $O_1, O_2, \ldots, O_s = P$ satisfying the conditions of Definition 14.5.1; we show that we are able to build a root tower

$$\mathbb{Q}[x_1, y_1, \ldots, x_n, y_n] =: K_0 \subseteq K_1 \subseteq \cdots \subseteq K_s \subseteq K_{s+1} \ni x + iy,$$

such that for each i either $K_i = K_{i-1}$ or $[K_i : K_{i-1}] = 2$. Inductively, let us assume that we have already built K_{i-1} and for each $j < i$

if O_j is a line or a circle its equation is a polynomial over $K_{i-1}[X, Y]$,
if $O_j = (\alpha_j, \beta_j)$ is a point, then $\alpha_j, \beta_j \in K_{i-1}$

and let us build K_i: if O_i is

(1) the line passing through two given points, then we set $K_i := K_{i-1}$ and the equation of O_i is a polynomial over $K_{i-1}[X, Y]$;
(2) the circle whose centre is a given point and which passes through another given point, then we set $K_i := K_{i-1}$ and the equation of O_i is a polynomial over $K_{i-1}[X, Y]$;
(3) the intersection $O_i = (\alpha_i, \beta_i)$ of two given lines,

$$aX + bY - c = dX + eY - f = 0,$$

$a, b, c, d, e, f \in K_{i-1}$, then we set $K_i := K_{i-1}$ since $\alpha_i, \beta_i \in K_{i-1}$;
(4) an intersection $O_i = (\alpha_i, \beta_i)$ of a given line with a given circle, i.e. a solution of

$$aX + bY - c = X^2 + Y^2 + dX + eY - f = 0,$$

$a, b, c, d, e, f \in K_{i-1}$; then, by expressing one variable in terms of the other via the linear equation and substituting it in the quadratic one, we obtain a quadratic equation in one variable, whose discriminant we will denote by $D \in K_{i-1}$. Then we set $K_i := K_{i-1}[\sqrt{D}]$ and $\alpha_i, \beta_i \in K_i$;
(5) an intersection $O_i = (\alpha_i, \beta_i)$ of two given circles, i.e. a solution of

$$X^2 + Y^2 + aX + bY - c = X^2 + Y^2 + dX + eY - f = 0,$$

$a, b, c, d, e, f \in K_{i-1}$; it is then sufficient to subtract the two equations to obtain a linear equation and reduce this to the previous case.

With this construction we finally obtain a field K_s such that $x, y \in K_s$; then we define $K_{s+1} := K_s[i]$ so that $x + iy \in K_{s+1}$.

(2) $\implies$ (1) Conversely, assuming we are given a root tower

$$\mathbb{Q}(x_1, y_1, \ldots, x_n, y_n) =: K_0 \subset K_1 \subset \cdots \subset K_{r-1} \subset K_r \ni x + iy$$

such that for all i, $[K_i : K_{i-1}] = 2$, i.e.

for all i, there exists $x_i \in K_i \setminus K_{i-1}, y_i \in K_{i-1}, : x_i^2 = y_i$, $K_i : K_{i-1}[x_i]$;

then, denoting $O_i := (x_i, 0)$ by Proposition 14.5.7, for all i, O_i is constructable in terms of $\mathcal{O} \cup \{O_j : j < i\}$; as a consequence P is constructable in terms of $\mathcal{O}$. ♄

Corollary 14.5.10. *If $\mathbb{F}$ is a field, $\mathbb{Q} \subseteq \mathbb{F} \subseteq \mathbb{C}$ and $P = (x, y)$ is constructable by ruler and compass in terms of $\mathcal{O} = \{(x_i, y_i), 1 \le i \le n\}$, with $x_i, y_i \in \mathbb{F}$, then $x + iy$ is algebraic over $\mathbb{F}$ with degree a power of 2.* ♄

To prove the converse of this corollary we need to recall another group theory fact:

Fact 14.5.11. *Let $\mathfrak{G}$ be a group such that $\#\mathfrak{G} = 2^s$; then there is a a chain of subgroups*

$$\mathfrak{G} = \mathfrak{G}_0 \supseteq \mathfrak{G}_1 \supseteq \cdots \supseteq \mathfrak{G}_{s-1} \supseteq \mathfrak{G}_s = \{1\},$$

where for each i, $\mathfrak{G}_i/\mathfrak{G}_{i+1}$ has order 2.

Corollary 14.5.12. *Given a set $\mathcal{O}$ of points $P_i := (x_i, y_i), 1 \le i \le n$, and a point $P := (x, y)$, let $\mathbb{F} := \mathbb{Q}(x_1, y_1, \ldots, x_n, y_n)$ and let $\mathbb{K} \supset \mathbb{F}$ be an algebraic extension such that $[\mathbb{K} : \mathbb{F}] = 2^s$ and $x + iy \in \mathbb{K}$. Then P is constructable by ruler and compass in terms of $\mathcal{O}$.*

Proof The claim follows from Fact 14.5.1, Theorem 14.3.16 and Proposition 14.3.12 ♄

Remark 14.5.13. The condition that $x + iy$ has degree 2^s over

$$\mathbb{F} := \mathbb{Q}(x_1, y_1, \ldots, x_n, y_n)$$

is *necessary* for $x + iy$ to be constructable by ruler and compass, but *not sufficient*: the polynomial $f(X) := X^4 + pX + p$, p a prime, is irreducible and

14.5 Constructions with Ruler and Compass

has two real roots (cf. Remark 14.4.6), so that if $\mathbb{F} := \mathbb{Q}$, α denotes a root of f and $\mathbb{K} = \mathbb{F}[\alpha]$ we have that

α has degree $4 = 2^2$ over $\mathbb{F}$, but
$G(\mathbb{K}/\mathbb{F}) = \mathsf{S}_4$,
$\#(G(\mathbb{K}/\mathbb{F})) = 24$, so that
α is not constructable by ruler and compass.

Remark 14.5.14. The theory developed in this section allows us to prove the insolvability of classical geometric constructions by ruler and compass:

Trisection of an angle: given a 'generic' angle ψ is it possible to construct by ruler and compass an angle ϕ such that $\psi = 3\phi$?
The answer is negative: in fact the ability to construct ψ is equivalent to the ability to construct $\cos(\psi)$ and the relation

$$\cos(3\phi) = 4\cos^3(\phi) - 3\cos(\phi),$$

holds so that, setting $\mathbb{F} := \mathbb{Q}[\cos(\psi)]$, with α a root of the irreducible polynomial $f(X) = 4X^3 - 3X - \cos(\psi)$ and $\mathbb{K} := \mathbb{F}[\alpha] = \mathbb{F}[X]/f(X)$, we deduce that α has degree 3 over $\mathbb{F}$ and so is not constructable[7].

Duplication of a cube: given a generic cube is it possible to construct by ruler and compass a cube whose volume is double that of the original one, i.e. given $c \in \mathbb{Q}$, $c > 0$, is it possible to construct by ruler and compass $d \in \mathbb{Q}$, $d > 0$, such that $d^3 = 2c^3$, i.e. $d = \sqrt[3]{2c}$? The answer is again negative since $\sqrt[3]{2}$ is a solution of the irreducible polynomial $X^3 - 2$ and so has degree 3.

The *rectification of a circumference*, i.e. the construction of π, and the *squaring of a circle*, i.e. the construction of $\sqrt{\pi}$, are not solvable for a more cogent reason: π is transcendental!

The *construction of regular polygons*, i.e. the construction of the primitive nth root of unity (n prime), which satisfies the polynomial $\sum_{i=0}^{n-1} X^i$ is solvable iff $n - 1$ is a power of 2, i.e. for the prime numbers $2^s + 1$; the solution was found by Gauss (Section 12.2).

---

[7] Note the important requirement that ψ is generic: in fact for specific values such as $\psi := \frac{\pi}{2}$ constructing $\phi := \frac{\pi}{6}$ is obvious. What is required is a construction (an algorithm) which yields a solution (an output) for each given angle (input).

Part two
Factorization

And when he had opened the second seal, I heard the second beast say, Come and see.
    And there went out another horse that was red; and power was given to him that sat thereon to take peace from the earth, and that they should kill one another: and there was given unto him a great sword.
Revelations

The things depending from Jove: blood, tin, sapphire, mint, deer, eagle, dolphin.
E.C. Agrippa, *De occulta philosophia*

And the heart of Allah bleeds for the wound of Mostar's bridge. And his rage is upon the offenders. The names of Mark Mammon and Rambo Satan are engraved upon his heart.
Hasan as-Sabah II, *The Hashishiyun Manifesto*

15
Prelude

15.1 A Computation

As an introduction to this part of the book, and as a conclusion to the previous one, I want to discuss the factorization of the polynomial

$$f(X) = X^{24} - 1 \in K[X],$$

where K is *any* field such that $\text{char}(K) = 0$.

Let $F \supseteq K$ be a splitting field of f, $\theta \in F$ denoting any primitive 24th root of unity, $\theta_i := \theta^i$, $i = 1, \ldots, 24$, and $\phi_i : F \mapsto F$ be the K-isomorphism defined by $\phi_i(\theta) = \theta^i$.

Since $f' = 24X^{23}$ and so $\gcd(f, f') = 1$, f is squarefree and

$$\theta_i \neq \theta_j \iff i \neq j.$$

We know that $F = K[\theta]$ and that

$$X^{24} - 1 = \prod_{j=1}^{24}(X - \theta_i) \tag{15.1}$$

is a factorization in $F[X]$.

First let us study the factorization of f over $\mathbb{Q}$: denoting $S := \mathbb{Z}_{24}$, we know that S can be partitioned into disjoint subsets as

$$S = S_1 \cup S_2 \cup S_3 \cup S_4 \cup S_6 \cup S_8 \cup S_{12} \cup S_{24},$$

where $R_i := \{\theta_j, j \in S_i\}$ is the set of the primitive ith roots of unity, so that:

$$\begin{aligned} S_1 &= \{0\}, \text{ and so } R_1 = \{1\}, \\ S_2 &= \{12\}, \text{ and so } R_1 = \{-1\}, \\ S_3 &= \{8, 16\}, \\ S_4 &= \{6, 18\}, \end{aligned}$$

$$S_6 = \{4, 20\},$$
$$S_8 = \{3, 9, 15, 21\},$$
$$S_{12} = \{2, 10, 14, 22\},$$
$$S_{24} = \{1, 5, 7, 11, 13, 17, 19, 23\},$$

and that f has the factorization

$$f(X) = \Phi_1(X)\Phi_2(X)\Phi_3(X)\Phi_4(X)\Phi_6(X)\Phi_8(X)\Phi_{12}(X)\Phi_{24}(X)$$

where

$$\Phi_i(X) := \prod_{j \in S_i}(X - \theta_i)$$

represents the irreducible ith cyclotomic polynomials f_i, which we can compute using the results of Section 7.6; denoting $g_i(X) := X^i - 1$, we have:

$$\begin{aligned}
\Phi_1(X) &:= g_1(X) & &= X - 1, \\
\Phi_2(X) &:= \tfrac{g_2(X)}{\Phi_1(X)} & &= X + 1, \\
\Phi_3(X) &:= \tfrac{g_3(X)}{\Phi_1(X)} & &= X^2 + X + 1, \\
\Phi_4(X) &:= \Phi_2(X^2) & &= X^2 + 1, \\
\Phi_6(X) &:= \tfrac{\Phi_2(X^3)}{\Phi_2(X)} & &= X^2 - X + 1, \\
\Phi_8(X) &:= \Phi_2(X^4) & &= X^4 + 1, \\
\Phi_{12}(X) &:= \Phi_6(X^2) & &= X^4 - X^2 + 1, \\
\Phi_{24}(X) &:= \Phi_6(X^4) & &= X^8 - X^4 + 1.
\end{aligned}$$

In order to factorize f over any field K, $\mathrm{char}(K) = 0$, we need to go back to Equation 15.1 and analyse it in $\mathbb{C}[X]$, where we know the value of a primitive root of unity

$$\theta := \cos\frac{\pi}{12} + i\sin\frac{\pi}{12} = \frac{\sqrt{6}+\sqrt{2}}{4} + i\frac{\sqrt{6}-\sqrt{2}}{4},$$

and so we can easily obtain all the 24th roots of unity:

$$\begin{aligned}
\theta_1 &= \tfrac{\sqrt{6}+\sqrt{2}}{4} + i\tfrac{\sqrt{6}-\sqrt{2}}{4}, & \theta_{13} &= \tfrac{-\sqrt{6}-\sqrt{2}}{4} + i\tfrac{-\sqrt{6}+\sqrt{2}}{4}, \\
\theta_2 &= \tfrac{\sqrt{3}}{2} + i\tfrac{1}{2}, & \theta_{14} &= \tfrac{-\sqrt{3}}{2} + i\tfrac{-1}{2}, \\
\theta_3 &= \tfrac{1}{\sqrt{2}} + i\tfrac{1}{\sqrt{2}}, & \theta_{15} &= \tfrac{-1}{\sqrt{2}} + i\tfrac{-1}{\sqrt{2}}, \\
\theta_4 &= \tfrac{1}{2} + i\tfrac{\sqrt{3}}{2}, & \theta_{16} &= \tfrac{-1}{2} + i\tfrac{-\sqrt{3}}{2}, \\
\theta_5 &= \tfrac{\sqrt{6}-\sqrt{2}}{4} + i\tfrac{\sqrt{6}+\sqrt{2}}{4}, & \theta_{17} &= \tfrac{-\sqrt{6}+\sqrt{2}}{4} + i\tfrac{-\sqrt{6}-\sqrt{2}}{4},
\end{aligned}$$

15.1 A Computation

$$\begin{aligned}
\theta_6 &= i, & \theta_{18} &= -i, \\
\theta_7 &= \tfrac{-\sqrt{6}+\sqrt{2}}{4} + i\tfrac{\sqrt{6}+\sqrt{2}}{4}, & \theta_{19} &= \tfrac{\sqrt{6}-\sqrt{2}}{4} + i\tfrac{-\sqrt{6}-\sqrt{2}}{4}, \\
\theta_8 &= \tfrac{-1}{2} + i\tfrac{\sqrt{3}}{2}, & \theta_{20} &= \tfrac{1}{2} + i\tfrac{-\sqrt{3}}{2}, \\
\theta_9 &= \tfrac{-1}{\sqrt{2}} + i\tfrac{1}{\sqrt{2}}, & \theta_{21} &= \tfrac{1}{\sqrt{2}} + i\tfrac{-1}{\sqrt{2}}, \\
\theta_{10} &= \tfrac{-\sqrt{3}}{2} + i\tfrac{1}{2}, & \theta_{22} &= \tfrac{\sqrt{3}}{2} + i\tfrac{-1}{2}, \\
\theta_{11} &= \tfrac{-\sqrt{6}-\sqrt{2}}{4} + i\tfrac{\sqrt{6}-\sqrt{2}}{4}, & \theta_{23} &= \tfrac{\sqrt{6}+\sqrt{2}}{4} + i\tfrac{-\sqrt{6}+\sqrt{2}}{4}, \\
\theta_{12} &= -1, & \theta_0 &= 1.
\end{aligned}$$

The above computations allow us to easily deduce

Lemma 15.1.1. *Let*

$$\theta := \frac{\sqrt{6}+\sqrt{2}}{4} + i\frac{\sqrt{6}-\sqrt{2}}{4} \in \mathbb{C},$$

and $\mathbb{F} = \mathbb{Q}(\theta)$. *Then:*

(1) $i = \theta^6$, $\sqrt{2} = \theta^{21} + \theta^3$, $\sqrt{3} = 2\theta^2 - \theta^6$;
(2) $\mathbb{F} = \mathbb{Q}(i, \sqrt{2}, \sqrt{3})$;
(3) $[\mathbb{F} : \mathbb{Q}] = 8$;
(4) Φ_{24} *is the minimal polynomial of* θ.

Proof.

(1) The results follow from the above values of θ_i. In fact:

the value of θ_6 implies $i = \theta^6$;
from the value of θ_3 we deduce

$$\sqrt{2}\theta^3 = 1 + i \implies \sqrt{2} = \theta^{21} + \theta^3;$$

from the value of θ_2 we deduce

$$\sqrt{3} = 2\theta_2 - i = 2\theta_2 - \theta_6.$$

(2) Obviously $\theta \in \mathbb{Q}(i, \sqrt{2}, \sqrt{3})$ and the previous result implies

$$\mathbb{Q}(i, \sqrt{2}, \sqrt{3}) \subseteq \mathbb{F}.$$

(3) Is then an obvious consequence.
(4) Although we know that Φ_{24} is irreducible by Proposition 7.6.10, we can simply deduce it from the facts that $\Phi_{24}(\theta) = 0$ and $\deg(\Phi_{24}) = 8 = [\mathbb{F} : \mathbb{Q}]$.

$\square$

Since X^8-X^4+1 is the minimal polynomial of θ, we can therefore represent each θ_i as an element in

$$\mathbb{F} = \mathbb{Q}[\theta] = \mathrm{Span}_{\mathbb{Q}}(\{1, \theta, \theta^2, \theta^3, \theta^4, \theta^5, \theta^6, \theta^7\})$$

as

$$\begin{array}{lll}
\theta_1 = \theta, & \theta_2 = \theta^2, & \theta_3 = \theta^3, \\
\theta_4 = \theta^4, & \theta_5 = \theta^5, & \theta_6 = \theta^6, \\
\theta_7 = \theta^7, & \theta_8 = \theta^4 - 1, & \theta_9 = \theta^5 - \theta, \\
\theta_{10} = \theta^6 - \theta^2, & \theta_{11} = \theta^7 - \theta^3, & \theta_{12} = -1, \\
\theta_{13} = -\theta, & \theta_{14} = -\theta^2, & \theta_{15} = -\theta^3, \\
\theta_{16} = -\theta^4, & \theta_{17} = -\theta^5, & \theta_{18} = -\theta^6, \\
\theta_{19} = -\theta^7, & \theta_{20} = -\theta^4 + 1, & \theta_{21} = -\theta^5 + \theta, \\
\theta_{22} = -\theta^6 + \theta^2, & \theta_{23} = -\theta^7 + \theta^3, & \theta_0 = 1;
\end{array}$$

from which we also get

$$\begin{array}{ll}
\sqrt{2} = -\theta^5 + \theta^3 + \theta, & \sqrt{-2} = \theta^5 + \theta^3 - \theta, \\
\sqrt{3} = -\theta^6 + 2\theta^2, & \sqrt{-3} = 2\theta^4 - 1, \\
\sqrt{6} = -2\theta^7 + \theta^5 + \theta^3 + \theta, & \sqrt{-6} = 2\theta^7 + \theta^5 - \theta^3 + \theta, \\
i = \theta^6.
\end{array}$$

Corollary 15.1.2.

The Galois group $\mathfrak{G}$ of $[\mathbb{F} : \mathbb{Q}]$ is

$$\mathfrak{G} = \{\phi_1, \phi_5, \phi_7, \phi_{11}, \phi_{13}, \phi_{17}, \phi_{19}, \phi_{23}\} \cong \mathbb{Z}_2 \times \mathbb{Z}_2 \times \mathbb{Z}_2.$$

The intermediate fields Δ such that $\mathbb{Q} \subseteq \Delta \subseteq \mathbb{F}$ and their associated subgroups $\mathfrak{g} \subseteq \mathfrak{G}$ are

$$\begin{array}{llll}
\Delta_1 = \mathbb{Q}, & \mathfrak{g}_1 = \mathfrak{G}, \\
\Delta_2 = \mathbb{Q}(i), & \mathfrak{g}_2 = \{\phi_1, \phi_5, \phi_{13}, \phi_{17}\}, \\
\Delta_3 = \mathbb{Q}(\sqrt{2}), & \mathfrak{g}_3 = \{\phi_1, \phi_7, \phi_{17}, \phi_{23}\}, \\
\Delta_4 = \mathbb{Q}(\sqrt{3}), & \mathfrak{g}_4 = \{\phi_1, \phi_{11}, \phi_{13}, \phi_{23}\}, \\
\Delta_5 = \mathbb{Q}(\sqrt{-2}), & \mathfrak{g}_5 = \{\phi_1, \phi_{11}, \phi_{17}, \phi_{19}\}, \\
\Delta_6 = \mathbb{Q}(\sqrt{-3}), & \mathfrak{g}_6 = \{\phi_1, \phi_7, \phi_{13}, \phi_{19}\}, \\
\Delta_7 = \mathbb{Q}(\sqrt{6}), & \mathfrak{g}_7 = \{\phi_1, \phi_5, \phi_{19}, \phi_{23}\}, \\
\Delta_8 = \mathbb{Q}(\sqrt{-6}), & \mathfrak{g}_8 = \{\phi_1, \phi_5, \phi_7, \phi_{11}\}, \\
\Delta_9 = \mathbb{Q}(i, \sqrt{6}), & \mathfrak{g}_9 = \{\phi_1, \phi_5\},
\end{array}$$

$$\Delta_{10} = \mathbb{Q}(\sqrt{2}, \sqrt{-3}), \qquad \mathfrak{g}_{10} = \{\phi_1, \phi_7\},$$
$$\Delta_{11} = \mathbb{Q}(\sqrt{-2}, \sqrt{3}), \qquad \mathfrak{g}_{11} = \{\phi_1, \phi_{11}\},$$
$$\Delta_{12} = \mathbb{Q}(i, \sqrt{3}), \qquad \mathfrak{g}_{12} = \{\phi_1, \phi_{13}\},$$
$$\Delta_{13} = \mathbb{Q}(i, \sqrt{2}), \qquad \mathfrak{g}_{13} = \{\phi_1, \phi_{17}\},$$
$$\Delta_{14} = \mathbb{Q}(\sqrt{-2}, \sqrt{-3}), \qquad \mathfrak{g}_{14} = \{\phi_1, \phi_{19}\},$$
$$\Delta_{15} = \mathbb{Q}(\sqrt{2}, \sqrt{3}), \qquad \mathfrak{g}_{15} = \{\phi_1, \phi_{23}\},$$
$$\Delta_{16} = \mathbb{F}, \qquad \mathfrak{g}_{16} = \{\phi_1\}.$$

Proof Since $\mathbb{F} = \mathbb{Q}(i, \sqrt{2}, \sqrt{3})$, it is clear that $\mathfrak{G}$ consists of the 8 $\mathbb{Q}$-isomorphisms ψ which satisfy

$$\psi(i) = \pm i, \quad \psi(\sqrt{2}) = \pm\sqrt{2}, \quad \psi(\sqrt{3}) = \pm\sqrt{3}.$$

On the other hand, the conjugate roots of θ are $\theta_i : i \in S_{24}$, so that $\mathfrak{G} = \{\phi_i, i \in S_{24}\}$.

To associate these two representations with each other, we only have to compute $\psi(\theta)$ and apply the fact that

$$\psi(\theta) = \theta_i \implies \psi = \phi_i;$$

for instance for the $\mathbb{Q}$-isomorphism ψ such that

$$\psi(i) = -i, \quad \psi(\sqrt{2}) = +\sqrt{2}, \quad \psi(\sqrt{3}) = +\sqrt{3},$$

we have $\psi(\theta) = \theta_{23}$ and $\psi = \phi_{23}$.

Then, listing all intermediate fields and their associated subgroups is just bookkeeping. ♃

It is then easy to compute the orbits of each S_j with respect to $\mathfrak{g}_i$:

Lemma 15.1.3.

S_1, S_2 are stable under $\mathfrak{g}_i$, for all i;

The orbits of S_3 are $\begin{cases} \{8\}, \{16\} & \text{if } i = 6, 10, 12, 14; \\ S_3 & \text{otherwise.} \end{cases}$

The orbits of S_4 are $\begin{cases} \{6\}, \{18\} & \text{if } i = 2, 9, 12, 13; \\ S_4 & \text{otherwise.} \end{cases}$

The orbits of S_6 are $\begin{cases} \{4\}, \{20\} & \text{if } i = 6, 10, 12, 14; \\ S_6 & \text{otherwise.} \end{cases}$

334   *Prelude*

The orbits of S_8 are $\begin{cases} \{3\}, \{9\}, \{15\}, \{21\} & \text{if } i = 13; \\ \{3, 9\}, \{15, 21\} & \text{if } i = 5, 11, 14; \\ \{3, 15\}, \{9, 21\} & \text{if } i = 2, 9, 12; \\ \{3, 21\}, \{9, 15\} & \text{if } i = 3, 10, 15; \\ S_8 & \text{otherwise.} \end{cases}$

The orbits of S_{12} are $\begin{cases} \{2\}, \{10\}, \{14\}, \{22\} & \text{if } i = 12; \\ \{2, 10\}, \{14, 22\} & \text{if } i = 2, 9, 13; \\ \{2, 14\}, \{10, 22\} & \text{if } i = 6, 10, 14; \\ \{2, 22\}, \{10, 14\} & \text{if } i = 4, 11, 15; \\ S_{12} & \text{otherwise.} \end{cases}$

Since $\mathfrak{G} = \{\phi_i : i \in S_{24}\}$, the orbits of S_{24} with respect to $\mathfrak{g}_i$ are the cosets of $\mathfrak{G}$ with respect to $\mathfrak{g}_i$.

Proof We only have to compute $\phi(\theta_k)$, for all $\phi \in \mathfrak{G}, k \in S$. The following computations

	ϕ_5	ϕ_7	ϕ_{11}	ϕ_{13}	ϕ_{17}	ϕ_{19}	ϕ_{23}
$k = 2$	10	14	22	2	10	14	22
$k = 3$	15	21	9	15	3	9	21
$k = 4$	20	4	20	4	20	4	20
$k = 6$	6	18	18	6	6	18	18
$k = 8$	16	8	16	8	16	8	16

are the relevant ones.   ♃

It is then obvious how to compute the factorization of f in $\Delta_i[X]$, for all i. For instance the factorization of f over $\Delta_{10} = \mathbb{Q}(\sqrt{2}, \sqrt{3})$ is

$$\begin{aligned} f(X) &= \Phi_1(X)\Phi_2(X)(X - \theta_8)(X - \theta_{16})\Phi_4(X)(X - \theta_4)(X - \theta_{20}) \\ &\quad \cdot h_8^{(a)}(X) h_8^{(b)}(X) h_{12}^{(a)}(X) h_{12}^{(b)}(X) \\ &\quad \cdot h_{24}^{(a)}(X) h_{24}^{(b)}(X) h_{24}^{(c)}(X) h_{24}^{(d)}(X), \end{aligned}$$

where

$$\begin{aligned} h_8^{(a)} &= (X - \theta_3)(X - \theta_{21}) = X^2 - \sqrt{2}X + 1, \\ h_8^{(b)} &= (X - \theta_9)(X - \theta_{15}) = X^2 + \sqrt{2}X + 1, \\ h_{12}^{(a)} &= (X - \theta_2)(X - \theta_{14}) = X^2 - \frac{1}{2} - \frac{\sqrt{-3}}{2}, \\ h_{12}^{(b)} &= (X - \theta_{10})(X - \theta_{22}) = X^2 - \frac{1}{2} + \frac{\sqrt{-3}}{2}, \end{aligned}$$

$$h_{24}^{(a)} = (X-\theta)(X-\theta_7) = X^2 - \frac{\sqrt{2}+\sqrt{-6}}{2}X + \frac{-1+\sqrt{-3}}{2},$$

$$h_{24}^{(b)} = (X-\theta_5)(X-\theta_{11}) = X^2 - \frac{-\sqrt{2}+\sqrt{-6}}{2}X + \frac{-1-\sqrt{-3}}{2},$$

$$h_{24}^{(c)} = (X-\theta_{13})(X-\theta_{19}) = X^2 - \frac{-\sqrt{2}-\sqrt{-6}}{2}X + \frac{-1+\sqrt{-3}}{2},$$

$$h_{24}^{(d)} = (X-\theta_{17})(X-\theta_{23}) = X^2 - \frac{+\sqrt{2}-\sqrt{-6}}{2}X + \frac{-1-\sqrt{-3}}{2}.$$

In fact, the following factorizations can be easily computed[1] and allow us to verify the validity of Galois Theory:

Corollary 15.1.4. *In Δ_i the factorization of Φ_j is:*

$$\Phi_3 = \begin{cases} \left(X - \frac{-1+\sqrt{-3}}{2}\right)\left(X - \frac{-1-\sqrt{-3}}{2}\right) & \text{if } i = 6, 10, 12, 14; \\ X^2 + X + 1 & \text{otherwise.} \end{cases}$$

$$\Phi_4 = \begin{cases} (X-i)(X+i) & \text{if } i = 2, 9, 12, 13; \\ X^2 + 1 & \text{otherwise.} \end{cases}$$

$$\Phi_6 = \begin{cases} \left(X - \frac{1+\sqrt{-3}}{2}\right)\left(X - \frac{1-\sqrt{-3}}{2}\right) & \text{if } i = 6, 10, 12, 14; \\ X^2 - X + 1 & \text{otherwise.} \end{cases}$$

$$\Phi_8 = \begin{cases} (X-\theta_3)(X-\theta_9)(X-\theta_{15})(X-\theta_{21}) & \text{if } i = 13; \\ (X^2 - \sqrt{-2}X - 1)(X^2 + \sqrt{-2}X - 1) & \text{if } i = 5, 11, 14; \\ (X^2 - i)((X^2 + i) & \text{if } i = 2, 9, 12; \\ (X^2 - \sqrt{2}X + 1)(X^2 + \sqrt{2}X + 1) & \text{if } i = 3, 10, 15; \\ X^4 + 1 & \text{otherwise.} \end{cases}$$

---

[1] The computations are obvious. For instance to factorize f_{24} in Δ_{14}, we have to note that the orbit of 1 in $S_{24} = \mathfrak{G}$ with respect to $\mathfrak{g}_{14}$ is $\{1, 19\}$. Therefore we have to compute

$$h(X) := (X-\theta_1)(X-\theta_{19}) = X^2 - (\theta_1 + \theta_{19})X + \theta_{20}$$

which is

$$h(X) = X^2 + \frac{-\sqrt{6}+\sqrt{-2}}{2}X + \frac{1-\sqrt{-3}}{2}.$$

Since $\Delta_{14} = \mathbb{Q}(\sqrt{-2}, \sqrt{-3})$ and, therefore, the four $\mathbb{Q}$-automorphisms ψ of Δ_{14} are those such that

$$\psi(\sqrt{-2}) = \pm\sqrt{-2}, \quad \psi(\sqrt{-3}) = \pm\sqrt{-3},$$

what we have to do is simply to express $h(X)$ in terms of $\sqrt{-2}$ and $\sqrt{-3}$ and apply the automorphisms ψ in order to get the four factors of f_{24} which are

$$X^2 + \frac{-(\pm\sqrt{-2})(\pm\sqrt{-3}) + (\pm\sqrt{-2})}{2}X + \frac{1-(\pm\sqrt{-3})}{2}.$$

In this way, it was easy for me to compute all the listed factorizations by hand.

$$\Phi_{12} = \begin{cases} (X - \theta_2)(X - \theta_{10})(X - \theta_{14})(X - \theta_{22}) & \text{if } i = 12; \\ (X^2 - iX - 1)(X^2 + iX - 1) & \text{if } i = 2, 9, 13; \\ \left(X^2 + \frac{-1-\sqrt{-3}}{2}\right)\left(X^2 + \frac{-1+\sqrt{-3}}{2}\right) & \text{if } i = 6, 10, 14; \\ \left(X^2 - \sqrt{3}X + 1\right)\left(X^2 - \sqrt{3}X + 1\right) & \text{if } i = 4, 11, 15; \\ X^4 - X^2 + 1 & \text{otherwise.} \end{cases}$$

$\qquad\qquad\qquad\qquad\qquad\qquad\qquad\qquad\qquad\qquad\qquad\qquad$ ⨆

Corollary 15.1.5. *In Δ_i the factorization of Φ_{24} is the one reported in Table 15.1.*

Remark 15.1.6. It is clear that the above results allow us to solve the problem of factorizing $f(X)$ over $K[X]$, where $\text{char}(K) = 0$: it is sufficient to compute the maximal value i such that $\Delta_i \subseteq K$ and read the corresponding factorization.

To solve the same problem when $\text{char}(K) \neq 0$ we can use Proposition 7.6.13.

We limit ourselves to discussing an example. Beforehand we have, of course, to separate the cases in which f is not squarefree, by computing $f' = 24X^{23}$ from which we conclude that

$$\begin{aligned} f \text{ is squarefree} &\iff \gcd(f, f') = 1 \\ &\iff f' \neq 0 \\ &\iff \text{char}(K) \neq 2, 3. \end{aligned}$$

Of course, when

$\text{char}(K) = 2$, we have $f = (X^3 - 1)^8 = (X - 1)^8(X^2 + X + 1)^8$.

Clearly $g(X) = X^2 + X + 1$, which while being irreducible in $\mathbb{Z}_2$, splits in $GF(4)$.

Using the representation $GF(4) = \mathbb{Z}_2[\alpha] = \mathbb{Z}_2[X]/g(X)$ we obtain $g(X) = (X + \alpha)(X + \alpha + 1)$.

Therefore in $K[X]$, $f(X)$ factorizes as

$$f(X) = \begin{cases} (X - 1)^8(X^2 + X + 1)^8 & \text{if } GF(4) \not\subset K, \\ (X - 1)^8(X + \alpha)^8(X + \alpha + 1)^8 & \text{if } GF(4) \subset K, \end{cases}$$

$\text{char}(K) = 3$, we have $f = (X^8 - 1)^3$. Since $GF(9)$ is the splitting field of $X^9 - X$, we can conclude that, representing $GF(9)$ as

$$GF(9) = \mathbb{Z}_3[\alpha] := \mathbb{Z}_3[X]/g(X)$$

where $g(X) = X^2 + X - 1$, in $K[X]$, $f(X)$ factorizes as

$$f(X) = \begin{cases} (X - 1)^3(X + 1)^3(X^2 + 1)^3 \\ \quad \cdot (X^2 - X - 1)^3(X^2 + X - 1)^3 & \text{if } GF(9) \not\subset K, \\ \prod_{j=1}^{8}(X - \alpha^j)^3 & \text{if } GF(9) \subset K, \end{cases}$$

15.1 A Computation

Table 15.1. *Factorization of Φ_{24} in $\Delta_i[X]$.*

$i =$	
1	$X^8 - X^4 + 1$
2	$(X^4 - iX^2 - 1)(X^4 + iX^2 - 1)$
3	$(X^4 - \sqrt{2}X^3 + X^2 - \sqrt{2}X + 1)$
	$\cdot (X^4 + \sqrt{2}X^3 + X^2 + \sqrt{2}X + 1)$
4	$(X^4 - \sqrt{3}X^2 + 1)((X^4 + \sqrt{3}X^2 + 1)$
5	$(X^4 + \sqrt{-2}X^3 - X^2 - \sqrt{-2}X + 1)$
	$\cdot (X^4 - \sqrt{-2}X^3 - X^2 + \sqrt{-2}X + 1)$
6	$(X^4 - \frac{\sqrt{-3}+1}{2})(X^4 - \frac{-\sqrt{-3}+1}{2})$
7	$(X^4 - \sqrt{6}X^3 + 3X^2 - \sqrt{6}X + 1)$
	$\cdot (X^4 + \sqrt{6}X^3 + 3X^2 + \sqrt{6}X + 1)$
8	$(X^4 - \sqrt{-6}X^3 - 3X^2 + \sqrt{-6}X + 1)$
	$\cdot (X^4 - \sqrt{-6}X^3 - 3X^2 + \sqrt{-6}X + 1)$
9	$(X^2 + \frac{-\sqrt{6}-\sqrt{-6}}{2}X + i)(X^2 + \frac{-\sqrt{6}+\sqrt{-6}}{2}X - i)$
	$\cdot (X^2 + \frac{+\sqrt{6}+\sqrt{-6}}{2}X + i)(X^2 + \frac{+\sqrt{6}-\sqrt{-6}}{2}X - i)$
10	$(X^2 + \frac{-\sqrt{2}-\sqrt{-6}}{2}X + \frac{-1+\sqrt{-3}}{2})(X^2 + \frac{+\sqrt{2}+\sqrt{-6}}{2}X + \frac{-1+\sqrt{-3}}{2})$
	$\cdot (X^2 + \frac{-\sqrt{2}+\sqrt{-6}}{2}X + \frac{-1-\sqrt{-3}}{2})(X^2 + \frac{+\sqrt{2}-\sqrt{-6}}{2}X + \frac{-1-\sqrt{-3}}{2})$
11	$(X^2 + \frac{-\sqrt{-6}+\sqrt{-2}}{2}X - 1)(X^2 + \frac{+\sqrt{-6}-\sqrt{-2}}{2}X - 1)$
	$\cdot (X^2 + \frac{+\sqrt{-6}+\sqrt{-2}}{2}X - 1)(X^2 + \frac{-\sqrt{-6}-\sqrt{-2}}{2}X - 1)$
12	$(X^2 + \frac{-\sqrt{3}-i}{2})(X^2 + \frac{-\sqrt{3}+i}{2})$
	$\cdot (X^2 + \frac{+\sqrt{3}-i}{2})(X^2 + \frac{+\sqrt{3}+i}{2})$
13	$(X^2 + \frac{-\sqrt{2}+\sqrt{-2}}{2}X - i)(X^2 + \frac{-\sqrt{2}-\sqrt{-2}}{2}X + i)$
	$\cdot (X^2 + \frac{+\sqrt{2}-\sqrt{-2}}{2}X - i)(X^2 + \frac{+\sqrt{2}+\sqrt{-2}}{2}X + i)$
14	$(X^2 + \frac{-\sqrt{6}+\sqrt{-2}}{2}X + \frac{1-\sqrt{-3}}{2})(X^2 + \frac{+\sqrt{6}-\sqrt{-2}}{2}X + \frac{1-\sqrt{-3}}{2})$
	$\cdot (X^2 + \frac{+\sqrt{6}+\sqrt{-2}}{2}X + \frac{1+\sqrt{-3}}{2})(X^2 + \frac{-\sqrt{6}-\sqrt{-2}}{2}X + \frac{1+\sqrt{-3}}{2})$
15	$(X^2 + \frac{-\sqrt{6}-\sqrt{2}}{2}X + 1)(X^2 + \frac{+\sqrt{6}+\sqrt{2}}{2}X + 1)$
	$\cdot (X^2 + \frac{+\sqrt{6}-\sqrt{2}}{2}X + 1)(X^2 + \frac{-\sqrt{6}+\sqrt{2}}{2}X + 1)$
16	$(X - \theta_1)(X - \theta_5)(X - \theta_7)(X - \theta_{11})$
	$\cdot (X - \theta_{13})(X - \theta_{17})(X - \theta_{19})(X - \theta_{23})$

Having eliminated the exceptional cases char$(K) = 2, 3$, here is the promised example:

Example 15.1.7. In the case char$(K) = 5$, since $24 = 5^2 - 1$,

$GF(25)$ is the splitting field of $f(X)$;
the multiplicative order of 5 mod 24 is 2, so that, representing $GF(25)$ as

$$GF(25) = \mathbb{Z}_5[\alpha] = \mathbb{Z}_5[X]/g(X), g(X) := X^2 + 2X - 2,$$

the orbits of conjugate roots and the corresponding factors in $GF(5)$ are:

$$
\begin{array}{rll}
\{\alpha, \alpha^5\}: & (X - \alpha)(X + \alpha + 2) & = X^2 + 2X - 2, \\
\{\alpha^2, \alpha^{10}\}: & (X + 2\alpha - 2)(X - 2\alpha - 1) & = X^2 + 2X - 1, \\
\{\alpha^3, \alpha^{15}\}: & (X - \alpha - 1)(X + \alpha + 1) & = X^2 + 2, \\
\{\alpha^4, \alpha^{20}\}: & (X + \alpha - 2)(X - \alpha + 1) & = X^2 - X + 1, \\
\{\alpha^6\}: & & X + 2, \\
\{\alpha^7, \alpha^{11}\}: & (X + 2\alpha)(X - 2\alpha + 1) & = X^2 + X + 2, \\
\{\alpha^8, \alpha^{16}\}: & (X + \alpha - 1)(X - \alpha + 2) & = X^2 + X + 1, \\
\{\alpha^9, \alpha^{21}\}: & (X + 2\alpha + 2)(X - 2\alpha - 2) & = X^2 - 2, \\
\{\alpha^{12}\}: & & X + 1, \\
\{\alpha^{13}, \alpha^{17}\}: & (X + \alpha)(X - \alpha - 2) & = X^2 - 2X - 2, \\
\{\alpha^{14}, \alpha^{22}\}: & (X - 2\alpha + 2)(X + 2\alpha + 1) & = X^2 - 2X - 1, \\
\{\alpha^{18}\}: & & = X - 2, \\
\{\alpha^{19}, \alpha^{23}\}: & (X - 2\alpha)(X + 2\alpha - 1) & = X^2 - X + 2.
\end{array}
$$

Remark 15.1.8. It is worthwhile noting that when $K = GF(5)$,

the orbits coincide with those of $\Delta_9 = \mathbb{Q}(i, \sqrt{6})$;
$GF(5)$ contains the roots of -1 and 6 since $6 = (\pm 1)^2$ and $-1 = -6 = (\pm 2)^2$, so in $\mathbb{Z}_5[X]$:
the factorizations on $GF(5)$ and on Δ_9 coincide if we substitute $\sqrt{6}$ with ± 1 and i with ± 2;
the $GF(5)$-automorphisms of $GF(25)$ which leaves $GF(5)$ invariant are the identity and the morphism $\psi(\alpha) = \alpha^5$.

15.2 An Exercise

Problem 15.2.1. Compute the factorization of the polynomial

$$f(X) = X^8 - 1 \in K[X],$$

15.2 An Exercise

where K is *any* field such that $\text{char}(K) = 0$, using the information that

$$\alpha = \frac{\sqrt{2} + \sqrt{-2}}{2}$$

is a primitive 8th root of unity, and proving that the following hold:

The fields Δ such that $\mathbb{Q} \subseteq \Delta \subseteq \mathbb{F} := \mathbb{Q}(\alpha)$ are

$$\begin{aligned}
\Delta_1 &:= \mathbb{Q}, \\
\Delta_2 &:= \mathbb{Q}(i), \\
\Delta_3 &:= \mathbb{Q}(\sqrt{2}), \\
\Delta_4 &:= \mathbb{Q}(\sqrt{-2}), \\
\Delta_5 &:= \mathbb{F}.
\end{aligned}$$

Let i be the maximal value i such that $\Delta_i \subseteq K$. The factorization of $f(X)$ in $K[X]$ is then:

$i =$ 1 $(X - 1)(X + 1)(X^2 + 1)(X^4 + 1)$,
      2    $(X - 1)(X + 1)(X - i)(X + i)(X^2 - i)(X^2 + i)$,
      3    $(X - 1)(X + 1)(X^2 + 1)(X^2 - \sqrt{2}X + 1)$
          $\cdot (X^2 + \sqrt{2}X + 1)$,
      4    $(X - 1)(X + 1)(X^2 + 1)(X^2 - \sqrt{-2}X - 1)$
          $\cdot (X^2 + \sqrt{-2}X - 1)$,
      5    $(X - 1)(X + 1)(X - i)(X + i)(X - \frac{+\sqrt{2}+\sqrt{-2}}{2})$
          $\cdot (X - \frac{-\sqrt{2}+\sqrt{-2}}{2})(X - \frac{+\sqrt{2}-\sqrt{-2}}{2})(X - \frac{-\sqrt{2}-\sqrt{-2}}{2})$.

Finally compute the Galois group $\mathfrak{G}$ of $[\mathbb{F} : \mathbb{Q}]$ and, for each i, the subgroup $\mathfrak{g}_i$ associated to Δ_i.

Solution: The first thing we have to do is to verify whether f is squarefree: since $f'(X) = 8X^7$ we can conclude that:

If $\text{char}(K) = 2$, then $f' = 0$; therefore by Proposition 4.6.1, we conclude that

$$f(X) = (X - 1)^8 \in \mathbb{Z}_2[X] \subseteq K[X].$$

Otherwise $f' \neq 0$, $\gcd(f, f') = 1$ and f is squarefree in $K[X]$.

The information that

$$\alpha = \frac{\sqrt{2} + \sqrt{-2}}{2} = \theta_3$$

is a primitive[2] 8th root of unity allows us to list the four roots of f which are

$$\alpha_i = \alpha^i = \theta_{3i}, \quad i \in \{1, 3, 5, 7\};$$

---
[2] Where θ is the primitive 24th root of unity introduced in the previous section.

and to partition $S := \mathbb{Z}_8$ into the disjoint subsets
$$S = S_1 \cup S_2 \cup S_4 \cup S_8$$
so that $R_i := \{\alpha_j, j \in S_i\}$ consists of the primitive ith roots of unity:
$$\begin{aligned} S_1 &= \{0\}, \\ S_2 &= \{4\}, \\ S_4 &= \{2, 6\}, \\ S_8 &= \{1, 3, 5, 7\}, \end{aligned}$$
and to obtain a partial factorization of $X^8 - 1$ in $\mathbb{Q}[X]$ as
$$X^8 - 1 = \Phi_1(X)\Phi_2(X)\Phi_4(X)\Phi_8(X),$$
where $\Phi_i(X) = \prod_{j \in S_i} X - \alpha_j$, i.e.
$$\begin{aligned} \Phi_1(X) &= X - 1, \\ \Phi_2(X) &= X + 1, \\ \Phi_4(X) &= X^2 + 1, \\ \Phi_8(X) &= X^4 + 1. \end{aligned}$$
Since $\alpha = \frac{\sqrt{2}+\sqrt{-2}}{2} = \frac{1+i}{\sqrt{2}}$, then $\mathbb{F} \subset \mathbb{Q}(i, \sqrt{2})$. The relations
$$i = \alpha_2, \quad \sqrt{2} = \alpha^{-1}(1+i) = \alpha_7 + \alpha,$$
allow us to conclude that

$\mathbb{F} = \mathbb{Q}(i, \sqrt{2})$;

$[\mathbb{F} : \mathbb{Q}] = 4$;

$f_8(X) = X^4 + 1$ is the minimal polynomial of α;

$\mathbb{F} = \mathrm{Span}_{\mathbb{Q}}(\{1, \alpha, \alpha^2, \alpha^3\})$;

$\mathfrak{G} := G(\mathbb{F}/\mathbb{Q}) \cong \mathbb{Z}_2 \times \mathbb{Z}_2$ and consists of the four $\mathbb{Q}$-isomorphisms ψ defined by
$$\psi(i) = \pm i, \quad \psi(\sqrt{2}) = \pm\sqrt{2};$$
to represent each α_i in $\mathrm{Span}_{\mathbb{Q}}(\{1, \alpha, \alpha^2, \alpha^3\})$ as
$$\begin{aligned} \alpha_1 &= \alpha = -\alpha_5, \\ \alpha_2 &= \alpha^2 = -\alpha_6, \\ \alpha_3 &= \alpha^3 = -\alpha_7, \\ \alpha_4 &= -1 = -\alpha_8, \end{aligned}$$

from which we deduce that
$$\begin{aligned} i &= \alpha^2, \\ \sqrt{2} &= \alpha - \alpha^3, \\ \sqrt{-2} &= \alpha + \alpha^3. \end{aligned}$$

Denoting $\phi_i : \mathbb{F} \mapsto \mathbb{F}$ the $\mathbb{Q}$-isomorphism such that $\phi_i(\alpha) = \alpha^i$ we can then easily obtain $\mathfrak{G}$ by computation:

$$\begin{aligned} \psi(i) &= +i, & \psi(\sqrt{2}) &= +\sqrt{2}, & \Longrightarrow \psi(\alpha) &= \alpha, & \psi &= \phi_1 \\ \psi(i) &= +i, & \psi(\sqrt{2}) &= -\sqrt{2}, & \Longrightarrow \psi(\alpha) &= \alpha_5, & \psi &= \phi_5 \\ \psi(i) &= -i, & \psi(\sqrt{2}) &= +\sqrt{2}, & \Longrightarrow \psi(\alpha) &= \alpha_7, & \psi &= \phi_7 \\ \psi(i) &= -i, & \psi(\sqrt{2}) &= -\sqrt{2}, & \Longrightarrow \psi(\alpha) &= \alpha_3, & \psi &= \phi_3 \end{aligned}$$

This allows us to list all the intermediate fields Δ_i such that $\mathbb{Q} \subseteq \mathbb{F}$ and the associated subgroups $\mathfrak{g}_i \subseteq \mathfrak{G}$, which are:

$$\begin{aligned} \Delta_1 &:= \mathbb{Q}, & \mathfrak{g}_1 &= \mathfrak{G}, \\ \Delta_2 &:= \mathbb{Q}(i), & \mathfrak{g}_2 &= \{\phi_1, \phi_5\}, \\ \Delta_3 &:= \mathbb{Q}(\sqrt{2}), & \mathfrak{g}_3 &= \{\phi_1, \phi_7\}, \\ \Delta_4 &:= \mathbb{Q}(\sqrt{-2}), & \mathfrak{g}_4 &= \{\phi_1, \phi_3\}, \\ \Delta_5 &:= \mathbb{F}, & \mathfrak{g}_5 &= \{\phi_1\}; \end{aligned}$$

so as to compute the orbits of each S_j with respect to $\mathfrak{g}_i$:

S_1, S_2 are stable under $\mathfrak{g}_i$;

the orbits of S_4 are $\begin{cases} \{2\}, \{6\} & \text{if } i = 2, 5, \\ \{2, 6\} & \text{otherwise}; \end{cases}$

the orbits of S_6 are $\begin{cases} \{1, 3, 5, 7\} & \text{if } i = 1, \\ \{1, 5\}, \{3, 7\} & \text{if } i = 2, \\ \{1, 7\}, \{3, 5\} & \text{if } i = 3, \\ \{1, 3\}, \{5, 7\} & \text{if } i = 4, \\ \{1\}, \{3\}, \{5\}, \{7\} & \text{if } i = 5; \end{cases}$

and to derive the claimed factorization. $\quad\boxed{2\!\!\!\downarrow}$

With this result, it is now easy to factorize $f(X) = X^8 - 1$ in $K[X]$ for any field K, without any restriction over its characteristic; we just need the following considerations:

Lemma 15.2.2. *Let K be a field,* $\text{char}(K) \neq 2$ *and*

$$\Sigma := \{-1, 2, -2\} \subset K;$$

then:

(1) If $\text{char}(K) \neq 0$, at least one element in Σ has a root in K.
(2) If two elements in Σ have a root in K, the same holds for the third.

Proof

(1) Consider the multiplicative group $K^* := K \setminus \{0\}$ and consider the group morphism $\sigma : K^* \mapsto K^*$ defined by $\sigma(a) := a^2$; since $\ker(\sigma) = \{1, -1\}$ we can deduce that
$$\text{Im}(\sigma) = \{a^2 : a \in K^*\}$$
is a subgroup of index 2. Therefore the map $\chi : K^* \mapsto \{1, -1\}$ defined by
$$\chi(a) = \begin{cases} 1 & \text{iff } a \in \text{Im}(\sigma); \\ -1 & \text{iff } a \notin \text{Im}(\sigma); \end{cases}$$
is a group morphism.
Therefore
$$\chi(-1) = \chi(2) = -1 \implies \chi(-2) = 1;$$
$$\chi(-1) = \chi(-2) = -1 \implies \chi(2) = 1;$$
$$\chi(2) = \chi(-2) = -1 \implies \chi(-1) = 1.$$

(2) If $a, b \in K$ are such that

$a^2 = -1, b^2 = 2$ then $c = ab$ is such that $c^2 = -2$;

$a^2 = -1, b^2 = -2$ then $c = ab$ is such that $c^2 = 2$;

$a^2 = 2, b^2 = -2$ then $c = \frac{ab}{2}$ is such that $c^2 = -1$.

♃

Now we can describe the factorization of $f(X) = X^8 + 1$ in $K[X]$ in terms of K:

Proposition 15.2.3. *Let K be a field, $S := \{-1, 2, -2\} \subset K$ and $f(X) = X^8 + 1 \in K[X]$. Then in $K[X]$ we have:*

if $\text{char}(K) = 2$, *then*
$$f(X) = (X + 1)^8;$$

if $\text{char}(K) \neq 2$ *and no one element in Σ is a square in K (e.g. $K = \mathbb{Q}$) then*
$$f(X) = (X - 1)(X + 1)(X^2 + 1)(X^4 + 1);$$

if $\text{char}(K) \neq 2$ *and -1 is the only element in Σ which has a square in K (e.g. $K = \mathbb{Q}(i), \mathbb{Z}_5$) then, denoting $\beta \in K$ such that $\beta^2 = -1$,*
$$f(X) = (X - 1)(X + 1)(X - \beta)(X + \beta)(X^2 - \beta)(X^2 + \beta);$$

15.2 An Exercise

if char$(K) \neq 2$ and 2 is the only element in Σ which has a square in K (e.g. $K = \mathbb{Q}(\sqrt{2}), \mathbb{R}, \mathbb{Z}_7$) then, denoting $\beta \in K$ such that $\beta^2 = 2$, in $K[X]$

$$f(X) = (X-1)(X+1)(X^2+1)(X^2-\beta X+1)(X^2+\beta X+1);$$

if char$(K) \neq 2$ and -2 is the only element in Σ which has a square in K (e.g. $K = \mathbb{Q}(\sqrt{-2}), \mathbb{Z}_{11}$) then, denoting $\beta \in K$ such that $\beta^2 = -2$,

$$f(X) = (X-1)(X+1)(X^2+1)(X^2-\beta X-1)(X^2+\beta X-1);$$

if char$(K) \neq 2$ and all elements in Σ have a square in K (e.g. $K = \mathbb{C}, \mathbb{Z}_{17}$) then, denoting $\beta, \gamma, \delta \in K$ such that $\beta^2 = 2, \gamma^2 = -2, \delta^2 = -1$,

$$f(X) = (X-1)(X+1)(X-\delta)(X+\delta)\left(X - \frac{\beta+\gamma}{2}\right)$$
$$\cdot \left(X - \frac{-\beta+\gamma}{2}\right)\left(X - \frac{\beta-\gamma}{2}\right)\left(X - \frac{-\beta-\gamma}{2}\right).$$

$\boxed{\text{4}}$

Example 15.2.4. It is worthwhile checking the factorization in some instances:

$K = \mathbb{Z}_5$: then $\beta := 2$ satisfies $\beta^2 = -1$ and we have

$$f(X) = (X-1)(X+1)(X-2)(X+2)(X^2-2)(X^2+2);$$

$K = \mathbb{Z}_7$: then $\beta := 3$ satisfies $\beta^2 = 2$ and we have

$$f(X) = (X-1)(X+1)(X^2+1)(X^2-3X+1)(X^2+3X+1);$$

$K = \mathbb{Z}_{11}$: then $\beta := 3$ satisfies $\beta^2 = -2$ and we have

$$f(X) = (X-1)(X+1)(X^2+1)(X^2-3X-1)(X^2+3X-1);$$

$K = \mathbb{Z}_{17}$: then $\beta := 6$ and $\gamma = 7$ satisfy $\beta^2 = 2, \gamma^2 = -2$, so that $\delta = \frac{\beta\gamma}{2} = 4$ satisfies $\delta^2 = -1$ and we have

$$f(X) = (X-1)(X+1)(X-4)(X+4)(X+2)(X+8)(X-2)(X-8).$$

Example 15.2.5. It is also worthwhile verifying what happens when $K = GF(9)$, which is the splitting field of $X^9 - X$ and so contains all the roots of $X^8 - 1$ and all the square roots of Σ. Let us choose, as a representation of $GF(9)$, $K = \mathbb{Z}_3[\epsilon]$ where $\epsilon^2 + \epsilon - 1 = 0$. It is easy to verify that[3]

$\gamma := 1$ satisfies $\gamma^2 = 1 = -2$,
$\beta := \epsilon - 1$ satisfies $\beta^2 = \epsilon^2 + \epsilon + 1 = 2$,
$\delta := \frac{\beta\gamma}{2} = -\beta = -\epsilon + 1$ satisfies $\delta^2 = -1$.

---

[3] Of course we will in fact verify that $\delta = -\beta$ and $\delta^2 = \beta^2 = 2 = -1$. But we are choosing the notation according to Proposition 15.2.3.

Therefore the roots are

$$\begin{aligned}
\alpha = \alpha_1 &= \tfrac{\beta}{2} + \tfrac{\gamma}{2} = \epsilon, \\
\alpha_2 &= \delta = -\epsilon + 1, \\
\alpha_3 &= -\tfrac{\beta}{2} + \tfrac{\gamma}{2} = -\epsilon - 1, \\
\alpha_4 &= -1, \\
\alpha_5 &= -\tfrac{\beta}{2} - \tfrac{\gamma}{2} = -\epsilon, \\
\alpha_6 &= -\delta = \epsilon - 1, \\
\alpha_7 &= \tfrac{\beta}{2} - \tfrac{\gamma}{2} = +\epsilon + 1, \\
\alpha_8 &= 1,
\end{aligned}$$

which in fact satisfy

$$\begin{aligned}
\alpha^2 &= \epsilon^2 = -\epsilon + 1 & &= \alpha_2, \\
\alpha^3 &= -\epsilon^2 + \epsilon = -\epsilon - 1 & &= \alpha_3, \\
\alpha^4 &= -\epsilon^2 - \epsilon = -1 & &= \alpha_4, \\
\alpha^5 &= -\epsilon & &= \alpha_5, \\
\alpha^6 &= -\epsilon^2 = +\epsilon - 1 & &= \alpha_6, \\
\alpha^7 &= \epsilon^2 - \epsilon = \epsilon + 1 & &= \alpha_7, \\
\alpha^8 &= \epsilon^2 + \epsilon = 1 & &= \alpha_8,
\end{aligned}$$

as expected.

The example allows some more analysis: in fact, there is a single subfield of $GF(9)$, i.e. $\mathbb{Z}_3$. Among the elements of Σ, the only one which has a root in $\mathbb{Z}_3$ is -2 whose roots are ± 1. We therefore have in $\mathbb{Z}_3$ the factorization

$$f(X) = (X - 1)(X + 1)(X^2 + 1)(X^2 - X - 1)(X^2 + X - 1).$$

Note that $\epsilon = \alpha$ has $X^2 + X - 1$ as its minimal polynomial. Therefore $GF(9) = \mathbb{Z}_3(\epsilon) = \mathbb{Z}_3(\alpha)$. Of course, the Galois group $\mathfrak{g} := G(GF(9)/GF(3))$ consists of two $GF(3)$-isomorphisms, the identity and the 'conjugation' ψ which satisfy $\psi(\epsilon) = -\epsilon$. The eight roots of the identity are then partitioned in the orbits with respect to $\mathfrak{g}$ as

$$\{\alpha_8\} \cup \{\alpha_4\} \cup \{\alpha_2, \alpha_6\} \cup \{\alpha_3, \alpha_5\} \cup \{\alpha_1, \alpha_7\}.$$

Remark 15.2.6. In fact the analysis contained in Proposition 15.2.3 can be concluded by noting that for each prime field $\mathbb{Z}_p$, $p \neq 2$, all square roots exist in $GF(p^2)$.

15.2 An Exercise

Therefore, if $\text{char}(K) = p \neq 0, 2$ either

- $K \supseteq GF(p^2)$, in which case f splits in linear factors according to Proposition 15.2.3;
- $K \not\supseteq GF(p^2)$, in which case the factorization of f depends on which elements in Σ have a square in $\mathbb{Z}_p$ in accordance with Proposition 15.2.3.

16
Kronecker III: factorization

The essence of a factorization tool for his model was completely clear to Kronecker, who wrote[1]

Die im Article 1 aufgestellte Definition der Irreductibilität entbehrt so lange einer sicheren Grundlage, als nicht eine Methode angegeben ist, mittels deren bei einer bestimmten, vorgelegten Function entschieden werden kann, ob dieselbe der aufgestellten Definition gemäss irreductibel ist oder nicht. [...] Desshalb soll hier eine neue Methode dargelegt werden, welche nur einfache, hier bereits verwendbare Hülfsmittel in Auspruch nimmt.
The definition of irreducibility enunciated in Article 1 lacks a firm foundation, the more so since no method is indicated by which, given a function, it can be decided whether or not its definition is irreducible. [...] A new method is presented, which will simply be applied here as a tool if needed.

and presented in the following pages tools which allowed him to factorize a polynomial

- in $\mathbb{Z}[X]$;
- in $k[X_1, \ldots, X_n][X]$ where k is a field for which there exists an algorithm for factorizing polynomials in $k[X]$;
- in $k(\alpha)[X]$ where α is an algebraic extension of k, a field for which there exists an algorithm for factorizing polynomials in $k[X]$.

Using the Gauss Lemma (Section 6.1), which strictly relates polynomial factorization over a unique factorization domain to that over its fraction field, his tools allowed him to factorize polynomial

- in $\mathbb{Q}[X]$, using the factorization algorithm in $\mathbb{Z}[X]$;

---

[1] L. Kronecker, Grundzüge einer Arithmetischen Theorie der Algebraischen Grössen, *Crelle's Journal*, **92** (1882), p. 11.

in $k(X_1, \ldots, X_n)[X]$ – where $k(X_1, \ldots, X_n)$ is a finite transcendental extension over a field k for which there exists a factorization algorithm –, using the factorization algorithm in $k[X_1, \ldots, X_n][X]$

and, therefore, a factorization algorithm for each finite extension field $k \supset \mathbb{Q}$, i.e. for any 0 characteristic field explicitly given in Kronecker's Model.

In the next three sections I will discuss Kronecker's proposals.

16.1 Von Schubert Factorization Algorithm over the Integers

The algorithm which we present here was proposed by von Schubert (1793) and rediscovered by Kronecker. It allows us to factorize polynomials $f(X)$ in $\mathbb{Z}[X]$ and, therefore, in $\mathbb{Q}[X]$ via the Gauss Lemma.

Algorithm 16.1.1 (von Schubert). Let us assume we are given a polynomial $f(X) \in \mathbb{Z}[X]$; through squarefree decomposition, we can moreover assume that it is squarefree.

Denoting $d := \deg(f)$, it is obvious that, if it is reducible, it must have a factor g such that $\deg(g) \leq \frac{d}{2}$.

We can therefore restrict our aim to finding a possible irreducible factor g of degree δ, $1 \leq \delta := \deg(g) \leq \frac{d}{2}$; if no such factor exists then we can conclude that f is irreducible; otherwise, we have just to reapply the algorithm to f/g; obviously, it is better to find g for increasing values of δ.

The tool we will use to find this g is nothing more than the obvious remark that, if g is a factor of f, then, for each $a \in \mathbb{Z}$, $g(a)$ divides $f(a)$.

Following the suggestion of this remark, in order to find a factor g of f with degree δ, let us

pick up $\delta + 1$ integers $a_0, a_1, \ldots, a_\delta \in \mathbb{Z}$;
for each i, evaluate $b_i := f(a_i)$ and factorize it;
pick up in all possible ways a sequence $c_1, \ldots, c_\delta$ such that for all i, c_i is a factor of b_i;
by the Lagrange Interpolation Formula, compute the single polynomial g satisfying

$$\deg(g) = \delta,$$
$$g(a_i) = c_i, \text{ for all } i;$$

verify whether g divides f.

The von Schubert Factorization Algorithm is drafted in Figure 16.1; the example in Example 16.1.2 should be sufficient to convince the reader of the

---
Fig. 16.1. Von Schubert Factorization Algorithm

$[g_1, \ldots, g_k] :=$ **Factorization**(f)
where
  $f \in \mathbb{Z}[X]$ is a squarefree polynomial
  $g_1, \ldots, g_k \in \mathbb{Z}[X]$ are the irreducible factors of f in $\mathbb{Z}[X]$
$d := \deg(f), L := [], \delta := 1$
While $\delta \leq \frac{d}{2}$ **do**
  Choose $a_0, a_1, \ldots a_\delta \in \mathbb{Z}$
  For $i = 0, \ldots, \delta$ **do**
    Factorize $b_i := f(a_i)$
  $S := \{(c_0, \ldots, c_\delta) : c_i \mid b_i, \forall i\}$,
  While $S \neq \emptyset$ **and** $\delta \leq \frac{d}{2}$ **do**
    Choose $(c_0, \ldots, c_\delta) \in S$
    $S := S \setminus \{(c_0, \ldots, c_\delta)\}$
    Compute $g(X) \in \mathbb{Z}[X]$ such that
      $\deg(g) = \delta$
      $g(a_i) = c_i, \forall i$
    If $g(X)$ divides $f(X)$ **then**
      $f := \frac{f}{g}$
      $d := \deg(f),$
      $L := [L, g]$
  $\delta := \delta + 1$
$[L, f]$

---

horrible complexity of the algorithm and the need of a better one even if expedients are applied such as:

first factorize f modulo some small primes, in order to avoid computations
  for impossible degrees δ of factors;
use small values for the $as \ldots$

Example 16.1.2. As an example let us try to factorize
$$f(X) = (X-2)(X+2)(X^2+4)(X^4+16) = (X^8 - 256).$$
For this example, factorization is, of course, a trivial task, but I chose it to allow easier verification: the reader is required to imagine that we are computing on a 'generic' polynomial of degree 8.

To find all factors of degree $\delta = 1$, we choose $a_0 := 1, a_1 := -1$, so that $b_0 = b_1 = -255$. The set of the factors of -255 is
$$F_1 := \{\pm 1, \pm 3, \pm 5, \pm 15, \pm 17, \pm 51, \pm 85, \pm 255\}$$
and therefore[2] card$(S) \leq 128$.

---
[2] In fact if $g(a_i) = c_i$, for all i, then $-g(a_i) = -c_i$, for all i; it is therefore useless to verify all tuples whose first element is negative.
For the same reason, it is also useless to verify the tuples such that $\gcd_i(c_i) \neq 1$, those such that c_i has more factors than $\frac{b_i}{c_i}$, for all i, *und so weiter*.

16.1 Von Schubert Factorization Algorithm over the Integers

We need to interpolate a linear polynomial g for each of the 128 tuples $(c_0, c_1) \in S$ as follows:

(1) $(1, 1)$; then $g(X) = 1$;
(2) $(1, -1)$; then $g(X) = X$ which does not divide f;
(3) $(1, 3)$; then $g(X) = 2 - X$ which divides f; so we get the factorization

$$f = (X - 2) f_1(X)$$

where

$$f_1(X) = X^7 + 2X^6 + 4X^5 + 8X^4 + 16X^3 + 32X^2 + 64X + 128; \quad (16.1)$$

(4) $(1, -3)$; then $g(X) = -1 + 2X$ which does not divide f_1;

...

(16) $(1, -255)$; then $g(X) = -126 + 127X$ which does not divide f_1;
(17) $(3, 1)$; then $g(X) = 2 + X$ which divides f_1; so we get the factorization

$$f = (X - 2)(X + 2) f_2(X)$$

where

$$f_2(X) = X^6 + 4X^4 + 16X^2 + 64.$$

While we have found the two linear factors of f, we still do not know that and therefore we have to go on, computing:

(18) $(3, 3)$; interpolation is useless;
(19) $(3, -3)$; interpolation is useless;
(20) $(3, 5)$; then $g(X) = 4 - X$ which does not divide f_2;

...

(113) $(255, -1)$; then $g(X) = 43 - 42X$ which does not divide f_2.

Since $\gcd(255, c_2) \neq 1$ for any $c_2 \in F \setminus \{\pm 1\}$, we have tested all the useful elements of S and we can therefore conclude that we have found all the linear factors of f.

So we move to finding all the factors of f_2 with degree $\delta = 2$: we choose $a_0 := 1, a_1 := -1, a_2 := 2$, so that $b_0 = b_1 = 85, b_2 = 256$. The set of the factors of 85 is

$$F_1 := \{\pm 1, \pm 5, \pm 17, \pm 85\}$$

and the set of those of 256 is

$$F_2 := \{\pm 1, \pm 2, \pm 4, \pm 8, \pm 16, \pm 32, \pm 64, \pm 128, \pm 256\}$$

and therefore card$(S) \leq 2^{16} = 65536$.

We interpolate a quadratic polynomial g for each of the 65536 tuples $(c_0, c_1, c_2) \in S$ as follows:

(1) $(1, 1, 1)$; then $g(X) = 1$;
(2) $(1, 1, -1)$; then $g(X) = -\frac{2}{3}X^2 + \frac{5}{3}$ which does not divide f_2;
(3) $(1, 1, 2)$; then $g(X) = \frac{1}{3}X^2 + \frac{2}{3}$ which does not divide f_2;
...
(180) $(5, -1, 256)$; then $g(X) = \frac{247}{3}X^2 + 4X - \frac{244}{3}$ which does not divide f_2;
(181) $(5, 5, 1)$; then $g(X) = -\frac{4}{3}X^2 + \frac{19}{3}$ which does not divide f_2;
...
(187) $(5, 5, 8)$; then $g(X) = X^2 + 4$ which divides f_2; so we get the factorization

$$f = (X - 2)(X + 2)(X^2 + 4)f_3(X)$$

where

$$f_3(X) = X^4 + 16;$$

(188) $(5, 5, -8)$; then $g(X) = -\frac{13}{3}X^2 + \frac{28}{3}$ which does not divide f_3;
...
(65536) $(85, -85, -256)$; then $g(X) = -87X^2 + 92$ which does not divide f_3.

Terminating this tour de force, we can conclude that there is no other quadratic factor and therefore that f_4 is irreducible and the factorization of f is

$$f = (X - 2)(X + 2)(X^2 + 4)(X^4 + 16).$$

16.2 Factorization of Multivariate Polynomials

Let k be a field on which we have a factorization algorithm for univariate polynomials and let us discuss Kronecker's technique for generalizing it to a factorization algorithm for *multivariate* polynomials.

Let us denote

$\mathcal{P} := k[X_1, \ldots, X_n]$;
$\deg_i(f)$ to be the degree of $f \in \mathcal{P}$ in the variable X_i;
$\mathcal{P}_d := \{f \in k[X_1, \ldots, X_n] : \deg_i(f) < d, \text{ for all } i\}$, for all $d \in \mathbb{N}$;
$\delta(d) := \sum_{i=1}^{n} d^i = d + d^2 + \cdots + d^n = d\frac{d^n-1}{d-1}$, for all $d \geq 2$;
$\chi_d : \mathcal{P} \mapsto k[X]$ to be the map defined by $\chi_d(X_i) = X^{d^{i-1}}$, for all i.

Lemma 16.2.1. *With the notation above:*

(1) The restriction $\overline{\chi}_d$ of the map χ_d to $\mathcal{P}_d$ is a k-vector space isomorphism between $\mathcal{P}_d$ and its image in $k[X]$ which is

$$\text{Im}(\overline{\chi}_d) := \{f \in k[X] : \deg(f) \leq \delta(d)\}.$$

16.2 Factorization of Multivariate Polynomials

(2) Let $f \in \mathcal{P}_d$, $g_1, g_2 \in k[X]$ be such that $g_1 g_2 = \chi_d(f)$; then $\deg(g_i) < \delta(d)$, for all i, and

$$\overline{\chi}_d^{-1}(g_1)\overline{\chi}_d^{-1}(g_2) = f.$$

(3) Let $f \in \mathcal{P}_d$; f is irreducible iff $\chi_d(f)$ is.

Proof

(1) For each term $t := X_1^{a_1} \cdots X_n^{a_n} \in \mathcal{P}_d$, we have $a_i < d$, for all i, and $\overline{\chi}_d(t) = X^\alpha$ where

$$\alpha = \sum_{i=1}^n a_i d^{i-1} < \sum_{i=1}^n d^i = \delta(d).$$

Conversely for each integer $0 \le \alpha < \delta$ there is a unique term $t := X_1^{a_1} \cdots X_n^{a_n} \in \mathcal{P}_d$ such that $\overline{\chi}_d(t) = X^\alpha$ and it can be computed by expressing α in its d-ary representation

$$\alpha = \sum_{i=1}^n a_i d^{i-1}, \; a_i < d \; \text{ for all } i.$$

(2) Since $\deg(\chi_d(f)) < \delta(d)$, the claim is obvious.
(3) It is an obvious consequence of the previous result. ⨆

Algorithm 16.2.2. On the basis of the above lemma, Kronecker's proposal to factorize a polynomial $F \in \mathcal{P}$, such that $\deg_i(F) < d$, for all i, consists of factorizing $f(X) := \chi_d(F) \in k[X]$ thus getting a factorization

$$f(X) = \prod_j g_j(X)$$

where $\deg(g_j) \le \deg(f) < \delta(d)$, so that

$$F = \prod_j \overline{\chi}_d(g_j)$$

is the required factorization.

Example 16.2.3. Let us consider the polynomial

$$F(X_1, X_2, X_3)] \in \mathbb{Q}[X_1, X_2, X_3]$$

defined by

$$F := X_1 X_2 X_3 + 16 X_1 X_2 + 4 X_1 X_3 + 2 X_2 X_3 + 64 X_1 + 32 X_2 + 8 X_3 + 128.$$

Since $\deg_i(F) = 1$, for all i, we can fix $d := 2$, so that $\delta(d) = 1 + 2 + 4 = 7$, and consider $\chi_2 : \mathcal{P}_2 \mapsto \{f \in k[X] : \deg(f) \leq 7\}$.

Note that we have

$$\begin{array}{rclcrcl}
\chi_2(1) & = & 1, & \chi_2(X_1) & = & X, \\
\chi_2(X_2) & = & X^2, & \chi_2(X_1 X_2) & = & X^3, \\
\chi_2(X_3) & = & X^4, & \chi_2(X_1 X_3) & = & X^5, \\
\chi_2(X_2 X_3) & = & X^6, & \chi_2(X_1 X_2 X_3) & = & X^7.
\end{array}$$

so that

$$\chi_2(F) = X^7 + 2X^6 + 4X^5 + 8X^4 + 16X^3 + 32X^2 + 64X + 128 = f_1(X)$$

where $f_1(X)$ is the polynomial introduced in Equation 16.1, of which we know the factorization in $\mathbb{Q}[X]$ which is

$$f_1 = (X+2)(X^2+4)(X^4+16)$$

so that, by applying $\overline{\chi}_2$, we deduce

$$F(X_1, X_2, X_3) = (X_1 + 2)(X_2 + 4)(X_3 + 16).$$

16.3 Factorization over a Simple Algebraic Extension

The last task for completing Kronecker's factorization approach is to describe an algorithm which allows us to factorize univariate polynomials over a field $k[\alpha]$ where k is a field such that $\operatorname{char}(k) = 0$ and we have a factorization algorithm over it.

Assuming we are given such a field in Kronecker's Model, we know therefore the minimal polynomial $m(X) \in k[X]$ of α. Putting $n := \deg(\alpha)$, we know that α has n conjugates which we will denote, if needed, as $\alpha = \alpha_1, \ldots, \alpha_n$; we will also denote the n k-isomorphisms from $k[\alpha]$ into $k[\alpha_i]$ by ψ_i.

With this notation, I will describe Kronecker's algorithm for factorization over $k[\alpha]$; but first I need some results.

Proposition 16.3.1. *If $f(X) \in k[\alpha][X]$ is irreducible, then $N_{k[\alpha]/k}(f)$ is the power of an irreducible polynomial in $k[X]$.*

Proof We know that $N_{k[\alpha]/k}(f) \in k[X]$.
Assume there are $g, h \in k[X]$ such that $N_{k[\alpha]/k}(f) = gh$, and $\gcd(g, h) = 1$. Since f divides $N_{k[\alpha]/k}(f)$, it divides either g or h but not both; let us say it divides g. Then $f_i := \psi_i(f)$ divides $\psi_i(g) = g$, for all i, and, since $\gcd(g, h) = 1$, then $\gcd(f_i, h) = 1$, for all i, so $\gcd(N_{k[\alpha]/k}(f), h) = 1$ and h is a constant. $\qed$

16.3 Factorization over a Simple Algebraic Extension

Proposition 16.3.2. *Let $f(X) \in k[\alpha][X]$ be such that $N_{k[\alpha]/k}(f)$ is squarefree.*

Let $N_{k[\alpha]/k}(f) = \prod_{i=1}^{r} g_i$ be a factorization of $N_{k[\alpha]/k}(f)$ into irreducible factors in $k[X]$ and let $h_i := \gcd_{k[\alpha]}(f, g_i)$. Then $f = \prod_{i=1}^{r} h_i$ is a factorization of f into irreducible factors in $k[\alpha][X]$.

Proof Let $f = \prod_{i=1}^{s} p_i^{e_i}$ be a factorization of f into irreducible factors in $k[\alpha][X]$.
Then $N_{k[\alpha]/k}(f) = \prod_{i=1}^{s} N_{k[\alpha]/k}(p_i)^{e_i}$.
Since $N_{k[\alpha]/k}(f)$ is squarefree, then $e_i = 1$ for all i and so f is squarefree.
Moreover since, by the above proposition, $N_{k[\alpha]/k}(p_i)$ is a power of an irreducible polynomial in $k[X]$, then it must be an irreducible polynomial and coincide with one of the g_js. Therefore $s = r$ and, after reindexing, $N_{k[\alpha]/k}(p_i) = g_i$; so, for all i, p_i divides g_i and divides also $h_i = \gcd(f, g_i)$.
Assuming p_i divides g_j for some $j \neq i$, we deduce the contradiction that $g_i = N_{k[\alpha]/k}(p_i)$ divides $N_{k[\alpha]/k}(g_j) = g_j^n$. This establishes the conclusion that
$$p_i = \gcd(f, g_i) = h_i, \text{ for all } i.$$
♃

Lemma 16.3.3. *Let $f(X) \in k[X]$ be a squarefree polynomial and let*
$$f_c(X) := f(X - c\alpha) \in k[\alpha][X], \text{ for all } c \in k.$$
There are only finitely many $c \in k$ such that $N_{k[\alpha]/k}(f_c)$ is not squarefree.

Proof Let $\beta_1, \ldots, \beta_m$ be the roots of f (in the splitting field of f over k). Then the roots of $\psi_i(f_c)$ are $\beta_j - c\alpha_i$, $1 \leq j \leq m$, so that the roots of $N_{k[\alpha]/k}(f_c)$ are
$$\beta_j - c\alpha_i, 1 \leq i \leq n, 1 \leq j \leq m.$$
Then $N_{k[\alpha]/k}(f_c)$ is not squarefree if and only if

there exists $i, j, k, l : \beta_j - c\alpha_i = \beta_l - c\alpha_k, \ k \neq i$ or $l \neq j$,

whence the claim. ♃

Lemma 16.3.4. *Let $f(X) \in k[\alpha][X]$ be a squarefree polynomial. Then there is a squarefree polynomial $g(X) \in k[X]$ such that f divides g.*

Fig. 16.2. Factorization over algebraic extensions

$[f_1, \ldots, f_s] := \textbf{AlgExtFactorization}(f)$
where
    $f(X) \in k[\alpha][X]$ is squarefree
    $f_i(X) \in k[\alpha][X]$ is irreducible
    $f = f_1 \ldots f_r$ is a factorization of f
$g := \text{Prim}(f)$
Repeat
    chooserandom $c \in k$
    $g := f(X - c\alpha)$
    $h := N_{k[\alpha]/k}(g)$
until h is squarefree
$[g_1, \ldots, g_r] := \textbf{Factorization}(h)$
For $i = 1, \ldots, r$ **do**
    $h_i := \gcd(g, g_i)$
    $g := \frac{g}{h_i}$
    $f_i := h_i(X + c\alpha)$
$[f_1, \ldots, f_r]$

Proof Let $G := N_{k[\alpha]/k}(f) \in k[X]$ and let $g \in k[X]$ be the squarefree associate of G. Since f is squarefree and divides G, then it divides g too. ⨆⨅

Proposition 16.3.5. *Let $f(X) \in k[\alpha][X]$ be a squarefree polynomial and, for $c \in k$, let*

$$f_c(X) := f(X - c\alpha) \in k[\alpha][X].$$

There are only finitely many $c \in k$ such that $N_{k[\alpha]/k}(f_c)$ is not squarefree.

Proof Let $g \in k[X]$ be as in Lemma 16.3.4. By Lemma 16.3.3 there are only finitely many $c \in k$ such that $N_{k[\alpha]/k}(g_c)$ is not squarefree, where

$$g_c(X) = g(X - c\alpha).$$

Since f divides g then $N_{k[\alpha]/k}(f_c)$ divides $N_{k[\alpha]/k}(g_c)$ and so it is squarefree, except for finitely many $c \in k$. ⨆⨅

Algorithm 16.3.6. We are now ready to describe (Figure 16.2) an algorithm for factorizing a polynomial $f(X) \in k[\alpha][X]$. Via computing squarefree associates or a distinct power factorization, we can assume f is squarefree.

16.3 Factorization over a Simple Algebraic Extension

Example 16.3.7. As an example let us consider the ring $\mathbb{Q}(\alpha)$ where α is a primitive 12th root of unity and the polynomial

$$\begin{aligned} f(X) &= X^4 + (-3\alpha - 1)X^3 + (-\alpha^2 + 3\alpha)X^2 \\ &\quad + (3\alpha^3 + \alpha^2)X - 3\alpha^3 \\ &= (X - \alpha)(X + \alpha)(X - 3\alpha)(X - 1) \in \mathbb{Q}(\alpha)[X] \end{aligned}$$

and let us try[3] to factorize f.

The computations we performed in Remark 7.6.7 inform us that the minimal polynomial of α is

$$m(X) = X^4 - X^2 + 1$$

and its conjugates are

$$\begin{aligned} \alpha_1 &= \alpha, \\ \alpha_2 &= \alpha^5 = \alpha^3 - \alpha, \\ \alpha_3 &= \alpha^7 = -\alpha, \\ \alpha_4 &= \alpha^{11} = -\alpha^3 + \alpha. \end{aligned}$$

By conjugation we get the four polynomials

$$\begin{aligned} f(X) &= X^4 + (-3\alpha - 1)X^3 + (-\alpha^2 + 3\alpha)X^2 \\ &\quad + (3\alpha^3 + \alpha^2)X - 3\alpha^3, \\ f(X, \alpha_2) &= X^4 + (-3\alpha^3 + 3\alpha - 1)X^3 + (3\alpha^3 + \alpha^2 - 3\alpha - 1)X^2 \\ &\quad + (3\alpha^3 - \alpha^2 + 1)X - 3\alpha^3, \\ f(X, \alpha_3) &= X^4 + (3\alpha - 1)X^3 + (-\alpha^2 - 3\alpha)X^2 \\ &\quad + (-3\alpha^3 + \alpha^2)X + 3\alpha^3, \\ f(X, \alpha_4) &= X^4 + (3\alpha^3 - 3\alpha - 1)X^3 + (-3\alpha^3 + \alpha^2 + 3\alpha - 1)X^2 \\ &\quad + (-3\alpha^3 - \alpha^2 + 1)X + 3\alpha^3, \end{aligned}$$

and we have

$$\begin{aligned} N_{k[\alpha]/k}(f) &= \prod_{i=1}^{4} f(X, \alpha_1) \\ &= X^{16} - 4X^{15} - 5X^{14} + 40X^{13} + 37X^{12} - 364X^{11} \\ &\quad + 410X^{10} + 356X^9 - 782X^8 - 284X^7 + 1210X^6 \\ &\quad - 364X^5 - 683X^4 + 360X^3 + 315X^2 - 324X + 81. \end{aligned}$$

Note that, by conjugation again, there are the factorizations

$$\begin{aligned} f(X, \alpha_1) &= (X - \alpha)(X + \alpha)(X - 3\alpha)(X - 1), \\ f(X, \alpha_2) &= (X - \alpha_2)(X + \alpha_2)(X - 3\alpha_2)(X - 1), \end{aligned}$$

---

[3] Of course, as usual, we assume that we do not know the factorization written above, but we use it to better understand the crucial point of Kronecker' proposal.

$$\begin{aligned}
f(X, \alpha_3) &= (X+\alpha)(X-\alpha)(X+3\alpha)(X-1), \\
f(X, \alpha_4) &= (X+\alpha_2)(X-\alpha_2)(X+3\alpha_2)(X-1), \\
N_{k[\alpha]/k}(f)(X) &= (X-1)^4(X-\alpha)^2(X-\alpha_2)^2(X+\alpha)^2(X+\alpha_2)^2 \\
&\quad \cdot (X-3\alpha)(X-3\alpha_2)(X+3\alpha)(X+3\alpha_2) \\
&= (X-1)^4(X^4-X^2+1)^2(X^4-9X^2+81),
\end{aligned}$$

since, in fact,

$$\begin{aligned}
X^4 - X^2 + 1 &= m(X) &= \prod_i (X - \alpha_i), \\
X^4 - 9X^2 + 81 &= 81 m\left(\tfrac{X}{3}\right) &= \prod_i (X - 3\alpha_i),
\end{aligned}$$

therefore $N_{k[\alpha]/k}(f)$ is not squarefree.

Let us now see what happens if we compute the norm of

$$g(X, \alpha) = f(X + 2\alpha);$$

we get

$$\begin{aligned}
g(X) &= X^4 + (5\alpha - 1)X^3 + (5\alpha^2 - 3\alpha)X^2 \\
&\quad + (-5\alpha^3 + \alpha^2)X + 3\alpha^3 - 6\alpha^2 + 6, \\
g(X, \alpha_2) &= X^4 + (5\alpha^3 - 5\alpha - 1)X^3 + (-3\alpha^3 - 5\alpha^2 + 3\alpha + 5)X^2 \\
&\quad + (-5\alpha^3 - \alpha^2 + 1)X + 3\alpha^3 + 6\alpha^2, \\
g(X, \alpha_3) &= X^4 + (-5\alpha - 1)X^3 + (5\alpha^2 + 3\alpha)X^2 \\
&\quad + (5\alpha^3 + \alpha^2)X - 3\alpha^3 - 6\alpha^2 + 6, \\
g(X, \alpha_4) &= X^4 + (-5\alpha^3 + 5\alpha - 1)X^3 + (3\alpha^3 - 5\alpha^2 - 3\alpha + 5)X^2 \\
&\quad + (5\alpha^3 - \alpha^2 + 1)X - 3\alpha^3 + 6\alpha^2, \\
N_{k[\alpha]/k}(g) &= \prod_{i=1}^4 g(X, \alpha_i) \\
&= X^{16} - 4X^{15} - 9X^{14} + 48X^{13} + 93X^{12} - 452X^{11} \\
&\quad - 130X^{10} + 1172X^9 + 1206X^8 - 1812X^7 - 2130X^6 \\
&\quad + 1732X^5 + 3145X^4 - 1008X^3 - 2061X^2 + 324X \\
&\quad + 1053,
\end{aligned}$$

and the factorizations

$$\begin{aligned}
g(X, \alpha_1) &= (X+\alpha)(X+3\alpha)(X-\alpha)(X-1+2\alpha), \\
g(X, \alpha_2) &= (X+\alpha_2)(X+3\alpha_2)(X-\alpha_2)(X-1+2\alpha_2), \\
g(X, \alpha_3) &= (X-\alpha)(X-3\alpha)(X+\alpha)(X-1-2\alpha), \\
g(X, \alpha_4) &= (X-\alpha_2)(X-3\alpha_2)(X+\alpha_2)(X-1-2\alpha_2),
\end{aligned}$$

16.3 Factorization over a Simple Algebraic Extension

$$
\begin{aligned}
N_{k[\alpha]/k}(g)(X) &= (X+\alpha)^2(X+\alpha_2)^2(X-\alpha)^2(X-\alpha_2)^2 \\
&\quad \cdot (X-3\alpha)(X-3\alpha_2)(X+3\alpha)(X+3\alpha_2) \\
&\quad \cdot (X-1+2\alpha)(X-1+2\alpha_2)(X-1-2\alpha) \\
&\quad \cdot (X-1-2\alpha_2) \\
&= (X^4 - X^2 + 1)^2(X^4 - 9X^2 + 81) \\
&\quad \cdot (X^4 - 4X^3 + 2X^2 + 4X + 13),
\end{aligned}
$$

where

$$
\begin{aligned}
X^4 - X^2 + 1 &= m(X) &&= \prod_i (X - \alpha_i), \\
X^4 - 9X^2 + 81 &= 81 m\left(\tfrac{X}{3}\right) &&= \prod_i (X - 3\alpha_i), \\
X^4 - 4X^3 + 2X^2 + 4X + 13 &= 16 m\left(\tfrac{-X+1}{2}\right) &&= \prod_i (X - 1 + 2\alpha_i),
\end{aligned}
$$

therefore $N_{k[\alpha]/k}(g)$ is not squarefree.

Let us now see what happens if we compute the norm of

$$g(X, \alpha) = f(X - 2\alpha);$$

we get

$$
\begin{aligned}
g(X) &= X^4 + (-11\alpha - 1)X^3 + (41\alpha^2 + 9\alpha)X^2 \\
&\quad + (-61\alpha^3 - 23\alpha^2)X + 15\alpha^3 + 30\alpha^2 - 30, \\
g(X, \alpha_2) &= X^4 + (-11\alpha^3 + 11\alpha - 1)X^3 \\
&\quad + (9\alpha^3 - 41\alpha^2 - 9\alpha + 41)X^2 \\
&\quad + (-61\alpha^3 + 23\alpha^2 - 23)X + 15\alpha^3 - 30\alpha^2, \\
g(X, \alpha_3) &= X^4 + (11\alpha - 1)X^3 + (41\alpha^2 - 9\alpha)X^2 \\
&\quad + (61\alpha^3 - 23\alpha^2)X - 15\alpha^3 + 30\alpha^2 - 30, \\
g(X, \alpha_4) &= X^4 + (11\alpha^3 - 11\alpha - 1)X^3 \\
&\quad + (-9\alpha^3 - 41\alpha^2 + 9\alpha + 41)X^2 \\
&\quad + (61\alpha^3 + 23\alpha^2 - 23)X - 15\alpha^3 - 30\alpha^2, \\
N_{k[\alpha]/k}(g) &= \prod_{i=1}^{4} g(X, \alpha_i) \\
&= X^{16} - 4X^{15} - 33X^{14} + 144X^{13} + 909X^{12} \\
&\quad - 4004X^{11} - 7138X^{10} + 38324X^9 \\
&\quad + 54534X^8 - 271284X^7 - 51858X^6 \\
&\quad + 469924X^5 + 703753X^4 - 435600X^3 \\
&\quad - 656325X^2 + 202500X + 658125,
\end{aligned}
$$

and the factorizations

$$
\begin{aligned}
g(X, \alpha_1) &= (X - 3\alpha)(X - \alpha)(X - 5\alpha)(X - 1 - 2\alpha), \\
g(X, \alpha_2) &= (X - 3\alpha_2)(X - \alpha_2)(X - 5\alpha_2)(X - 1 - 2\alpha_2), \\
g(X, \alpha_3) &= (X + 3\alpha)(X + \alpha)(X + 5\alpha)(X - 1 + 2\alpha),
\end{aligned}
$$

$$g(X, \alpha_4) = (X + 3\alpha_2)(X + \alpha_2)(X + 5\alpha_2)(X - 1 + 2\alpha_2),$$
$$N_{k[\alpha]/k}(g)(X) = (X - 3\alpha)(X - 3\alpha_2)(X + 3\alpha)(X + 3\alpha_2)$$
$$\cdot (X - \alpha)(X - \alpha_2)(X + \alpha)(X + \alpha_2)$$
$$\cdot (X - 5\alpha)(X - 5\alpha_2)(X + 5\alpha)(X + 5\alpha_2)$$
$$\cdot (X - 1 - 2\alpha)(X - 1 - 2\alpha_2)(X - 1 + 2\alpha)$$
$$\cdot (X - 1 + 2\alpha_2)$$
$$= (X^4 - 9X^2 + 81)(X^4 - X^2 + 1)$$
$$\cdot (X^4 + 4X^3 + 2X^2 - 4X + 13)(X^4 - 25X^2 + 625),$$

where

$$\begin{array}{rcll}
X^4 - 9X^2 + 81 & = & 81m\left(\frac{-X}{3}\right) & = \prod_i (X - 3\alpha_i), \\
X^4 - X^2 + 1 & = & m(X) & = \prod_i (X - \alpha_i), \\
X^4 + 4X^3 + 2X^2 - 4X + 13 & = & 16m\left(\frac{X+1}{2}\right) & = \prod_i (X - 1 - 2\alpha_i), \\
X^4 - 25X^2 + 625 & = & 625m\left(\frac{-X}{5}\right) & = \prod_i (X + 5\alpha_i).
\end{array}$$

Therefore $N_{k[\alpha]/k}(g)$ is squarefree and we have to compute

$$\begin{array}{rcl}
\gcd(g, X^4 - 9X^2 + 81) & = & X + 3\alpha, \\
\gcd(g, X^4 - X^2 + 1) & = & X - \alpha, \\
\gcd(g, X^4 - 4X^3 + 2X^2 + 4X + 13) & = & X - 1 - 2\alpha, \\
\gcd(g, X^4 - 25X^2 + 625) & = & X + 5\alpha,
\end{array}$$

to deduce the factorizations

$$f(X - 2\alpha) = (X - 3\alpha)(X - \alpha)(X - 1 - 2\alpha)(X - 5\alpha),$$
$$f(X) = (X - \alpha)(X + \alpha)(X - 1)(X - 3\alpha).$$

Remark 16.3.8. The reader will have noted the strict connection between this factorization algorithm and the computation of primitive elements (Lemma 8.4.2). In fact, $N_{k[\alpha]/k}(f_c)$ is a squarefree polynomial whose roots are $\xi_{ij} = \beta_i + c\alpha_j$, for all i, j, so that

$$f_c(X) = f(X - c\alpha) = \prod_i (X - c\alpha - \beta_i) = \prod_i (X - \xi_{i1});$$

also, denoting $m_i(X) := c^n m\left(\frac{X - \beta_i}{c}\right)$, for all i, we have

$$m_i(X) = c^n \prod_j \left(\frac{(X - \beta_i)}{c} - \alpha_j\right) = \prod_j (X - \beta_i - c\alpha_j) = \prod_j (X - \xi_{ij}).$$

In conclusion we get

$$N_{k[\alpha]/k}(f_c)(X) = \prod_i \prod_j (X - \xi_{ij})$$

16.3 Factorization over a Simple Algebraic Extension

$$= \prod_j \phi_j(f_c)$$
$$= \prod_i m_i(X).$$

This formula is of course related to the ones discussed in Proposition 6.7.1 which, for polynomials

$$F(Y) := \prod_{j=1}^n (Y - \mathbf{a}_j), \quad G(Y) := \prod_{i=1}^n (Y - \mathbf{b}_i),$$

give

$$\prod_j G(\mathbf{a}_j) = \prod_i \prod_j (\mathbf{a}_j - \mathbf{b}_i) = (-1)^{nm} \prod_i F(\mathbf{b}_i);$$

in fact, setting $G = f_c, \mathbf{b}_i = \beta_i, \mathbf{a}_j = X - c\alpha_j$, we get

$$\prod_i \prod_j (X - \beta_i - c\alpha_j) = \prod_j \phi_j(f_c),$$

while, setting $F = m_1, \mathbf{b}_i = \frac{X - \beta_i}{c}, \mathbf{a}_j = \alpha_j$, we get

$$\prod_i \prod_j (X - \beta_i - c\alpha_j) = \prod_i m_i(X).$$

Example 16.3.9. Let us consider the example $k := \mathbb{Q}, \alpha := \sqrt{3}$ so that $m(X) = X^2 - 3, \beta := \sqrt{2}$ so that $f(X) = X^2 - 2$ and $c := 1$ (cf. Example 8.4.3).

Then, denoting $\phi : \mathbb{Q}[\sqrt{3}] \mapsto \mathbb{Q}[\sqrt{3}]$ to be the conjugate $\mathbb{Q}$-isomorphism such that $\phi(\sqrt{3}) = -\sqrt{3}$ and setting $\xi := \sqrt{2} + \sqrt{3}$ we have the four conjugates

$$\xi_{++} := +\sqrt{2} + \sqrt{3}, \qquad \xi_{+-} := +\sqrt{2} - \sqrt{3},$$
$$\xi_{-+} := -\sqrt{2} + \sqrt{3}, \qquad \xi_{--} := -\sqrt{2} - \sqrt{3},$$

whose minimal polynomial

$$h(X) := X^4 - 10X^2 + 1$$

over $\mathbb{Q}[X]$ can be factorized as

$$h(X) = f_c \phi(f_c) = m(X - \sqrt{2})m(X + \sqrt{2})$$

where

$$\begin{aligned} f_c &= X^2 + 2\sqrt{3}X + 1 = (X - \xi_{++})(X - \xi_{-+}), \\ f_c &= X^2 - 2\sqrt{3}X + 1 = (X - \xi_{+-})(X - \xi_{--}), \end{aligned}$$

$$\begin{aligned} m(X+\sqrt{2}) &= X^2 + 2\sqrt{2}X - 1 = (X - \xi_{++})(X - \xi_{+-}), \\ m(X-\sqrt{2}) &= X^2 - 2\sqrt{2}X - 1 = (X - \xi_{-+})(X - \xi_{--}). \end{aligned}$$

Remark 16.3.10. The algorithm implicitly required that k be an infinite field, in order to guarantee that there is an element $c \in k$ such that $N_{k[\alpha]/k}(f_c)$ is squarefree. This requirement is satisfied if char$(k) = 0$ – the case which interested Kronecker – or k contains a transcendental element if char$(k) \neq 0$.

This algorithm, therefore, fails if k is an algebraic extension of a prime field $\mathbb{Z}_p$, which is, then, a finite field. On the other hand, Kronecker's proposal does not give any hint of how to factorize over a finite field, and even factorization over a field k, char$(k) = p \neq 0$, which contains a transcendental element, requires a factorization algorithm over the finite field $\mathbb{Z}_p$.

A factorization algorithm over a finite field is the subject of the next chapter.

17
Berlekamp

As we remarked in the previous chapter, Kronecker's factorization proposal did not solve the problem of factorization over a finite field.

This problem became interesting in the 1960s due to applications of polynomials over finite fields in computer science (feedback shift register sequences and error correcting codes) and was solved by Berlekamp in 1967 (eighty-five years after Kronecker's Algorithm!). A different, probabilistic, algorithm was then proposed by Cantor and Zassenhaus.

Both algorithms are presented in this chapter.

17.1 Berlekamp's Algorithm

Let F be a finite field of characteristic p and cardinality $q = p^n$.

Lemma 17.1.1. *Let $g(X) \in F[X]$. Then*

$$g^q - g = \prod_{\alpha \in F}(g - \alpha).$$

Proof This follows from the obvious factorization $Y^q - Y = \prod_{\alpha \in F}(Y - \alpha)$. ♃

Lemma 17.1.2. *Let $t(X) \in F[X]$ be a power of an irreducible polynomial,*

$$t(X) = s(X)^e,$$

and let $g(X) \in F[X]$.

Then $g^q - g$ is a multiple of $t \iff$ there exists $\alpha \in F$ such that $g \equiv \alpha \pmod{t}$.

Proof If $g^q - g$ is a multiple of t, then it is a multiple of s; since s is irreducible and $g^q - g = \prod_{\alpha \in F}(g - \alpha)$, this implies that there is $\alpha \in F$ such that $g - \alpha$ is a multiple of s.

Since for $\beta \neq \alpha$, $g - \beta$ and $g - \alpha$ are relatively prime, then s does not divide $g - \beta$, for all $\beta \neq \alpha$, and therefore $g - \alpha$ is a multiple of t.

Conversely if $g \equiv \alpha \pmod{t}$ then $g - \alpha$ is a multiple of t and so is $g^q - g$. ♃

Let us now fix a polynomial $f(X) \in F[X]$ and let $d := \deg(f)$.

Proposition 17.1.3. *Let $g(X) \in F[X]$ be such that*

$0 < \deg(g) < d$,
f *divides* $g^q - g$.

Then

$$f = \prod_{\alpha \in F} \gcd(f, g - \alpha)$$

is a non-trivial factorization of f.

Proof In fact

$$f = \gcd(f, g^q - g) = \gcd\left(f, \prod_{\alpha \in F} g - \alpha\right) = \prod_{\alpha \in F} \gcd(f, g - \alpha),$$

the last equality following from the fact that $g - \alpha$ and $g - \beta$ are relatively prime if $\beta \neq \alpha$.
Since

$$\deg(\gcd(f, g - \alpha)) \leq \deg(g - \alpha) < d = \deg(f)$$

the factorization is non-trivial. ♃

Let us denote by $F_d[X]$ the F-vector subspace of $F[X]$ consisting of the polynomials in $F[X]$ of degree less than d:

$$F_d[X] := \{g \in F[X] : \deg(g) < d\};$$

recall that $F_d[X]$ is endowed with a ring structure whose product is given by

$$g_1 \times g_2 = \mathbf{Rem}(g_1 g_2, f)$$

and that there is a ring isomorphism $\sigma : F_d[X] \mapsto F[X]/f(X)$ which associates to $g \in F_d[X]$ its residue class mod f.

17.1 Berlekamp's Algorithm

Let $f = \prod_{i=1}^{k} s_i^{e_i}$ be a factorization of f into irreducible factors and let $t_i = s_i^{e_i}$.

We recall that, by the Chinese Remainder Theorem, there is a ring isomorphism

$$\tau : F[X]/f(X) \mapsto F[X]/t_1(X) \oplus \cdots \oplus F[X]/t_k(X).$$

Under these isomorphisms if $g \in F_d[X]$, $\tau\sigma(g) = (g_1, \ldots, g_k)$ where g_i is the residue class of g mod t_i.

Note that the ring $F[X]/t_1(X) \oplus \cdots \oplus F[X]/t_k(X)$ contains

$$\tau\sigma(F) = \{(\alpha, \ldots, \alpha) : \alpha \in F\}$$

as an isomorphic copy of F, so that the restriction of its product turns it into an F-vector space and both σ and τ are F-vector space isomorphisms.

Lemma 17.1.4. *The morphism*

$$\epsilon : F_d[X] \mapsto F_d[X]$$

defined by

$$\epsilon(g) = g^q$$

is a ring morphism, so that in particular it is linear.

Proof Since $q = \text{card}(F)$, we have

$$g^q + h^q = (g + h)^q, \text{ for all } g, h \in F_d[X]$$

and

$$\alpha^q = \alpha, \text{ for all } \alpha \in F,$$

showing the linearity of ϵ. It is moreover compatible with multiplication and $\epsilon(1) = 1$, completing the proof. ♃

Corollary 17.1.5. *The subset*

$$V(f) := \{g \in F_d[X] : g^q \equiv g \bmod f\} \subseteq F_d[X]$$

is a subring and an F-vector space. ♃

Lemma 17.1.6. $\tau\sigma(V(f)) = \{(\alpha_1, \ldots, \alpha_k) : \alpha_i \in F, \text{ for all } i\}.$

Proof

$$g \in V(f) \iff g^q - g \text{ is a multiple of } f$$
$$\iff g^q - g \text{ is a multiple of } t_i, \forall i$$
$$\iff \text{for all } i, \text{ there exists } \alpha_i \in F : g - \alpha_i \text{ is a multiple of } t_i$$
$$\iff \text{for all } i, \text{ there exists } \alpha_i \in F : \tau\sigma(g) = (\alpha_1, \ldots, \alpha_n).$$

♃

Corollary 17.1.7. $\dim_F(V(f))$ *is the number of irreducible factors in the factorization of* f.

Proof The above lemma informs us that $V(f)$ is isomorphic under $\tau\sigma$ to the subvector space

$$\{(\alpha_1, \ldots, \alpha_k) : \alpha_i \in F, \text{ for all } i\}$$

of $F[X]/t_1(X) \oplus \cdots \oplus F[X]/t_k(X)$ whose dimension is k.

♃

Proposition 17.1.8. *Let* $\{g_1, \ldots, g_k\}$ *be a basis of* $V(f)$.
According to Lemma 17.1.2,

$$\text{for all } i, 1 \leq i \leq k, \text{ for all } j, 1 \leq j \leq k, \exists \, \alpha_{ij} \in F :$$
$$t_j \text{ divides } g_i - \alpha_{ij}.$$

Then:

(1) *for all* i, $\tau\sigma(g_i) = (\alpha_{i1}, \ldots, \alpha_{ik})$;
(2) *for all* j, l, *there is* i *such that* $\alpha_{ij} \neq \alpha_{il}$.

Proof

(1) It follows immediately from the description of $\tau\sigma$ implied by the proof of Lemma 17.1.6.
(2) Since, for $g = \sum_i c_i g_i$ we have

$$\tau\sigma(g) = \left(\sum_i c_i \alpha_{i1}, \ldots, \sum_i c_i \alpha_{ik} \right),$$

the assumption that there are j, l such that for all i, $\alpha_{ij} = \alpha_{il}$, implies the contradiction that

$$\tau\sigma(V(f)) \subseteq \{(\alpha_1, \ldots, \alpha_k) : \alpha_i \in F : \alpha_j = \alpha_l\}.$$

♃

17.1 Berlekamp's Algorithm

Example 17.1.9. To illustrate this result let us consider $F := \mathbb{Z}_3$ and $f := (X^9 - X)/(X^3 - X)$, which (by Theorem 7.2.2) is the product of all irreducible quadratic polynomials in $F[X]$, so that

$$f = X^6 + X^4 + X^2 + 1 = (X^2 + 1)(X^2 + X - 1)(X^2 - X - 1).$$

Denoting

$$t_1 := X^2 + 1, t_2 := X^2 + X - 1, t_3 := X^2 - X - 1,$$

the morphism τ is defined by

$$\tau(a_0 + a_1 X + a_2 X^2 + a_3 X^3 + a_4 X^4 + a_5 X^5) = (\tau_1, \tau_2, \tau_3)$$

where

$$\begin{aligned}
\tau_1 &:= (a_0 - a_2 + a_4) + (a_1 - a_3 + a_5)X, \\
\tau_2 &:= (a_0 + a_2 - a_3 - a_4) + (a_1 - a_2 - a_3 - a_5)X, \\
\tau_3 &:= (a_0 + a_2 + a_3 - a_4) + (a_1 + a_2 - a_3 - a_5)X.
\end{aligned}$$

As we will see in Example 17.1.12, a basis of

$$V(f) := \{g \in F_9[X] : g^9 \equiv g \bmod f\}$$

is $\{1, X^3 + X, X^4\}$ and so we have

$$\tau\sigma(1) = (1, 1, 1), \tau\sigma(X^3 + X) = (0, -1, 1), \tau\sigma(X^4) = (1, -1, -1).$$

We are now in a position to devise an algorithm for computing a complete factorization of f into powers of irreducible polynomials.

What we have to do is:

compute a basis $\{g_1, \ldots, g_k\}$ of $V(f)$;
pick up a polynomial g in this basis; and
compute $\gcd(f, g - \alpha)$, for all $\alpha \in F$, obtaining a non-trivial factorization of f, because of Proposition 17.1.3.
Repeatedly we can pick up in turn further polynomials g' in the basis and compute $\gcd(h, g' - \alpha)$, for each $\alpha \in F$ and for each h in the current non-trivial factorization of f.

We are guaranteed in this way to obtain a complete factorization because of Proposition 17.1.8: in fact, if two factors are not separated in this way, then for all i, there exists α_i such that both of them divide $g_i - \alpha_i$, which is contrary to Proposition 17.1.8.

Example 17.1.10. Continuing the computation of Example 17.1.12, let us choose $g := X^4$ and compute

$$\gcd(X^4, X^6 + X^4 + X^2 + 1) := 1$$
$$\gcd(X^4 - 1, X^6 + X^4 + X^2 + 1) := X^2 + 1$$
$$\gcd(X^4 + 1, X^6 + X^4 + X^2 + 1) := X^4 + 1.$$

Note that, since $X^4 + 1 = t_2 t_3$ and $\tau\sigma(X^4) = (1, -1, -1)$, the computation confirms Proposition 17.1.8.

To complete the factorization we take

$$g := X^3 + X, \quad h := X^4 + 1$$

and compute

$$\gcd(X^3 + X, X^4 + 1) := 1$$
$$\gcd(X^3 + X - 1, X^4 + 1) := X^2 - X - 1$$
$$\gcd(X^3 + X + 1, X^4 + 1) := X^2 + X - 1$$

as we expected from $\tau\sigma(X^3 + X) = (0, -1, 1)$.

In this way we get the factorization of f.

The only thing we have left to do is to discuss how to compute a basis of $V(f)$. This can be easily performed by linear algebra, if we represent the elements of $F_d[X]$ by d-tuples of elements of F and we are able to express ϵ in matrix form.

How to represent the elements of $F_d[X]$ by d-tuples is obvious: we use the isomorphism $\rho : F_d[X] \mapsto F^d$ defined by

$$\rho\left(\sum_{i=0}^{d-1} a_i X^i\right) = (a_0, \ldots, a_{d-1}).$$

For the second problem, it is again obvious that ϵ is given by the matrix Q whose jth column is $\rho\epsilon(X^{j-1}) = \rho(X^{q(j-1)})$, so that:

Proposition 17.1.11. *Let $g(X) \in F[X]$ be such that $0 < \deg(g) < d$. Then*

$$g \in V(f) \iff \rho(g) \in \ker(Q - I).$$

Proof $g^q - g$ is a multiple of f iff

$$0 = \rho(g^q - g) = \rho(g^q) - \rho(g) = (Q - I)\rho(g).$$

$\square$

17.1 Berlekamp's Algorithm

Denoting $r_j := \mathbf{Rem}(X^{qj}, f)$, to compute Q we can use the following inductive approach:

$$r_j = \mathbf{Rem}(r_{j-1}r_1, f).$$

Example 17.1.12. To complete the computation of Example 17.1.9 we have to compute $\mathbf{Rem}(X^{qj}, f)$, for all $j < 6$, and we do this as follows:

$$\begin{aligned}
r_0 &:= && 1, \\
r_1 &:= && X^3, \\
r_2 &:= \mathbf{Rem}(X^6, f(X)) &=& -X^4 - X^2 - 1, \\
r_3 &:= \mathbf{Rem}(-X^7 - X^5 - X^3, f(X)) &=& X, \\
r_4 &:= \mathbf{Rem}(X^4, f(X)) &=& X^4, \\
r_5 &:= \mathbf{Rem}(X^7, f(X)) &=& -X^5 - X^3 - X,
\end{aligned}$$

which gives us the matrices

$$Q = \begin{pmatrix} 1 & 0 & -1 & 0 & 0 & 0 \\ 0 & 0 & 0 & 1 & 0 & -1 \\ 0 & 0 & -1 & 0 & 0 & 0 \\ 0 & 1 & 0 & 0 & 0 & -1 \\ 0 & 0 & -1 & 0 & 1 & 0 \\ 0 & 0 & 0 & 0 & 0 & -1 \end{pmatrix}$$

and

$$Q - I = \begin{pmatrix} 0 & 0 & -1 & 0 & 0 & 0 \\ 0 & -1 & 0 & 1 & 0 & -1 \\ 0 & 0 & 1 & 0 & 0 & 0 \\ 0 & 1 & 0 & -1 & 0 & -1 \\ 0 & 0 & -1 & 0 & 0 & 0 \\ 0 & 0 & 0 & 0 & 0 & 1 \end{pmatrix}$$

from which we obtain the solution $\{1, X^3 + X, X^4\}$.

Corollary 17.1.13. *The following conditions are equivalent:*

(1) f is a power of an irreducible polynomial;
(2) $\dim_F(\ker(Q - I)) = 1$;
(3) $\ker(Q - I)$ is generated by $\sigma(1)$.

Corollary 17.1.14. *The following conditions are equivalent:*

(1) f is irreducible;
(2) $\dim_F(\ker(Q - I)) = 1$ and $\gcd(f, f') = 1$.

Fig. 17.1. Berlekamp's Algorithm

$[t_1, \ldots, t_k] := \mathbf{Factorization}(f, F)$
where
    $F = \{\alpha_1, \ldots, \alpha_q\}$ is a finite field
    $q := \text{card}(F)$
    $f(X) \in F[X]$
    $t_1, \ldots, t_k \in F[X]$ are powers of irreducible polynomials
    $f(X) = \prod_{i=1}^{k} t_i(X)$
$d := \deg(f), r_0 := 1, r_1 := \mathbf{Rem}(X^q, f)$
For $j = 2, \ldots, d - 1$ **do**
    $r_j := \mathbf{Rem}(r_1 r_{j-1}, f)$
Let Q be the matrix whose jth column is $\rho(r_{j-1})$
$k := \dim_F(\ker(Q - I))$
Let g_i, $2 \leq i \leq k$, be such that
    $\{\rho(1), \rho(g_2), \ldots, \rho(g_k)\}$ is a basis of $\ker(Q - I)$.
$L_0 := [\,], L := [f], i := 1$
While $\text{card}(L) + \text{card}(L_0) < k$ **do**
    $i := i + 1, j := 0$
    While $j < q$ **and** $\text{card}(L) + \text{card}(L_0) < k$ **do**
        $j := j + 1, L_0 := L, L := [\,]$
        While $L_0 \neq \emptyset$ **and** $\text{card}(L) + \text{card}(L_0) < k$ **do**
            $h := \mathbf{First}(L_0), L_0 := \mathbf{Rest}(L_0)$
            $h_0 := \gcd(g_i - \alpha_j, h)$
            If $0 < \deg(h_0) < \deg(h)$ **then**
                $L := L \cup [h_0, h/h_0]$
            else
                $L := L \cup [h]$
$L \cup L_0$

Algorithm 17.1.15 (Berlekamp's Algorithm). We now have all the tools we need to describe Berlekamp's Algorithm for polynomial factorization over a finite field F. It returns a factorization of a polynomial f into powers of irreducible polynomials, so that, if f is squarefree, it returns a complete factorization of f: cf. Figure 17.1.

To obtain a complete factorization, we would need an algorithm which given $t(X)$, a power of an irreducible polynomial in $F[X]$, $t(X) = s(X)^e$, computes its irreducible factor $s(X)$ and its multiplicity e. For completeness, we next describe (Figure 17.2) such an algorithm.

In practice, however, it is advisable to perform first a squarefree decomposition of f, and apply Berlekamp's Algorithm to each factor in the squarefree decomposition.

If not only a squarefree decomposition of f, but, subsequently, also a distinct degree decomposition is performed on each squarefree factor (as has been

Fig. 17.2. Irreducible Factor

$(s, e) := \mathbf{Irr}(t)$
where
    $s(X) \in F[X]$ is an irreducible polynomial
    $e \in \mathbb{N}$
    $t(X) \in F[X]$ is the power of an irreducible polynomial
    $t = s^e$
$e := 1, s := t$
Repeat
    While $s' = 0$ **do**
        let s_1 be s.t. $s(X) = s_1(X^p)$
        $e := ep, s := s_1$
    $s_1 := \gcd(s, s')$
    If $s_1 \neq 1$ **then**
        $s_2 := \frac{s}{s_1}$
        $e_1 := \frac{\deg(s)}{\deg(s_2)}$
        $e := ee_1$
        $s := s_2$
until $\gcd(s, s') = 1$
(s, e)

---

proposed, since this improves the performance of the algorithm), Berlekamp's Algorithm is simplified since the degree of its irreducible factors is known in advance: cf. Figure 17.3.

17.2 The Cantor–Zassenhaus Algorithm

To present the alternative probabilistic algorithm proposed by Cantor and Zassenhaus, I will use the same notation as in the previous section; so we have a field F, $\text{char}(F) = p$, $\text{card}(F) = q = p^n$, and a polynomial $f(X) \in F[X]$; unlike in Berlekamp's Algorithm, we will assume that f is squarefree and the product of k irreducible factors t_i which have the same degree δ:

$$f = \prod_{i=1}^{k} t_i;$$

therefore $\deg(f) := d = \delta k$.

The Cantor–Zassenhaus Algorithm is mainly based on the application of the isomorphism

$$\tau : F[X]/f(X) \mapsto F[X]/t_1(X) \oplus \cdots \oplus F[X]/t_k(X);$$

Fig. 17.3. Advanced Berlekamp Algorithm

$[t_1, \ldots, t_k] := \textbf{Factorization}(f, F)$
where
  $F = \{\alpha_1, \ldots, \alpha_q\}$ is a finite field
  $q := \text{card}(F)$
  $f(X) \in F[X]$ is a squarefree polynomial, all of which factors have degree $n < \deg(f)$
  $t_1, \ldots, t_k \in F[X]$ are the irreducible factors of f in $F[X]$
$d := \deg(f), r_0 := 1, r_1 := \textbf{Rem}(X^q, f)$
For $j = 2, \ldots, d-1$ **do**
  $r_j := \textbf{Rem}(r_1 r_{j-1}, f)$
Let Q be the matrix whose jth column is $\rho(r_{j-1})$
$k := \dim_F(\ker(Q - I))$
Let g_i, $2 \le i \le k$ be s.t.
  $\{\rho(1), \rho(g_2), \ldots, \rho(g_k)\}$ is a basis of $\ker(Q - I)$.
$L := [\,], L_1 := [f], i := 1$
While $\text{card}(L) < k$ **do**
  $i := i + 1, j := 0$
  While $j < q$ **and** $\text{card}(L) < k$ **do**
    $j := j + 1, L_0 := L_1, L_1 := [\,]$
    While $L_0 \ne \emptyset$ **and** $\text{card}(L) < k$ **do**
      $h := \textbf{First}(L_0), L_0 := \textbf{Rest}(L_0)$
      $h_0 := \gcd(g_i - \alpha_j, h)$
      If $\deg(h_0) = n$ **then**
        $L := L \cup [h_0]$
      else
        $L_1 := L_1 \cup [h_0]$
      If $\deg(h/h_0) = n$ **then**
        $L := L \cup [h/h_0]$
      else
        $L_1 := L_1 \cup [h/h_0]$
L

related with that, we will again use the ring

$$F_d[X] := \{g \in F[X] : \deg(g) < d\},$$

whose product is given by $g_1 \times g_2 = \textbf{Rem}(g_1 g_2, f)$, and the ring isomorphism

$$\sigma : F_d[X] \mapsto F[X]/f(X)$$

which associates to $g \in F_d[X]$ its residue class mod f.

Since the analysis requires the use of the idempotents of $F[X]/f(X)$, it is better to introduce them more explicitly than in Berlekamp's Algorithm (where, of course, they play an implicit role): we denote c_j to be the idempotent such that

$$c_i \equiv \begin{cases} 1 & \text{if } i = j \\ 0 & \text{otherwise} \end{cases} \pmod{t_j}.$$

17.2 The Cantor–Zassenhaus Algorithm

Lemma 17.2.1. *Let us assume* $g(X) \in F_d[X]$ *is a polynomial which, denoting* $(\alpha_1, \ldots, \alpha_k) := \tau\sigma(g)$, *satisfies*

$g \not\equiv 0, \pm 1 (\mod f)$;
$\alpha_i \in \{-1, 0, 1\}$, *for all* i.

Denoting $S_0 := \{i : \alpha_i = 0\}$, $S_1 := \{i : \alpha_i = 1\}$, *at least one between*

$$\gcd(f, g) = \prod_{i \in S_0} t_i$$

and

$$\gcd(f, g - 1) = \prod_{i \in S_1} t_i$$

is a proper factor of f.

Proof Denote, for $j \in \{-1, 0, 1\}$, $S_j := \{i : \alpha_i = j\}$ and

$$f_j := \gcd(f, g - j) = \prod_{i \in S_j} t_i.$$

Remarking that, for any $j \in \{-1, 0, 1\}$,

$$\begin{aligned} S_j = \{1, \ldots, k\} &\iff t_i \text{ divides } g - j, \text{ for all } i \\ &\iff f \text{ divides } g - j \\ &\iff \gcd(f, g - j) = f \\ &\iff g \equiv j (\mod f), \end{aligned}$$

and

$$\begin{aligned} S_j = \emptyset &\iff t_i \text{ does not divide } g - j, \text{ for all } i \\ &\iff \gcd(f, g - j) = 1, \end{aligned}$$

we can conclude that, either:

- $S_0 \neq \emptyset$, in which case $S_0 \neq \{1, \ldots, k\}$ – since $g \not\equiv 0 (\mod f)$ – and so f_0 is a proper factor;
- $S_0 = \emptyset$, in which case $S_j \neq \{1, \ldots, k\}$ for $j = \pm 1$ – since $g \not\equiv \pm 1 (\mod f)$ – and so f_1 and f_{-1} are proper factors. ♃

On the basis of this lemma, the aim is to produce an element $g(X) \in F_d[X]$ such that, denoting $(\alpha_1, \ldots, \alpha_k) := \tau\sigma(g)$, we have

$g \neq 0, \pm 1$;
$\alpha_i \in \{-1, 0, 1\}$, for all i.

To obtain such a g, Cantor and Zassenhaus proposed to generate it by producing a random non-constant polynomial $h(X) \in F_d[X] \setminus F$.

Restricting ourselves to the case $p \neq 2$ and setting

$$m := \frac{(q^\delta - 1)}{2}$$

in fact we find:

Proposition 17.2.2. *Let $h(X) \in F_d[X] \setminus F$ and*

$$g(X) := \mathbf{Rem}(h^m, f).$$

Then, denoting $(\alpha_1, \ldots, \alpha_k) := \tau\sigma(g)$,

$$\alpha_i \in \{-1, 0, 1\}, \text{ for all } i.$$

Proof Setting $(\beta_1, \ldots, \beta_k) := \tau(h)$ we have

$$h \equiv \sum_i \beta_i c_i \pmod{f},$$

from which we get

$$g \equiv h^m \equiv \sum_i \beta_i^m c_i \pmod{f},$$

and so $\alpha_i = \beta_i^m$ in $F[X]/t_i(X)$, for all i.
Therefore we have

$$\alpha_i^2 = \beta_i^{2m} = \beta_i^{q^\delta - 1}$$

in $F[X]/t_i(X) = GF(q^\delta)$, so that either

$\beta_i = 0$ and so $\alpha_i = 0 \pmod{t_i}$, or
$\beta_i \neq 0$ and so $\alpha_i^2 = \beta_i^{q^\delta - 1} = 1$ and $\alpha = \pm 1$.

♃

It is then sufficient to repeatedly write a random polynomial $h(X) \in F_d[X] \setminus F$ and compute $g(X) := \mathbf{Rem}(h^m, f)$ until $g \neq 0, \pm 1$.

Algorithm 17.2.3 (Cantor–Zassenhaus Algorithm).

How to devise a factorization algorithm is quite clear, on the basis of the above results. It is presented in Figure 17.4.

The analysis of such an algorithm requires us, of course, to analyse the probability that a random polynomial $h(X) \in F_d[X] \setminus F$ is such that $\mathbf{Rem}(h^m, f) \neq 0, \pm 1$:

17.2 The Cantor–Zassenhaus Algorithm

Fig. 17.4. Cantor–Zassenhaus Algorithm

$[t_1, \ldots, t_k] :=$ **Factorization**(f, F)
where
  F is a finite field
  $q := \text{card}(F)$
  $f(X) \in F[X]$ is a squarefree polynomial whose factors have the same degree δ
  $t_1, \ldots, t_k \in F[X]$ are irreducible polynomials
  $f(X) = \prod_{i=1}^{k} t_i(X)$
$L := \emptyset, L_0 := [f]$
$d := \deg(f), m := m := \frac{(q^\delta - 1)}{2}$
While $L_0 \neq \emptyset$ **do**
  Choose $h \in F[X] \setminus F$ such that $\deg(h) \leq d$
  $g := \textbf{Rem}(h^m, f)$
  If $g \neq 0, \pm 1$ **then**
    $L_1 := L_0, L_0 := []$
    Repeat
      $p := \textbf{First}(L_1), L_1 := \textbf{Rest}(L_1)$
      $p_0 := \gcd(p, g), p_1 := \gcd(p, g - 1), p_{-1} := \frac{p}{p_0 p_1}$
      For $i \in \{-1, 0, 1\}$ **do**
        If $\deg(p_i) = \delta$ **then** $L := L \cup [p_i]$
        If $\deg(p_i) > \delta$ **then** $L_0 := L_0 \cup [p_i]$
    until $L_1 \neq \emptyset$
L

Lemma 17.2.4. *There are $2m^k - q + 1$ polynomials h in $F_d[X] \setminus F$ which satisfy $h^m \equiv 0, \pm 1 (\bmod f)$.*

Proof Since h is chosen to be a non-constant in F, we have $h^m \not\equiv 0 \pmod{f}$ because

$$h \not\equiv 0 (\bmod f) \Rightarrow \text{ there exists } i : h \not\equiv 0 (\bmod t_i)$$
$$\Rightarrow \text{ there exists } i : h^m \not\equiv 0 (\bmod t_i)$$
$$\Rightarrow h^m \not\equiv 0 (\bmod f).$$

Since for each of the $q^\delta - 1$ polynomials $\beta_i \in F[X]/t_i$, $\beta_i \neq 0$, we have $(\beta^m)^2 = 1$, then m elements satisfy $\beta^m = 1$ while the other m elements satisfy $\beta^m = -1$.
Therefore among the elements $h = \sum_i \beta_i c_i \in F_d[X]$ there are

m^k of them satisfying $\beta_i^m = 1$, for all i, and so $h^m \equiv 1 (\bmod t_i)$, for all i,
  $h^m \equiv 1 (\bmod f)$,
and m^k satisfying $\beta_i^m = -1$, for all i and so $h^m \equiv -1 (\bmod f)$,

giving a total of $2m^k$ elements $h \in F_d[X]$ satisfying $h^m \equiv \pm 1 \pmod{f}$, among which the $q - 1$ non-zero constants are included. ♃

Proposition 17.2.5. *The probability of randomly choosing a polynomial h among the $q^d - q$ ones in $F_d[X] \setminus F$ which satisfy $h^m \equiv 0, \pm 1 \pmod{f}$ is*

$$\frac{2m^k - q + 1}{q^d - q} < \frac{1}{2}.$$

Proof The evaluation of the probability being obvious, we only have to remark that

$$\frac{2m^k - q + 1}{q^d - q} < \frac{2m^k}{q^d - q} = \frac{2\left(\frac{q^\delta - 1}{2}\right)^k}{q^d - q} = \frac{1}{2^{k-1}} \frac{(q^\delta - 1)^k}{q^d - q} \leq \frac{1}{2^{k-1}} \leq \frac{1}{2}.$$

♃

Example 17.2.6. As an example to show the behaviour and the probabilistic distribution aspects of the Cantor–Zassenhaus result, we fix $F := \mathbb{Z}_3$ and we recall that the irreducible polynomials of degree 2 are

$$t_1 := X^2 + 1, \quad t_2 := X^2 + X - 1, \quad t_3 := X^2 - X - 1.$$

We want to show the behaviour of the generic polynomial $h \in (\mathbb{Z}_3)_4[X]$ with respect to the polynomials

$$\tau_1 := t_2 t_3, \quad \tau_2 := t_1 t_3, \quad \tau_3 := t_1 t_2.$$

To do so we have to compute[1] for all $h \in (\mathbb{Z}_3)_4[X]$

$h_j := h^4 \pmod{t_j}, j = 1, 2, 3,$
$\chi_j := h^4 \pmod{\tau_j}, j = 1, 2, 3.$

The result is contained in the tables in Figure 17.5, from which an easy count allows us to verify that there are exactly $m^2 = 4^2 = 16$ elements such that $\chi_j = 1$ (and the same number such that $\chi_j = -1$).

Note that there are 8 non-zero elements h in $\mathbb{Z}_3\lfloor X \rfloor_2$ (which represent

---

[1] Since, on the basis of Bezout, we have

$$\begin{aligned} 1 &= (-X+1)t_2 + (X+1)t_3, \\ 1 &= Xt_1 - (X+1)t_3, \\ 1 &= (-X)t_1 + (X-1)t_2; \end{aligned}$$

to get χ_i we have to compute

$$\begin{aligned} \chi_1 &:= h_3(-X+1)t_2 + h_2(X+1)t_3, \\ \chi_2 &:= h_3 X t_1 - h_1(X+1)t_3, \end{aligned}$$

17.2 The Cantor–Zassenhaus Algorithm

Fig. 17.5. Illustration of the structure of Example 17.2.1

h	h_1	h_2	h_3	χ_1	χ_2	χ_3
$-X^3 - X^2 - X - 1$	0	1	-1	$-X^3 - X$	$-X^3 - X$	$-X^3 - X$
$-X^3 - X^2 - X$	1	-1	-1	-1	$X^3 + X + 1$	$-X^3 - X + 1$
$-X^3 - X^2 - X + 1$	1	-1	1	$X^3 + X$	1	$-X^3 - X + 1$
$-X^3 - X^2 - 1$	1	-1	0	$-X^3 - X + 1$	$-X^3 - X + 1$	$-X^3 - X + 1$
$-X^3 - X^2$	-1	-1	1	$X^3 + X$	$-X^3 - X - 1$	-1
$-X^3 - X^2 + 1$	1	1	1	1	$-X^3 - X - 1$	$X^3 + X - 1$
$-X^3 - X^2 + X - 1$	1	1	-1	$-X^3 - X$	$X^3 + X + 1$	1
$-X^3 - X^2 + X$	-1	0	1	$-X^3 - X - 1$	$-X^3 - X - 1$	$-X^3 - X - 1$
$-X^3 - X^2 + X + 1$	-1	1	-1	$-X^3 - X$	-1	$X^3 + X - 1$
$-X^3 - X - 1$	1	0	1	$-X^3 - X - 1$	1	$X^3 + X + 1$
$-X^3 - X$	0	1	1	1	$X^3 + X$	$-X^3 - X$
$-X^3 - X + 1$	1	1	0	$X^3 + X - 1$	$-X^3 - X + 1$	1
$-X^3 - 1$	-1	-1	1	$X^3 + X$	$-X^3 - X - 1$	-1
$-X^3$	1	-1	-1	-1	$X^3 + X + 1$	$-X^3 - X + 1$
$-X^3 + 1$	-1	1	-1	$-X^3 - X$	-1	$X^3 + X - 1$
$-X^3 + X - 1$	-1	-1	-1	-1	-1	-1
$-X^3 + X$	1	1	1	1	1	1
$-X^3 + X + 1$	-1	-1	-1	-1	-1	-1
$-X^3 + X^2 - X - 1$	1	1	-1	$-X^3 - X$	$X^3 + X + 1$	1
$-X^3 + X^2 - X$	1	-1	-1	-1	$X^3 + X + 1$	$-X^3 - X + 1$
$-X^3 + X^2 - X + 1$	0	-1	1	$X^3 + X$	$X^3 + X$	$X^3 + X$
$-X^3 + X^2 - 1$	-1	1	1	1	$-X^3 - X - 1$	$X^3 + X - 1$
$-X^3 + X^2$	-1	1	-1	$-X^3 - X$	-1	$X^3 + X - 1$
$-X^3 + X^2 + 1$	1	0	-1	$X^3 + X + 1$	$X^3 + X + 1$	$X^3 + X + 1$
$-X^3 + X^2 + X - 1$	-1	-1	1	$X^3 + X$	$-X^3 - X - 1$	-1
$-X^3 + X^2 + X$	-1	1	0	$X^3 + X - 1$	$X^3 + X - 1$	$X^3 + X - 1$
$-X^3 + X^2 + X + 1$	1	-1	1	$X^3 + X$	1	$-X^3 - X + 1$
$-X^2 - X - 1$	1	1	1	1	1	1
$-X^2 - X$	-1	1	-1	$-X^3 - X$	-1	$X^3 + X - 1$
$-X^2 - X + 1$	-1	0	-1	$X^3 + X + 1$	-1	$-X^3 - X - 1$
$-X^2 - 1$	0	-1	-1	-1	$-X^3 - X$	$X^3 + X$
$-X^2$	1	1	1	1	1	1
$-X^2 + 1$	1	-1	-1	-1	$X^3 + X + 1$	$-X^3 - X + 1$
$-X^2 + X - 1$	1	1	1	1	1	1
$-X^2 + X$	-1	-1	1	$X^3 + X$	$-X^3 - X - 1$	-1
$-X^2 + X + 1$	-1	-1	0	$-X^3 - X + 1$	$X^3 + X - 1$	-1
$-X - 1$	-1	-1	1	$X^3 + X$	$-X^3 - X - 1$	-1
$-X$	1	-1	-1	-1	$X^3 + X + 1$	$-X^3 - X + 1$
$-X + 1$	-1	1	-1	$-X^3 - X$	-1	$X^3 + X - 1$
-1	1	1	1	1	1	1
0	0	0	0	0	0	0
1	1	1	1	1	1	1
$X - 1$	-1	1	-1	$-X^3 - X$	-1	$X^3 + X - 1$
X	1	-1	-1	-1	$X^3 + X + 1$	$-X^3 - X + 1$
$X + 1$	-1	-1	1	$X^3 + X$	$-X^3 - X - 1$	-1
$X^2 - X - 1$	-1	-1	0	$-X^3 - X + 1$	$X^3 + X - 1$	-1
$X^2 - X$	-1	-1	1	$X^3 + X$	$-X^3 - X - 1$	-1
$X^2 - X + 1$	1	1	1	1	1	1
$X^2 - 1$	1	-1	-1	-1	$X^3 + X + 1$	$-X^3 - X + 1$
X^2	1	1	1	1	1	1
$X^2 + 1$	0	-1	-1	-1	$-X^3 - X$	$X^3 + X$
$X^2 + X - 1$	-1	0	-1	$X^3 + X + 1$	-1	$-X^3 - X - 1$
$X^2 + X$	-1	1	-1	$-X^3 - X$	-1	$X^3 + X - 1$
$X^2 + X + 1$	1	1	1	1	1	1

Fig. 17.5. (cont.)

h	h_1	h_2	h_3	χ_1	χ_2	χ_3
$X^3 - X^2 - X - 1$	1	-1	1	$X^3 + X$	1	$-X^3 - X + 1$
$X^3 - X^2 - X$	-1	1	0	$X^3 + X - 1$	$X^3 + X - 1$	$X^3 + X - 1$
$X^3 - X^2 - X + 1$	-1	-1	1	$X^3 + X$	$-X^3 - X - 1$	-1
$X^3 - X^2 - 1$	1	0	-1	$X^3 + X + 1$	$X^3 + X + 1$	$X^3 + X + 1$
$X^3 - X^2$	-1	1	-1	$-X^3 - X$	-1	$X^3 + X - 1$
$X^3 - X^2 + 1$	-1	1	1	1	$-X^3 - X - 1$	$X^3 + X - 1$
$X^3 - X^2 + X - 1$	0	-1	1	$X^3 + X$	$X^3 + X$	$X^3 + X$
$X^3 - X^2 + X$	1	-1	-1	-1	$X^3 + X + 1$	$-X^3 - X + 1$
$X^3 - X^2 + X + 1$	1	1	-1	$-X^3 - X$	$X^3 + X + 1$	1
$X^3 - X - 1$	-1	-1	-1	-1	-1	-1
$X^3 - X$	1	1	1	1	1	1
$X^3 - X + 1$	-1	-1	-1	-1	-1	-1
$X^3 - 1$	-1	1	-1	$-X^3 - X$	-1	$X^3 + X - 1$
X^3	1	-1	-1	-1	$X^3 + X + 1$	$-X^3 - X + 1$
$X^3 + 1$	-1	-1	1	$X^3 + X$	$-X^3 - X - 1$	-1
$X^3 + X - 1$	1	1	0	$X^3 + X - 1$	$-X^3 - X + 1$	1
$X^3 + X$	0	1	1	1	$X^3 + X$	$-X^3 - X$
$X^3 + X + 1$	1	0	1	$-X^3 - X - 1$	1	
$X^3 + X^2 - X - 1$	-1	1	-1	$-X^3 - X$	-1	$X^3 + X - 1$
$X^3 + X^2 - X$	-1	0	1	$-X^3 - X - 1$	$-X^3 - X - 1$	$-X^3 - X - 1$
$X^3 + X^2 - X + 1$	1	1	-1	$-X^3 - X$	$X^3 + X + 1$	1
$X^3 + X^2 - 1$	-1	1	1	1	$-X^3 - X - 1$	$X^3 + X - 1$
$X^3 + X^2$	-1	-1	1	$X^3 + X$	$-X^3 - X - 1$	-1
$X^3 + X^2 + 1$	1	-1	0	$-X^3 - X + 1$	$-X^3 - X + 1$	$-X^3 - X + 1$
$X^3 + X^2 + X - 1$	1	-1	1	$X^3 + X$	1	$-X^3 - X + 1$
$X^3 + X^2 + X$	1	-1	-1	-1	$X^3 + X + 1$	$-X^3 - X + 1$
$X^3 + X^2 + X + 1$	0	1	-1	$-X^3 - X$	$-X^3 - X$	$-X^3 - X$

$\mathbb{Z}_3[X]/t_j(X))$, for all j, 4 of which satisfy $h^4 \equiv 1 \pmod{t_j}$ while the others satisfy $h^4 \equiv -1 \pmod{t_j}$. The 16 elements such that $\chi_1 = 1$ are those which are obtained by Chinese remaindering with respect to the decomposition

$$\mathbb{Z}_3[X]/\tau_1(X) \cong \mathbb{Z}_3[X]/t_2(X) \oplus \mathbb{Z}_3[X]/t_3(X)$$

as solutions of the system

$$h \equiv \begin{cases} h_2 \pmod{t_2} \\ h_3 \pmod{t_3} \end{cases}$$

where each h_j runs over the 4 elements satisfying $h_j^4 \equiv 1 \pmod{t_j}$.

Remark 17.2.7. Recall that in our analysis we restricted ourselves to the case $p \neq 2$. The case when q is even, is treated in a similar way yielding the results below, for which we use the same notation as before.

$$\chi_3 := h_2(-X)t_1 + h_1(X-1)t_2.$$

17.2 The Cantor–Zassenhaus Algorithm

Lemma 17.2.8. *The polynomial* $M(X) := X^2 + X + 1 \in \mathbb{Z}_2[X]$ *is irreducible over* $\mathbb{Z}_2$ *and therefore*

$$GF(4) = \mathbb{Z}_2[X]/M(X) = \mathbb{Z}_2[\alpha] = \{0, 1, \alpha, \alpha + 1\}.$$

♃

Lemma 17.2.9. *Let* K *be a finite field such that* $p := \operatorname{char}(K) = 2, q := \operatorname{card}(K)$ *and*

$$K \supseteq GF(4) = \mathbb{Z}_2[\alpha].$$

Let $h(X) \in K_d[X]/K$, $m := \frac{q^\delta - 1}{3}$, *and* $g(X) := \mathbf{Rem}(h^m, f)$.
Then, if $g \notin GF(4)$, *the factorization*

$$f(X) = \gcd(f, g) \gcd(f, g - 1) \gcd(f, g - \alpha) \gcd(f, g - \alpha - 1)$$

is non-trivial over K.

Proof The argument is similar to that of Lemma 17.2.1 and is left to the reader.

♃

Lemma 17.2.10. *Let* F *be a field such that* $p := \operatorname{char}(F) = 2$ *and* $q := \operatorname{card}(F) = 2^n$.

If n *is even and so* $q \equiv 1 \pmod{3}$ *then* $F \supseteq GF(4)$.
If n *is odd and so* $q \equiv -1 \pmod{3}$ *then*

$$K := F[X]/M(X) = F[\alpha] \supseteq GF(4).$$

♃

Algorithm 17.2.11. According to the lemmas above, how to modify the algorithm of Figure 17.4 then becomes clear:

If n is even, we compute a factorization over F and we only have to

modify the definition of m;
compute the factors $p_\beta := \gcd(f, g - \beta)$, for all $\beta \in GF(4)$, and distribute the non-constant ones between L and L_0 according to their degree, i.e. whether they are irreducible or not,

If n is odd, we compute a factorization over $K := F[\alpha]$ and then we combine the factors which are conjugate over F.

Example 17.2.12. As an example let us try to compute all the monic irreducible polynomials in $\mathbb{Z}_2[X]$ whose degree is 4.

Since we know that $X^{2^d} - X$ is the product of all the monic irreducible polynomials in $\mathbb{Z}_2[X]$ whose degree divides d, it is clear that our task is to factorize

$$f(X) = \frac{X^{16} - X}{X^4 - X} = X^{12} + X^9 + X^6 + X^3 + 1$$

over $GF(4)$, where we will find quadratic factors. We choose the random polynomial

$$h(X) = X^{12} + \alpha X^{11} + \alpha X^{10} + \alpha^2 X^9 + X^8 + \alpha X^7 + X^6 + \alpha^2 X^5 + X^4 + \alpha X^3 + \alpha X^2,$$

therefore, since $m = \frac{4^2-1}{3} = 5$, we compute

$$g(X) := h^5(X) \equiv \alpha^2(X^{10} + X^8 + X^5 + X^2 + 1) \pmod{f(X)}$$

and we have

$$\begin{aligned}
\gcd(f, g) &= 1, \\
\gcd(f, g - 0) &= 1, \\
p_\alpha := \gcd(f, g - \alpha) &= X^6 + X^5 + \alpha X^4 + X^3 + \alpha^2 X^2 + \alpha^2, \\
p_{\alpha^2} := \gcd(f, g - \alpha^2) &= X^6 + X^5 + \alpha^2 X^4 + X^3 + \alpha X^2 + \alpha.
\end{aligned}$$

Then we choose another random polynomial

$$h(X) = \alpha X^{12} + \alpha X^{11} + X^7 + \alpha X^6 + \alpha X^5 + X^4 + \alpha X^2 + X$$

and we obtain

$$g(X) := h^5(X) \equiv X^9 + X^8 + X^6 + \alpha^2 X^5 + \alpha^2 X^4 + X^2 + \alpha^2 X + \alpha \pmod{f(X)}.$$

The non-trivial gcds that we obtain are:

$$\begin{aligned}
\gcd(g - 1, p_\alpha) &= X^2 + \alpha X + 1, \\
\gcd(g - \alpha, p_{\alpha^2}) &= X^4 + X + \alpha + 1, \\
\gcd(g - \alpha^2, p_\alpha) &= X^2 + X + \alpha, \\
\gcd(g - 1, p_{\alpha^2}) &= X^2 + X + \alpha^2, \\
\gcd(g - \alpha, p_\alpha) &= X^2 + \alpha X + \alpha.
\end{aligned}$$

We do not need a third polynomial to complete the factorization, since, by conjugation, we can deduce that the two missing factors are the conjugate ones of $X^2 + \alpha X + 1$ and $X^2 + \alpha X + \alpha$, so that

$$X^4 + X + \alpha + 1 = (X^2 + \alpha^2 X + 1)(X^2 + \alpha^2 X + \alpha^2).$$

17.2 The Cantor–Zassenhaus Algorithm

Therefore we obtain the factorization

$$\begin{aligned} f(X) &= (X^2 + \alpha X + 1)(X^2 + \alpha^2 X + 1)(X^2 + \alpha^2 X + \alpha^2)(X^2 + X + \alpha) \\ &\quad \times (X^2 + X + \alpha^2)(X^2 + \alpha X + \alpha) \end{aligned}$$

in $GF(4)[X]$; then conjugation gives us the factors in $\mathbb{Z}_2[X]$:

$$\begin{aligned} (X^2 + \alpha^2 X + 1)(X^2 + \alpha X + 1) &= X^4 + X^3 + X^2 + X + 1, \\ (X^2 + X + \alpha)(X^2 + X + \alpha^2) &= X^4 + X + 1, \\ (X^2 + \alpha X + \alpha)(X^2 + \alpha^2 X + \alpha^2) &= X^4 + X^3 + 1. \end{aligned}$$

To conclude, we just need to evaluate the probability of picking up a polynomial $h(X) \in K_d[X]/K$ such that $g(X) := \mathbf{Rem}(h^m, f) \notin GF(4)$:

Proposition 17.2.13. *Let K be a finite field such that $p := \mathrm{char}(K) = 2$, $q := \mathrm{card}(K)$ and*

$$K \supseteq GF(4) = \mathbb{Z}_2[\alpha],$$

and let $m = \frac{q^\delta - 1}{3}$.

The probability of randomly choosing a polynomial h among the $q^d - q$ ones in $K_d[X] \setminus K$ which satisfy $\mathbf{Rem}(h^m, f) \notin GF(4)$ is

$$\frac{3m^k - q + 1}{q^d - q} < \frac{1}{3}.$$

Proof We only need to adjust the argument of Lemma 17.2.4:
For each of the $q^\delta - 1$ polynomials $\beta_i \in K[X]/t_i$, $\beta_i \neq 0$ we have $(\beta^m)^3 = 1$.
Then there are m elements satisfying $\beta^m = \alpha^j$, for all $j \in \mathbb{Z}_3$.
Therefore, among the elements $h = \sum_i \beta_i c_i \in F_d[X]$, for each $j \in \{1, 2, 3\}$, there are m^k of them satisfying $\beta_i^m = \alpha^j$, for all i and so $h^m \equiv \alpha^j \pmod{f}$.
This gives the probability

$$\frac{3m^k - q + 1}{q^d - q} < \frac{3m^k}{q^d - q} = \frac{3\left(\frac{q^\delta - 1}{3}\right)^k}{q^d - q} = \frac{1}{3^{k-1}} \frac{(q^\delta - 1)^k}{q^d - q} \leq \frac{1}{3^{k-1}} \leq \frac{1}{3}.$$

♃

18
Zassenhaus

The impractical nature of von Schubert's Algorithm for factorization over $\mathbb{Z}$ is evident even from the example I have presented. The absence of a 'reasonable' factorization algorithm for polynomials over the integers was one of the major weaknesses of Kronecker's Model.

Berlekamp's Algorithm mended this flaw: in fact in 1969 Zassenhaus suggested substituting von Schubert's Algorithm with an application of Berlekamp's Algorithm and a lemma by Hensel.

Hensel's Lemma gives an algorithm which allows us to 'lift' a factorization over D/p to one over D/p^n where D is a principal ideal domain and $p \in D$ is irreducible.

Zassenhaus proposed computing a factorization of a polynomial f over D, based on a factorization algorithm over D/p, by the following approach:

factorize the image of f over D/p;
lift, via Hensel, this factorization to one over D/p^n for a 'suitably' large n – the 'suitability' of n is based on the ability to recover all the coefficients of the factors of f over D – and
obtain the factors over D, by combining the ones over D_{p^n} and checking if they divide f.

In this chapter I will first introduce Hensel's Lemma (Section 18.1) and then I will discuss Zassenhaus' proposal (Section 18.2) through its application to the cases in which the principal ideal domain is either

$K[Y]$ (Section 18.3), where we assume we have a factorization algorithm over K and, by iteration, we obtain an algorithm for the multivariate factorization of polynomials in $K[X_1, \ldots, X_n]$;

ℤ (Section 18.5), in which case the auxiliary algorithm is Berlekamp's and the 'suitability' of n is based on the classical analysis of the bounds relating coefficients and roots of a polynomial (Section 18.4).

Even if the Berlekamp–Hensel–Zassenhaus Algorithm for factorization over ℤ gives an incredible advantage with respect to von Schubert, its complexity is exponential in the degree of the polynomial to be factorized, as is proved by a worst case class of polynomials introduced by Swinnerton-Dyer (Section 18.6).

A more recent algorithm, that of Lenstra–Lenstra–Lovász (L^3), allows us to factorize polynomials over ℤ with polynomial complexity; a sketch of this result is the content of Section 18.7

18.1 Hensel's Lemma

In this section D is a principal ideal domain (and so a unique factorization domain), $p \in D$ is an irreducible element. For $n \in \mathbb{N}$ we denote $D_n := D/p^n$ (so that $D_1 = D/p$ is a field), $\pi_n : D[X] \mapsto D_n[X]$ the canonical projection and $\pi := \pi_1$.

Example 18.1.1. We suggest that the reader mainly consider the two following examples:

$D := \mathbb{Z}$, p a prime so that $D_n = \mathbb{Z}_{p^n}$,
$D := \mathbb{Q}[X]$, $p := X - \alpha$ so that $D_n = \mathbb{Q}[X]/(X - \alpha)^n$.

Theorem 18.1.2 (Hensel's Lemma). *Let $f(X) \in D[X]$ be such that*

$$\deg(f) = \deg(\pi(f)).$$

Let $g_1, h_1 \in D[X]$ be such that

(1) $f \equiv g_1 h_1 \pmod{p}$,
(2) $\deg(f) = \deg(g_1) + \deg(h_1)$,
(3) $\gcd(\pi(g_1), \pi(h_1)) = 1$.

Then for each $n \in \mathbb{N}$, there are $g_n, h_n \in D[X]$ such that

(1) $f \equiv g_n h_n \pmod{p^n}$,
(2) $g_n \equiv g_1 \pmod{p}$, $h_n \equiv g_1 \pmod{p}$,
(3) $\deg(g_n) = \deg(g_1)$, $\deg(h_n) = \deg(h_1)$.

Moreover if g_1 is monic, then, for all n, g_n is monic.

Proof Since

$$\begin{aligned}\deg(f) &= \deg(g_1) + \deg(h_1) \\ &\geq \deg(\pi(g_1)) + \deg(\pi(h_1)) \\ &= \deg(\pi(f)) \\ &= \deg(f),\end{aligned}$$

then $\deg(g_1) = \deg(\pi(g_1))$ and $\deg(h_1) = \deg(\pi(h_1))$.
Since $\gcd(\pi(g_1), \pi(h_1)) = 1$, by the Bezout Identity – note that $D_1 = D/p$ is a field –, there are $s, t \in D_1[X]$ such that

$$s\pi(g_1) + t\pi(h_1) = 1, \ \deg(s) < \deg(h_1), \ \deg(t) < \deg(g_1).$$

Given these crucial preliminaries, we can attack the proof, which is by induction on n, the case $n = 1$ being true by hypothesis.
So we can assume the result true for n and we set $q := p^n$.
Let $U \in D[X]$ be such that

$$f - g_n h_n = qU,$$

so that $\deg(U) \leq \deg(f)$ and let

$$u := \pi(U).$$

Let

$$b := \mathbf{Rem}(ut, \pi(g_1)), \ c := \mathbf{Quot}(ut, \pi(g_1)),$$

and let

$$a := us + c\pi(h_1).$$

Then in $D_1[X]$,

$$\begin{aligned}u &= us\pi(g_1) + ut\pi(h_1) \\ &= us\pi(g_1) + c\pi(g_1)\pi(h_1) + b\pi(h_1) \\ &= a\pi(g_1) + b\pi(h_1)\end{aligned}$$

and $\deg(b) < \deg(g_1)$.
As a consequence, since $\deg(u) \leq \deg(f)$ and

$$\deg(b\pi(h_1)) < \deg(g_1 h_1) = \deg(f),$$

we have $\deg(a\pi(g_1)) \leq \deg(f)$, $\deg(a) \leq \deg(h_1)$.
Let now $A, B \in D[X]$ be such that

$$\pi(A) = a, \ \pi(B) = b, \ \deg(A) = \deg(a), \ \deg(B) = \deg(b),$$

18.1 Hensel's Lemma

so that

$$U = Ag_n + Bh_n + pC$$

for some $C \in D[X]$.

Let

$$g_{n+1} := g_n + qB, h_{n+1} := h_n + qA.$$

We claim that g_{n+1}, h_{n+1} satisfy the inductive assumptions.
In fact:

(1) Since

$$\begin{aligned} f - g_{n+1}h_{n+1} &= f - g_n h_n - qAg_n - qBh_n - q^2AB \\ &= f - g_n h_n - qU + pqC - q^2AB \\ &= pqC - q^2AB \\ &= p^{n+1}(C - p^{n-1}AB), \end{aligned}$$

we have $f \equiv g_{n+1}h_{n+1} \pmod{p^{n+1}}$.

(2) Is obviously true by construction.

(3) Since $\deg(B) < \deg(g_1)$, then $\deg(g_{n+1}) = \deg(g_1)$.
Also $\deg(h_{n+1}) = \deg(h_1)$ holds, since

$$\deg(A) \le \deg(h_1) = \deg(h_n),$$

and $\operatorname{lc}(h_n) \equiv \operatorname{lc}(h_1) \pmod{p}$ ensures that $\operatorname{lc}(h_n)$ is not a multiple of q.

♃

Algorithm 18.1.3. This proof is constructive and can be directly translated into the algorithm, described in Figure 18.1, under suitable effective assumptions[1] on D.

---

[1] We must require that D is an effective principal ideal domain, i.e. that

  D is an effective ring;
  given $a, b \in D$, it is possible to check whether b divides a, and, in this case, to explicitly compute c such that $a = cb$ (this implies that it is possible, for each irreducible $p \in D$, to decide whether $a = 0$ in the field D/p);
  given $a, b \in D$, it is possible to compute $d = \gcd(a, b)$ and $s, t \in D$ such that $d = as + bt$ (this allows us to compute inverses in the ring D/b).

Moreover, for each $a \in D_n$ we must be able to compute $b \in D$ such that $\pi_n(b) = a$. All these requirements are satisfied if D possesses an Euclidean Algorithm (e.g. if D is $\mathbb{Z}$ or $k[X]$).

Fig. 18.1. Hensel Lifting

$(g_n, h_n) := \textbf{HenselLifting}(f, g_1, h_1, p, n)$
where
  $f, g_1, h_1 \in D[X]$
  $p \in D$ is irreducible
  $n \in \mathbb{N}$
  f, g_1, h_1 satisfy the assumptions of Theorem 18.1.2
  g_n, h_n satisfy the thesis of Theorem 18.1.2
$(d, s, t) := \textbf{ExtGCD}(\pi(g_1), \pi(h_1))$
$q := 1$
For $i = 1..n - 1$, **do**
  $q := qp(= p^i)$
  $U := \frac{f - g_i h_i}{q}$
  $u := \pi(U)$
  $(c, b) := \textbf{PolynomialDivision}(ut, \pi(g_1))$
  $a := us + c\pi(h_1)$
  Choose $A, B \in D[X]$ such that $a = \pi(A), b = \pi(B)$
  $g_{i+1} := g_i + qB, h_{i+1} := h_i + qA$

Example 18.1.4. Let us choose $D := \mathbb{Z}$, $p = 3$ and

$$\begin{aligned} f &= 4X^6 + 27X^5 - 38X^4 - 6X^3 + 70X^2 - 105X + 49 \\ &= (4X^4 - X^3 - 3X^2 + 8X - 7)(X^2 + 7X - 7). \end{aligned}$$

The polynomials

$$g_1 := X^4 - X^3 - X - 1, \; h_1 := X^2 + X - 1$$

satisfy the assumptions of Theorem 18.1.2 since

$$g_1 h_1 = X^6 - 2X^4 - 2X^2 + 1 \equiv f \pmod{3},$$

and $\gcd(\pi(g_1), \pi(h_1)) = 1$ – in fact $\pi(g_1) = (X^2 + 1)(X^2 - X - 1)$.
The Euclidean Algorithm in $\mathbb{Z}_3[X]$ allows us to compute

$$s(X) := -1, \; t(X) := X^2 + X,$$

which satisfy $1 = s\pi(g_1) + t\pi(X)$.

Since

$$f - g_1 h_1 = 3(X^6 + 9X^5 - 12X^4 - 2X^3 + 24X^2 - 35X + 16),$$

we fix

$$U(X) := X^6 + 9X^5 - 12X^4 - 2X^3 + 24X^2 - 35X + 16,$$

$$u := X^6 + X^3 + X + 1$$

18.1 Hensel's Lemma

and, by the Division Algorithm of

$$ut := X^8 + X^7 + X^5 + X^4 + X^3 - X^2 + X$$

by g_1, we get

$$b = X^3 - X^2 + X - 1, \ c = X^4 - X^3 - X^2 + X - 1, \ a = X^2$$

obtaining

$$g_2 := X^4 + 2X^3 - 3X^2 + 2X - 4, \ h_2 := 4X^2 + X - 1$$

which satisfy the assumptions of Theorem 18.1.2, since

$$g_2 h_2 - f = 18X^5 - 27X^4 - 9X^3 + 81X^2 - 99X + 45 \equiv 0 (\text{mod } 9).$$

Iteratively we fix

$$U(X) := 2X^5 - 3X^4 - X^3 + 9X^2 - 11X + 5,$$

$$u = -X^5 - X^3 + X - 1$$

and, by the Division Algorithm of

$$ut := -X^7 - X^6 - X^5 - X^4 + X^3 - X$$

by g_1, we get

$$b = -X^3 + X^2 + 1, \ c = -X^3 + X^2 + 1, \ a = 0$$

obtaining

$$g_3 := X^4 - 7X^3 + 6X^2 + 2X + 5, \quad h_3 := 4X^2 + X - 1$$

which satisfy the assumptions of Theorem 18.1.2, since

$$g_3 h_3 - f = 54X^5 - 54X^4 - 27X^3 + 54X^2 - 108X + 54 \equiv 0 (\text{mod } 27).$$

Fixing

$$U := 2X^5 - 2X^4 - X^3 + 2X^2 - 4X + 2,$$

$$u := -X^5 + X^4 - X^3 - X^2 - X - 1$$

the Division Algorithm of

$$ut := -X^7 + X^4 + X^3 + X^2 - X$$

by g_1 gives us

$$b = X^3 - X^2 - 1, \ c = -X^3 - X^2 - X - 1, \ a = X - 1$$

from which we have
$$g_4 := X^4 + 20X^3 - 21X^2 + 2X - 22, \ h_4 := 4X^2 + 28X - 28$$
which satisfy
$$g_4 h_4 - f = 81(-X^5 - 6X^4 + 14X^3 - 6X^2 + 7X - 7).$$

Remark 18.1.5. Let $\psi : R \mapsto S$ be a surjective ring morphism and let us denote by ψ its polynomial extension $\psi : R[X] \mapsto S[X]$. If $f(X) \in R[X]$ is such that $\psi(f)$ is irreducible (respectively squarefree), then so is f. The converse is not, in general, true. Consider e.g.,

- $R := \mathbb{Z}$, $S := \mathbb{Z}_5$, ϕ the canonical projection and $f := X^2 + 1$ where $\psi(f) = (X - 2)(X + 2)$;
- $R := \mathbb{Z}$, $S := \mathbb{Z}_2$, ϕ the canonical projection and $f := X^2 + 1$ where $\psi(f) = (X + 1)^2$.

Let $\rho_n : D_n[X] \mapsto D[X]$ be a map such that

$\pi_n \rho_n(a) = a$, for all $a \in D_n$,
$\rho_n(0) = 0, \rho_n(1) = 1, \rho_n(X) = X$,

and denote $\rho := \rho_1$.

Proposition 18.1.6. *Let $g(X) \in D[X]$ be a primitive polynomial such that*

lc(g) *is not a multiple of p,*
$\pi(g)$ *is a squarefree polynomial.*

Let $g_1, \ldots, g_r$ be the monic irreducible factors of $\pi(g)$, so that

$$\pi(g) = \pi(\mathrm{lc}(g)) \prod_{i=1}^{r} g_i.$$

Then, for all $n \in \mathbb{N}$, there are monic polynomials $G_1, \ldots, G_r \in D[X]$ such that

(1) *for all i, $\pi(G_i) = g_i$;*
(2) $\pi_n(g) = \pi_n(\mathrm{lc}(g)) \prod_{i=1}^{r} \pi_n(G_i)$;
(3) *for all i, $\pi_n(G_i)$ is irreducible.*

If D is an effective principal ideal domain, then, given $g_1, \ldots, g_r$, it is possible to compute such $G_1, \ldots, G_r$.

18.1 Hensel's Lemma

Proof There are several schemes for performing this; the easiest to describe (not necessarily the most efficient) is as follows:

Let us apply the Hensel Lifting Algorithm, to compute

$$(G_1, H_1) := \textbf{HenselLifting}\left(g, \rho(g_1), \text{lc}(g)\rho\left(\prod_{i=2}^{r} g_i\right), p, n\right),$$

so that

$\pi(G_1) = g_1,$
$\pi(H_1) = \pi(\text{lc}(g)) \prod_{i=2}^{r} g_i,$
$\pi_n(g) = \pi_n(G_1)\pi_n(H_1).$

Iteratively assume we have computed $G_1, \ldots, G_k, H_k \in D[X]$ such that

$\pi(G_i) = g_i,$ for all $i \leq k,$
$\pi(H_k) = \pi(\text{lc}(g)) \prod_{i=k+1}^{r} g_i,$
$\pi_n(g) = \prod_{i=1}^{k} \pi_n(G_i)\pi_n(H_k),$

we then apply the Hensel Lifting Algorithm, to compute

$$(G_{k+1}, H_{k+1}) := \textbf{HenselLifting}\left(H_k, \rho(g_{k+1}), \text{lc}(g)\rho\left(\prod_{i=k+2}^{r} g_i\right), p, n\right),$$

so that

$\pi(G_{k+1}) = g_{k+1},$
$\pi(H_{k+1}) = \pi(\text{lc}(g)) \prod_{i=k+2}^{r} g_i,$
$\pi_n(H_k) = \pi_n(G_{k+1})\pi_n(H_{k+1}),$

and therefore

$$\pi_n(g) = \prod_{i=1}^{k+1} \pi_n(G_i)\pi_n(H_{k+1}).$$

After $r - 1$ iterations we have computed $G_1, \ldots, G_{r-1}, H_{r-1} \in D[X]$ such that

$\pi(G_i) = g_i,$ for all $i < r,$
$\pi(H_{r-1}) = \pi(\text{lc}(g)) g_r,$
$\pi_n(g) = \prod_{i=1}^{r-1} \pi_n(G_i)\pi_n(H_{r-1}).$

Let then h be the monic associate of $\pi_n(H_{r-1})$ and let $G_r := \rho_n(h)$, so that (1) and (2) hold immediately and (3) holds because $\pi_n(G_i)$ are then irreducible since their homomorphic images in $D_1[X]$, g_i, are. □

Example 18.1.7. Let us try to compute a factorization of

$$g = 4X^6 + 27X^5 - 38X^4 - 6X^3 + 70X^2 - 105X + 49$$

over $\mathbb{Z}_{3^4}$ using the computations performed in Example 18.1.4.

We have over $\mathbb{Z}_3$ the factorization

$$f = (X^2 + 1)(X^2 - X - 1)(X^2 + X - 1).$$

So, setting

$$g_3 := X^2 + 1, \quad g_2 := X^2 - X - 1, \quad g_1 := X^2 + X - 1,$$

we first apply the Hensel Lifting Algorithm, to compute

$$(G_1, H_1) := \textbf{HenselLifting}(f, \rho(g_1), \text{lc}(g)\rho(g_2 g_3), 3, 4);$$

the computation has already been done in Example 18.1.4 which returned

$$\begin{aligned} G_1 &:= 4X^2 + 28X - 28, \\ H_1 &:= X^4 + 20X^3 - 21X^2 + 2X - 22; \end{aligned}$$

so we have to apply

$$(G_2, H_2) := \textbf{HenselLifting}(H_1, \rho(g_2), \text{lc}(g)\rho(g_3), 3, 4):$$

after having computed $s = -X - 1$, $t = X$, which satisfy $sg_2 + tg_3 = 1$, Hensel returns the factorizations:

$$\begin{aligned} H_1 = \quad & (X^2 - X - 1)(X^2 + 1) \pmod{3}, \\ & (X^2 + 2X - 4)(X^2 + 1) \pmod{9}, \\ & (X^2 + 2X - 13)(X^2 - 9X + 10) \pmod{27}, \\ & (X^2 - 25X - 13)(X^2 - 36X - 17) \pmod{81}, \end{aligned}$$

so that we obtain the factorization

$$f = (X^2 + 7X - 7)(X^2 - 25X - 13)(X^2 - 36X - 17).$$

Algorithm 18.1.8. We will refer by

$$(G_1, \ldots, G_r) := \textbf{HenselLifting}(g, g_1, \ldots, g_r, p, n)$$

an algorithm which, either by the scheme of the proof of Proposition 18.1.6 or by a similar one, allows us to compute such G_is from our knowledge of the g_is.

18.2 The Zassenhaus Algorithm

Let D be a principal ideal domain and Q its fraction field; we will assume that D is an effective principal domain, so that Q is an effective field.

Let $f(X) \in Q[X]$; by computing either the squarefree associate or the distinct power factorization of f, we will assume w.l.o.g. that f is squarefree. Therefore $g(X) := \text{Prim}(f) \in D[X]$ is squarefree too; by the Gauss Lemma (Theorem 6.1.9) factorizing f in $Q[X]$ is equivalent to factorizing g in $D[X]$.

The Zassenhaus Algorithm reduces factorization in $D[X]$ to one in $D/p[X]$, for a suitable irreducible p, which can be lifted, via Hensel's Lemma (Theorem 18.1), to the one in $D/p^n[X]$ and then reinterpreted in $D[X]$.

The first thing we need to do is to understand what 'suitability' means for irreducible $p \in D$: if, given a polynomial $g \in D[X]$, we want to lift a factorization of g mod p to one mod p^n by the Hensel Lemma, we need the hypotheses to be satisfied; we need in particular that

$\deg(g) = \deg(\pi(g))$, i.e. p does not divide $\text{lc}(g)$, and
for any factorization $\pi(g) = \pi(g_1)\pi(h_1)$ in $D/p[X]$, we have $\gcd(\pi(g_1), \pi(h_1)) = 1$, which holds iff $\pi(g)$ is squarefree.

Lemma 18.2.1. *Let D be an infinite domain and let $g(X) \in D[X]$ be squarefree. Then there are infinitely many irreducible elements $p \in D$ such that*

(1) p does not divide $\text{lc}(g)$, and
(2) $\pi(g)$ is squarefree, where $\pi : D \mapsto D/p$ denotes the canonical projection.

Proof Recall (Proposition 6.5.4) that $\pi(g)$ is not squarefree if and only if p divides the discriminant of g. Therefore, there are only finitely many irreducible elements in D which divide both $\text{lc}(g)$ and the discriminant of g. Since D is infinite the claim follows. $\boxed{2\!\!\!\downarrow}$

If D is infinite – which is the case with $D = \mathbb{Z}$ or $D = K[X]$, for any field K –, if we successively try several random irreducible $p \in D$ which do not divide $\text{lc}(g)$ and check by the squarefree test whether $\pi(g)$ is squarefree, after finitely many trials we find a 'suitable' irreducible p, i.e. one such that p does not divide $\text{lc}(g)$ and $\pi(g)$ is squarefree.

Let p be such an irreducible, and let us use the same notation as Section 18.1 so that for all $n \in \mathbb{N}$,

$D_n := D/p^n$,
$\pi_n : D[X] \mapsto D_n[X]$ is the canonical projection,
$\rho_n : D_n[X] \mapsto D[X]$ is a map such that

$\pi_n \rho_n(a) = a$, for all $a \in D_n$,
$\rho_n(0) = 0, \rho_n(1) = 1, \rho_n(X) = X$,

and ρ and π will denote respectively ρ_1 and π_1.

Assume we are able to obtain a factorization $\pi(g) = \mathrm{lc}(\pi(g)) \prod_{j=1}^r g_j$ into monic irreducible factors $g_j \in D_1[X]$; since $\pi(g)$ is squarefree, we have

$$\gcd(g_i, g_j) = 1, \ i \neq j.$$

Computing

$$(G_1, \ldots, G_r) := \textbf{HenselLifting}(g, g_1, \ldots, g_r, p, n)$$

(cf. Algorithm 18.1.8), we obtain monic polynomials $G_1, \ldots, G_r \in D[X]$ such that

for all j, $\pi(G_j) = g_j$;
$\pi_n(g) = \pi_n(\mathrm{lc}(g)) \prod_{j=1}^r \pi_n(G_j)$;
for all j, $\pi_n(G_j)$ is irreducible.

Let now $g = \prod_{i=1}^s h_i$ be a factorization into distinct irreducible primitive polynomials in $D[X]$; the relation between the G_js and the h_is is given by the following:

Lemma 18.2.2. *There is a partition into s disjoint subsets $I_1, \ldots, I_s$ of $\{1, \ldots, r\}$ such that $\forall i$:*

$$\pi_n(h_i) = \pi_n \left(\mathrm{lc}(h_i) \prod_{j \in I_i} G_j \right)$$

Proof In fact

$$\pi_n(\mathrm{lc}(g)) \prod_{j=1}^r \pi_n(G_j) = \pi_n(g) = \prod_{i=1}^s \pi_n(h_i);$$

therefore for all i, $\pi_n(h_i)$ is associate to the product of some of the G_js. ♃

While the content of this section gives some hints about how to factorize polynomials in $Q[X]$ where Q is the fraction field of an effective principal domain, it falls well short of its goal. In fact:

it requires a factorization algorithm for polynomials over fields D/p;
it does not give any hint of how to compute the partition $\{1, \ldots, r\} = \bigcup_{j=1}^{s} I_j$, nor of how to choose which among the polynomials h_i is such that

$$\pi_n(h_i) = \pi_n \left(\mathrm{lc}(h_i) \prod_{j \in I_i} G_j \right)$$

is a factor of g;
it does not give any hint of how to choose n, nor of how to determine the leading coefficient $\mathrm{lc}(h_i)$ of a factor h_i of g.

In the rest of this chapter, we will show how to solve the above problems if

$D = K[X], Q = K(X)$, assuming a factorization algorithm is given in $K[X]$;
$D = \mathbb{Z}, Q = \mathbb{Q}$.

18.3 Factorization Over a Simple Transcendental Extension

The discussion of the problems posed by Zassenhaus' proposal is much simpler in the case $D = K[X]$ than in the setting of the original proposal by Zassenhaus, $D = \mathbb{Z}$.

Let us therefore assume that an effective field K, $\mathrm{char}(K) = 0$, is given such that there is a factorization algorithm in $K[X]$ and we intend to give a factorization algorithm in $K(Y)[X]$, where $K(Y)$ is a rational function field or, equivalently, a transcendental extension.

Therefore, with specific notation for this chapter, we have $D := K[Y]$, $Q := K(Y)$, and the primitive elements of D are the linear factors $(Y - \alpha)$ where $\alpha \in K$.

For a fixed $\alpha \in K$, we have

$$D_p = K[Y]/(Y - \alpha) \equiv K, \quad D_{p^n} = K[Y]/(Y - \alpha)^n;$$

$$\pi_n : K[Y] \mapsto K[Y]/(Y - \alpha)^n$$

is the canonical projection, and

$$\rho_n : K[Y]/(Y - \alpha)^n \mapsto K[Y]$$

the map which sends $f \in K[Y]/(Y-\alpha)^n$ into the unique polynomial $g \in K[Y]$ such that $\deg(g) < n$ and $\pi_n(g) = f$.

So let us be given a primitive squarefree polynomial

$$g(Y, X) \in K[Y, X] = K[Y][X];$$

by random choices, we can obtain $\alpha \in K$ which is not a root of both $\text{lc}(g) \in K[Y]$ and the discriminant of g in $K[Y]$, so that $\text{lc}(g)(\alpha) \neq 0$ and $\pi(g)$ is squarefree. Then, by whatever factorization algorithm is available in $K[X]$, we obtain a factorization $\pi(g) = \pi(\text{lc}(g))\prod_{j=1}^{r} g_j$ into irreducible monic factors, and then, for each n, we can apply the Hensel Lifting Algorithm to compute monic polynomials $G_j \in K[Y]/(Y-\alpha)^n[X]$ such that

for all j, $\pi_n(G_j) = g_j$;

$\pi_n(g) = \pi_n(\text{lc}(g))\prod_{j=1}^{r} \pi_n(G_j)$;

for all j, $\pi_n(G_j)$ is irreducible;

if $g = \prod_{i=1}^{s} h_i$ is a factorization into distinct irreducible primitive polynomials in $K[Y][X]$, then there is a partition of $\{1, \ldots, r\}$ into s disjoint subsets $I_1, \ldots, I_s$ such that

$$\pi_n(h_i) = \pi_n\left(\text{lc}(h_i)\prod_{j \in I_i} G_j\right).$$

Remark 18.3.1. In this setting, most of the questions posed at the end of the previous chapter are quite easy:

we can choose $n = \deg_Y(g)$;

since $\text{lc}(h_i)$ divides $\text{lc}(g)$, it would be sufficient to compute $\frac{\text{lc}(g)}{\text{lc}(h_i)}h_i$ instead of h_i, the advantage being that we know the leading coefficient of $\frac{\text{lc}(g)}{\text{lc}(h_i)}h_i$: it is nothing more than $\text{lc}(g)$!

The only difficult problem is how to compute the partition $\cup_{i=1}^{s} I_i = \{1, \ldots, r\}$.

In order to solve this question, let us remark that:

Lemma 18.3.2. *Let $g, \alpha, g_j, n, G_j, h_i, I_i$ be as above and assume*

$$n \geq \deg_Y(g).$$

Then, for all i, $1 \leq i \leq s$, we have

$\frac{\text{lc}(g)}{\text{lc}(h_i)}h_i = \rho_n\pi_n\left(\text{lc}(g)\prod_{j \in I_i} G_j\right)$ *divides* $\text{lc}(g)g$.

18.3 Factorization Over a Simple Transcendental Extension

Proof Both are polynomials whose degree in Y is bound by n and their image under π_n is the same, so they are equal.

Moreover, let h_{i1} be such that $g = h_i h_{i1}$; then

$$\left(\frac{\mathrm{lc}(g)}{\mathrm{lc}(h_i)} h_i\right)\left(\frac{\mathrm{lc}(g)}{\mathrm{lc}(h_{i1})} h_{i1}\right) = \mathrm{lc}(g) g.$$

♃

This lemma suggests the hypothesis that, in order to check whether $I \subseteq \{1, \ldots, r\}$ is one of the subsets I_i giving the factors h_i, we could check whether

$$\rho_n \pi_n \left(\mathrm{lc}(g) \prod_{j \in I} G_j\right)$$

divides $\mathrm{lc}(g) g$. The hypothesis is correct as proved by the next

Proposition 18.3.3. *Let $g, \alpha, g_j, n, G_j, h_i, I_i$ be as above and assume*

$$n \geq \deg_Y(g).$$

Let I be a minimal subset of $\{1, \ldots, r\}$ such that

$$H := \rho_n \pi_n \left(\mathrm{lc}(g) \prod_{j \in I} G_j\right)$$

divides $\mathrm{lc}(g) g$; then $\mathrm{Prim}(H)$ is an irreducible factor of g.

Proof We have $\mathrm{lc}(g) g = H H_1$ for some $H_1 \in K[Y][X]$, and so

$$g = \mathrm{Prim}(H) \mathrm{Prim}(H_1).$$

Assume $\mathrm{Prim}(H)$ is not irreducible, so that $\mathrm{Prim}(H) = H_{11} H_{12}$. Then

$$\pi_n(H_{11}) \pi_n(H_{12}) = \pi_n(\mathrm{Prim}(H))$$

is associated to $\prod_{j \in I} G_j$. So there is $J \subset I$ such that $\pi_n(H_{11})$ is associated to $\prod_{j \in J} G_j$ and then

$$\rho_n \pi_n \left(\mathrm{lc}(g) \prod_{j \in J} G_j\right) = \frac{\mathrm{lc}(g)}{\mathrm{lc}(H_{11})} H_{11}$$

divides $\mathrm{lc}(g) g$, in contradiction to the minimality of I.

♃

Fig. 18.2. Zassenhaus' Algorithm

$[f_1, \ldots, f_s] :=$ **Zassenhaus – Factorization**(f)
where
 $f(Y, X) \in K(Y)[X]$ is a squarefree polynomial
 $f_i(Y, X) \in K(Y)[X]$ are monic irreducible polynomials
 $f = f_1, \ldots, f_r$
$g := \text{Prim}(f)$
Repeat
 chooserandom $\alpha \in K$
until $\text{lc}(g)(\alpha) \neq 0$ **and** $\pi(g)$ is squarefree
$[g_1, \ldots, g_r] :=$ **Factorization**$(g(\alpha, X), K)$
$n := \deg_Y(g)$
$q := p^n$
$(G_1, \ldots, G_r) :=$ **HenselLifting**$(g, g_1, \ldots, g_r, (Y - \alpha), n)$
$S := \{I : I \subseteq \{1, \ldots, r\}\}, T := \{1, \ldots, r\}, h := g, \text{List} := []$
While $\#(\min_< S) \leq \frac{\#(T)}{2}$ **do**
 $I := \min_< S$
 $S := S \setminus \{I\}$
 $H := \rho_n \pi_n \left(\text{lc}(g) \prod_{j \in I} G_j \right)$
 If H divides $lc(h)h$ **then**
  $T := T \setminus I$
  $h := h/H$
  $\text{List} := \text{List} \cup [\text{lc}(H)^{-1} H]$
$\text{List} \cup [h]$

Algorithm 18.3.4. The above proposition answers the question of how to find the partition elements I_i: they are the minimal subsets satisfying the above property.

Finding the factors h_i of g by combining the G_js is then just book-keeping.

Zassenhaus' Algorithm for factorization over $K(Y)$ can be now described in Figure 18.2.

We will denote by $<$ some total ordering on the subsets of $\{1, \ldots, r\}$ such that

$$I \subset J \implies I < J;$$

in the description below we order the subsets so that subsets of lesser cardinality are lesser; another choice is to order them so that the lesser subsets I are those such that $\sum_{j \in I} \deg(G_j)$ is lesser.

Remark 18.3.5. Let us remark that:

 by an iterative application of this algorithm, we obtain factorization in $K(Y_1, \ldots, Y_r)[X]$ and an algorithm to factorize primitive polynomials in $K[Y_1, \ldots, Y_r][X]$;

therefore, in order to factorize

$$f \in K[X_1, \ldots, X_n] = K[X_1, \ldots, X_{n-1}][X_n],$$

we compute, by gcd, $\text{Cont}(f) \in K[X_1, \ldots, X_{n-1}]$ and $\text{Prim}(f) \in K[X_1, \ldots, X_{n-1}][X_n]$. $\text{Prim}(f)$ is then factorized by the algorithm above, while the same algorithm is recursively applied to factorize $\text{Cont}(f)$.

The scheme discussed above is a simplification of the original Zassenhaus Algorithm for factorizing over $\mathbb{Z}$. The original tool is essentially Proposition 18.3.3. But when $D := \mathbb{Z}$, the choice of n requires the ability to bound the coefficients of the factors of a polynomial f in terms of its own coefficients; this ability goes back to Cauchy whose results will be discussed in the next section.

18.4 Cauchy Bounds

For $f(X) := \sum_{i=0}^{d} a_i X^i \in \mathbb{C}[X]$, we will denote

$$|f| := \max\{|a_i| : 0 \le i \le d\}.$$

In order to get a hint of how to choose n for a factorization algorithm in $\mathbb{Q}$ we need to evaluate the absolute value of the roots of f in terms of its coefficients, and more precisely in terms of $|f|$:

Proposition 18.4.1 (Cauchy). *Let $f(X) := \sum_{i=0}^{d} a_i X^i \in \mathbb{C}[X]$ and let $\alpha \in \mathbb{C}$ be a root of f. Then:*

(1) $|\alpha| \le 1 + \frac{|f|}{|a_d|}$.

(2) *Let* $R_i := \sqrt[i]{\frac{d|a_{d-i}|}{|a_d|}}$, $R := \max_i \{R_i\}$. *Then* $|\alpha| \le R$.

(3) *Let* $S_i := \sqrt[i]{\frac{|a_{d-i}|}{|a_d|}}$, $S := \max_i S_i$. *Then* $|\alpha| < 2S$.

Proof

(1) If $|\alpha| \le 1$, the claim is obviously true. So let us assume $|\alpha| > 1$. Then, since $f(\alpha) = 0$,

$$a_d \alpha^d = -\sum_{i=0}^{d-1} a_i \alpha^i,$$

and therefore:

$$|a_d||\alpha|^d \le \sum_{i=0}^{d-1} |a_i||\alpha|^i$$

$$\leq |f| \sum_{i=0}^{d-1} |\alpha|^i$$
$$= |f| \left(|\alpha|^d - 1\right) (|\alpha| - 1)^{-1}$$
$$\leq |f||\alpha|^d (|\alpha| - 1)^{-1}$$

so that
$$|a_d|(|\alpha| - 1) \leq |f|,$$
whence the claim.

(2) Let j be such that
$$|a_j||\alpha|^j \geq |a_i||\alpha|^i, \text{ for all } i.$$

As above we have
$$|a_d||\alpha|^d \leq \sum_{i=0}^{d-1} |a_i||\alpha|^i \leq d|a_j||\alpha|^j$$

so that
$$|\alpha| \leq \sqrt[d-j]{\frac{d|a_j|}{|a_d|}}.$$

(3) If $|\alpha| \leq S$, there is nothing to prove, so we can assume $|\alpha| > S$. Then since
$$\frac{|a_{d-i}|}{|a_d|} \leq S^i,$$
we have
$$|\alpha|^d \leq \sum_{i=0}^{d-1} \frac{|a_i||\alpha|^i}{|a_d|} \leq \sum_{i=0}^{d-1} S^{d-i}|\alpha|^i,$$

from which, setting
$$\beta := \frac{|\alpha|}{S},$$
and dividing by S^d, we have
$$\beta^d \leq \sum_{i=0}^{d-1} \beta^i;$$
since $\beta > 1$, then
$$(\beta - 1)\beta^d \leq \beta^d - 1 < \beta^d$$
and therefore $\beta - 1 < 1$, $\beta < 2$, $|\alpha| < 2S$.

□

18.4 Cauchy Bounds

The following result shows that (2) and (3) are nearly optimal:

Proposition 18.4.2. *Let* $f(X) := \sum_{i=0}^{d} a_i X^i \in \mathbb{C}[X]$ *and let* S_i, S *be as in Proposition 18.4.1.*
Then there is $\alpha \in \mathbb{C}$ *such that* $f(\alpha) = 0$ *and* $|\alpha| \geq \frac{S}{d}$.

Proof Let us denote by $\alpha_1, \ldots, \alpha_d$ the d roots of f and let us order them so that
$$|\alpha_1| \geq |\alpha_2| \geq \cdots \geq |\alpha_d|.$$
Let j be such that $S_j = S$. Since $a_{d-j} = (-1)^j a_d \sigma_j(\alpha_1, \ldots, \alpha_d)$, then
$$|a_d| \binom{d}{j} |\alpha_1 \cdots \alpha_j| \geq |a_{d-j}|.$$
Therefore
$$d^j |\alpha_1|^j \geq \binom{d}{j} |\alpha_1 \cdots \alpha_j| \geq \frac{|a_{d-j}|}{|a_d|}$$
and so $d|\alpha_1| \geq \sqrt[j]{\frac{|a_{d-j}|}{|a_d|}} = S_j = S$, $|\alpha_1| \geq \frac{S}{d}$. □

A sort of vice-versa of the results of Proposition 18.4.1 is much easier to obtain through Newton results (Section 6.2):

Proposition 18.4.3. *Let* $f(X) \in \mathbb{C}[X]$ *be a polynomial and let* $g(X) \in \mathbb{C}[X]$ *be a monic factor of* f *such that* $\deg(g) = d$.
Let $R := \max\{|\alpha| : \alpha \in \mathbb{C}, f(\alpha) = 0\}$.
Then $|g| \leq \max\{\binom{d}{k} R^k : k \leq d\}$.

Proof Let $\alpha_1, \ldots, \alpha_d$ be the roots of $g(X) = X^d + \sum_{i=0}^{d-1} a_i X^i$. Then, since $a_k = (-1)^k \sigma_k(\alpha_1, \ldots, \alpha_d)$, we get $|a_k| \leq \binom{d}{k} R^k$. □

Remark 18.4.4. It is clear that the results of Proposition 18.4.1 and 18.4.3 allow us, given a polynomial $f(X) \in \mathbb{C}[X]$, to evaluate $|g|$ in terms of $|f|$ for any factor g of f. A stronger result, but one not so easy to prove is Landau Mignotte Inequality[2]:

---

[2] M. Mignotte, Some useful bounds in B. Buchberger, G.E. Collins, R. Loos (eds.) *Computer Algebra, System and Algebraic Computation*, Springer, 1982

Proposition 18.4.5 (Landau–Mignotte Inequality). *Let*

$$Q := \sum_{i=0}^{q} b_i X^i, \quad P := \sum_{i=0}^{d} a_i X^i \in \mathbb{Z}[X].$$

If Q divides P, then, denoting

$$\|P\| := \sqrt{\sum_{i=0}^{d} |a_i|^2},$$

we have

$$\sum_{i=0}^{q} b_i \leq 2^q \left|\frac{b_q}{a_d}\right| \|P\|.$$

$\boxed{2\!\!\!\!4}$

Later we will also need another result of Mignotte:

Proposition 18.4.6. *Let $Q, P \in \mathbb{Z}[X]$ be such that Q divides P. Then, setting $m := \deg(Q)$*

$$\|Q\| \leq \binom{2m}{m}^{\frac{1}{2}} \|P\|.$$

$\boxed{2\!\!\!\!4}$

18.5 Factorization over the Rationals

We are now able to discuss the original Zassenhaus proposal for using Berlekamp's Algorithm and the Hensel Lemma to obtain factorization over $\mathbb{Q}$ and, by the Gauss Lemma, over $\mathbb{Z}$.

Therefore, specializing the notation of Section 18.1, we have $D := \mathbb{Z}$, $Q := \mathbb{Q}$ and the primitive elements of D are the integer primes p.

For a fixed prime $p \in \mathbb{Z}$, we have $D_p = \mathbb{Z}_p$, $D_{p^n} = \mathbb{Z}_{p^n}$;

$$\pi_n : \mathbb{Z} \mapsto \mathbb{Z}_{p^n}$$

is the canonical projection, and

$$\rho_n : \mathbb{Z}_{p^n} \mapsto \mathbb{Z}$$

the map which sends $a \in \mathbb{Z}_{p^n}$ into the unique integer $\rho_n(a)$ such that

$\pi_n(\rho_n(a)) = a,$
$-\frac{p^n}{2} < a \leq \frac{p^n}{2}.$

18.5 Factorization over the Rationals

With an abuse of notation
$$\pi_n : \mathbb{Z}[X] \mapsto \mathbb{Z}_{p^n}[X]$$
is the canonical projection, and
$$\rho_n : \mathbb{Z}_{p^n}[X] \mapsto \mathbb{Z}[X]$$
denotes the polynomial extension.

So let us be given a primitive squarefree polynomial $g(X) \in \mathbb{Z}[X]$; by random choices, we can obtain a prime $p \in \mathbb{Z}$ which divides neither $\mathrm{lc}(g)$ nor the discriminant of g, so that $\pi(g)$ is squarefree.

Then, by Berlekamp and Hensel, we obtain, for a suitable n, monic polynomials
$$G_1, \ldots, G_r \in \mathbb{Z}_{p^n}[X]$$
such that

$\pi_n(g) = \pi_n(\mathrm{lc}(g)) \prod_{j=1}^{r} \pi_n(G_j)$;

for all j, $\pi_n(G_j)$ is irreducible;

if $g = \prod_{i=1}^{s} h_i$ is a factorization into distinct irreducible primitive polynomials in $\mathbb{Z}[X]$, then there is a partition of $\{1, \ldots, r\}$ into s disjoint subsets $I_1, \ldots, I_s$ such that

$$\pi_n(h_i) = \pi_n\left(\mathrm{lc}(h_i) \prod_{j \in I_i} G_j\right).$$

Remark 18.5.1. The questions posed at the end of Section 18.1 can be now solved easily:

choosing the 'suitable' n requires just an application of the results of the previous section;

the leading coefficient of h_i can be assumed to be $\mathrm{lc}(g)$, just by evaluating, instead of h_i, its associate $h_0 := \frac{\mathrm{lc}(g)}{\mathrm{lc}(h_i)} h_i$ (to evaluate $|h_0|$ see Lemma 18.5.2);

computing the partition of $\{1, \ldots, r\}$ could be obtained as in Section 18.3, provided that Proposition 18.3.3 can be generalized to this setting.

Therefore, our task now is to prove Proposition 18.5.4 below.

Lemma 18.5.2. *Let $g(X) \in \mathbb{Z}[X]$ be a primitive polynomial and let R be such that $R \geq |h_1|$ for each monic irreducible factor h_1 of $g \in \mathbb{Q}[X]$.*
For each primitive irreducible factor h of g, we have

$\mathrm{lc}(h)$ divides $\mathrm{lc}(g)$;

$|h| \leq |\mathrm{lc}(h)| R$;

let $h_0 := (\mathrm{lc}(g)/\mathrm{lc}(h))h \in \mathbb{Z}[X]$. Then

$\mathrm{lc}(h_0) = \mathrm{lc}(g)$ and

$|h_0| \le |\mathrm{lc}(g)|R$.

♃

Lemma 18.5.3. *Let $g, p, g_j, n, G_j, h_i, I_i$ be as above, $q = p^n$ and assume*

$$\left|\frac{\mathrm{lc}(g)}{\mathrm{lc}(h_i)}\right| |h_i| < \frac{q}{2}.$$

Then for all i, denoting

$$H_i := \rho_n \pi_n \left(\mathrm{lc}(g) \prod_{j \in I_i} G_j\right),$$

we have $\frac{\mathrm{lc}(g)}{\mathrm{lc}(h_i)} h_i = H_i$.

Proof In fact by Lemma 18.2.2, we know that

$$\pi_n(h_i) = \pi_n\left(\mathrm{lc}(h_i) \prod_{j \in I_i} G_j\right).$$

Therefore

$$\frac{\mathrm{lc}(g)}{\mathrm{lc}(h_i)} h_i \equiv \rho_n \pi_n\left(\mathrm{lc}(g) \prod_{j \in I_i} G_j\right) \pmod{q}.$$

Since both $|H_i|$ and $\left|\frac{\mathrm{lc}(g)}{\mathrm{lc}(h_i)}\right| |h_i|$ are less than $\frac{q}{2}$, then $h_i = H_i$.

♃

Proposition 18.5.4. *Under the same notation:*

(1) $\frac{\mathrm{lc}(g)}{\mathrm{lc}(h_i)} h_i = H_i$ divides $\mathrm{lc}(g)g$.
(2) Let I be a minimal subset of $\{1, \dots, r\}$ such that

$$H := \rho_n \pi_n\left(\mathrm{lc}(g) \prod_{j \in I} G_j\right)$$

divides $\mathrm{lc}(g)g$; then $\mathrm{Prim}(H)$ is an irreducible factor of g.

Proof

(1) Let h_{i1} be such that $g = h_i h_{i1}$; then

$$\left(\frac{\mathrm{lc}(g)}{\mathrm{lc}(h_i)} h_i\right)\left(\frac{\mathrm{lc}(g)}{\mathrm{lc}(h_{i1})} h_{i1}\right) = \mathrm{lc}(g)g.$$

Fig. 18.3. Berlekamp–Hensel–Zassenhaus's Algorithm

$[f_1, \ldots, f_s] := \mathbf{BHZ} - \mathbf{Factorization}(f)$
where
 $f(X) \in \mathbb{Q}[X]$ is a squarefree polynomial
 $f_i(X) \in \mathbb{Q}[X]$ are monic irreducible polynomials
 $f = f_1, \ldots, f_r$
$g := \mathrm{Prim}(f)$
Compute R such that $R \geq |h_1|$, for all h_1 monic irreducible factors of f in $\mathbb{Q}[X]$.
Repeat
  chooserandom $p \in \mathbb{Z}$ prime
until p does not divide $\mathrm{lc}(g)$ and $(g \bmod p)$ is squarefree
$[g_1, \ldots, g_r] := \mathbf{Factorization}(g, \mathbb{Z}_p)$
Compute n such that $\frac{p^n}{2} \geq \left|\mathrm{lc}(g)\right| R$
$(G_1, \ldots, G_r) := \mathbf{HenselLifting}(g, g_1, \ldots, g_r, p, n)$
$S := \{I : I \subseteq \{1, \ldots, r\}\}, T := \{1, \ldots, r\}, h := g, \mathrm{List} := []$
While $\#(\min_< S) \leq \frac{\#(T)}{2}$ **do**
  $I := \min_< S$
  $S := S \setminus \{I\}$
  $H := \rho_n \pi_n \left(\mathrm{lc}(g) \prod_{j \in I} G_j \right)$
  If H divides $\mathrm{lc}(h)h$ **then**
   $T := T \setminus I$
   $h := \frac{h}{H}$
   $\mathrm{List} := \mathrm{List} \cup [\mathrm{lc}(H)^{-1} H]$
$\mathrm{List} \cup [h]$

(2) We have $\mathrm{lc}(g)g = H H_1$ for some $H_1 \in \mathbb{Z}[X]$, and so

$$g = \mathrm{Prim}(H)\mathrm{Prim}(H_1).$$

Assume $\mathrm{Prim}(H)$ is not irreducible, so $\mathrm{Prim}(H) = H_{11} H_{12}$. Then $\pi_n(H_{11})\pi_n(H_{12}) = \pi_n(\mathrm{Prim}(H))$ is associated to $\prod_{j \in I} G_j$. So there is $J \subset I$ such that $\pi_n(H_{11})$ is associated to $\prod_{j \in J} G_j$ and then

$$H_{11} \equiv \rho_n \pi_n \left(\mathrm{lc}(g) \prod_{j \in J} G_j \right) = \frac{\mathrm{lc}(g)}{\mathrm{lc}(H_{11})} H_{11} (\bmod q)$$

divides $\mathrm{lc}(g)g$, in contradiction to the minimality of I. ♃

Algorithm 18.5.5 (Berlekamp–Hensel–Zassenhaus). On the basis of these results, we can now present the Berlekamp–Hensel–Zassenhaus Factorization Algorithm for $\mathbb{Q}[X]$ (Figure 18.3).

Example 18.5.6. An elementary illustration of the algorithm is given by Example 18.1.4 and Example 18.1.7, where we discussed the factorization of

$$\begin{aligned} f &= 4X^6 + 27X^5 - 38X^4 - 6X^3 + 70X^2 - 105X + 49 \\ &= (4X^4 - X^3 - 3X^2 + 8X - 7)(X^2 + 7X - 7) \end{aligned}$$

and, through Hensel we obtained the factorization in $\mathbb{Z}_{81}$

$$f = (X^2 + 7X - 7)(X^2 - 25X - 13)(X^2 - 36X - 17).$$

Since $4|h| \leq 40$ for the factors h of f, we could apply Zassenhaus' proposal and we just note that

$$\rho_4\left(\pi_4\left(4(X^2 + 7X - 7)\right)\right) \text{ and}$$
$$\rho_4\left(\pi_4\left(4(X^2 - 25X - 13)(X^2 - 36X - 17)\right)\right) = 4X^4 - X^3 - 3X^2 + 8X - 7$$

divide, in fact, $4f$.

18.6 Swinnerton-Dyer Polynomials

Remark 18.6.1. From the computations of Chapter 15 we can deduce that, if $\alpha \in \mathbb{C}$ is an 8th root of unity,

$\mathbb{Q}(\alpha) = \mathbb{Q}(i, \sqrt{2})$,

the cyclotomic polynomial $f_8 = X^4 + 1$ factorizes into quadratic or linear factors in any finite field.

This curious behaviour is shared by a series of polynomials introduced by Swinnerton-Dyer (1969).

Definition 18.6.2. *Let*

$$\mathcal{P} := \{p \in \mathbb{N} : p \text{ is prime }\} \cup \{-1\}$$

and for each finite subset $P := \{p_1, \ldots, p_n\} \subset \mathcal{P}$ *let*

$$\alpha_P := \sqrt{p_1} + \sqrt{p_2} + \cdots + \sqrt{p_n}.$$

The minimal polynomial $\mathfrak{s}_P$ of α_P is the Swinnerton-Dyer polynomial *produced by P.*

Our aim is to prove that, for any prime $q \in \mathbb{N}$, $\mathfrak{s}_P$ factorizes in $\mathbb{Z}_q$ into linear and quadratic factors.

Let us fix $P := \{p_1, \ldots, p_n\} \subset \mathcal{P}$; since our argument is by induction, we will also denote for all i, $1 \leq i \leq n$,

$P_i := \{p_1, \ldots, p_i\}$,
$\mathbb{K}_i := \mathbb{Q}(\sqrt{p_1}, \sqrt{p_2}, \ldots, \sqrt{p_i})$,

18.6 Swinnerton-Dyer Polynomials

$\alpha_i := \alpha_{P_i} = \sqrt{p_1} + \sqrt{p_2} + \cdots + \sqrt{p_i}$,

$\mathfrak{s}_i := \mathfrak{s}_{P_i}$, the minimal polynomial of α_i.

Lemma 18.6.3. *With the notation above, for all i:*

(1) $\mathfrak{s}_i \in \mathbb{Z}[X]$;
(2) *A $\mathbb{Q}$-vector basis of $\mathbb{K}_i$ is*

$$V_i := \left\{ \sqrt{\prod_{j \in S} p_j} : S \subseteq \{1, \ldots, i\} \right\};$$

(3) $[\mathbb{K}_i : \mathbb{Q}] = 2^i$;
(4) $\mathbb{K}_i = \mathbb{Q}[\alpha_i]$.

Proof Since the claim is trivial when $i = 1$, let us assume it is true for $i = n-1$ and let us prove it for $i = n$.

(1) By the definition we have

$$\mathfrak{s}_n[X] = \mathfrak{s}_{n-1}(X + \sqrt{p_n})\mathfrak{s}_{n-1}(X - \sqrt{p_n})$$

and so, by conjugation, $\mathfrak{s}_{n-1} \in \mathbb{Z}[X]$ implies $\mathfrak{s}_n \in \mathbb{Z}[X]$.

(2) We have just to prove that $\sqrt{p_n} \notin \mathrm{Span}_{\mathbb{Q}}(V_{n-1})$: otherwise we would have a relation

$$\sqrt{p_n} = \sum_{S \subseteq \{1,\ldots,n-1\}} c_S \sqrt{\prod_{j \in S} p_j}, \quad c_S \in \mathbb{Q}, \tag{18.1}$$

in which, since p_n is a prime different from those in P_{n-1},

> at least two coefficients c_{S_1}, c_{S_0} are non-zero, and
> there exists k, $1 \leq k \leq n-1$, which occurs in one of S_0 and S_1 but not in the other.

To fix the notation let us say $k = n - 1 \in S_1$, so that we can rewrite Equation 18.1 as

$$\sqrt{p_n} = d_0 + d_1\sqrt{p_{n-1}}, \quad d_0, d_1 \in \mathbb{K}_{n-2}, \quad d_0 \neq 0, \quad d_1 \neq 0$$

from which, by squaring, we deduce

$$\sqrt{p_{n-1}} = \frac{p_n - d_0^2 - d_1^2 p_{n-1}}{2d_0 d_1} \in \mathbb{K}_{n-2}$$

getting the hoped-for contradiction.
(3) Obvious.

(4) To compute a primitive element of $\mathbb{K}_n = \mathbb{Q}(\alpha_{n-1}, \sqrt{p_n})$ we only have to apply the construction implicit in Theorem 8.4.5: we have to consider the polynomials $f := \mathfrak{s}_{n-1}$ and $g := X^2 - p_n$. To prove that $\alpha_n = \alpha_{n-1} + \sqrt{p_n}$ is a primitive element of $\mathbb{K}_n[X]$ we have to prove that the only common root of $h(X) = f(\alpha_{n-1} + \sqrt{p_n} - X)$ and $g(X)$ is $\sqrt{p_n}$, which is tantamount to proving that $-\sqrt{p_n}$ is not a root of $\mathfrak{s}_{n-1}(\alpha_{n-1} + \sqrt{p_n} - X)$: this is true, since, otherwise, there would be a root β of $\mathfrak{s}_{n-1}$ such that

$$\beta - \alpha_{n-1} = 2\sqrt{p_n}$$

while the left-hand side is in $\mathbb{K}_{n-1}$ and the right-hand side is not.

♃

Corollary 18.6.4. *The irreducible polynomial $\mathfrak{s}_P$ is such that* $\deg(\mathfrak{s}_P) = 2^n$ *and its roots are*

$$\pm\sqrt{p_1} \pm \sqrt{p_2} + \cdots \pm \sqrt{p_n}.$$

♃

Theorem 18.6.5. *For each $P \subset \mathcal{P}$, and for each prime $q \in \mathbb{N}$, the projection of the irreducible polynomial $\mathfrak{s}_P \in \mathbb{Z}[X]$ in $\mathbb{Z}_q[X]$ factorizes there into linear and quadratic factors.*

Proof By the above results we know that $\mathfrak{s}_P$ is irreducible in $\mathbb{Z}[X]$.
On the other hand, since $GF(q^2)$ contains the square roots of each integer modulo q, the projection of $\mathfrak{s}_P$ splits into $GF(q^2)$, and therefore, by conjugation, its factors are either linear or quadratic in $GF(q)$.

♃

Remark 18.6.6. Kaltofen, Musser and Saunder[3] generalized the Swinnerton-Dyer polynomials, to the minimal polynomials whose root is

$$\beta_{r,P} := \sqrt[r]{p_1} + \sqrt[r]{p_2} + \cdots + \sqrt[r]{p_n}$$

where r is a prime; such polynomials have a similar property to Swinnerton-Dyer's : for any prime $q \in \mathbb{N}$, such polynomials factorize in $\mathbb{Z}_q$ into factors of degree at most r.

The proof presented here of the Swinnerton-Dyer property essentially applies also to their more general situation.

---

[3] In A generalized class of polynomials that are hard to factor, *SIAM J. Comp.* **12** (1983), 473–479.

Remark 18.6.7. Assume we are applying the Berlekamp–Hensel–Zassenhaus Algorithm to a Swinnerton-Dyer polynomial $\mathfrak{s}$ whose degree is $d := \deg(\mathfrak{s}) = 2^r$. Whatever prime we choose, the algorithm gives us at least $\frac{d}{2} = 2^{r-1}$ lifted polynomials to combine together to obtain a potential factor. Since the polynomial is irreducible all the $2^{\frac{d}{2}}$ combinations will fail, before we find out that $\mathfrak{s}$ is irreducible.

The complexity of the Berlekamp–Hensel–Zassenhaus Algorithm is therefore *exponential* in the degree! A factorization algorithm whose complexity is *polynomial* was proposed in 1982 by Lenstra–Lenstra–Lovász.

18.7 L^3 Algorithm

To describe the Lenstra–Lenstra–Lovász (L^3) factorization algorithm[4] for polynomials in $\mathbb{Z}[X]$ I will apply the notation of Section 18.2 to the case $D = \mathbb{Z}$.

Therefore let us assume we are given a primitive squarefree polynomial $g(X) \in \mathbb{Z}[X]$ whose factorization $g = \prod_{i=1}^{s} f_i$ into distinct irreducible primitive polynomials in $\mathbb{Z}[X]$ we aim to compute. Let us denote $d := \deg(g)$ and fix a prime p such that p does not divide $lc(g)$ and the discriminant of g.

Then for all $n \in \mathbb{N}$,

$\pi_n : \mathbb{Z}[X] \mapsto \mathbb{Z}_{p^n}[X]$ is the canonical projection,

$\rho_n : \mathbb{Z}_{p^n}[X] \mapsto \mathbb{Z}[X]$ is the map such that for each polynomial $f(X) \in \mathbb{Z}_{p^n}[X]$, $\rho_n(f) := \sum_{i=0}^{m} a_i X^i \in \mathbb{Z}[X]$ satisfies

$\pi_n \rho_n(f) = f$,
$-\frac{p^n}{2} < a_i \leq \frac{p^n}{2}$ for all i,

and ρ and π will denote respectively ρ_1 and π_1.

Let $\pi(g) = lc(\pi(g)) \prod_{j=1}^{r} g_j$ be its factorization into monic irreducible factors $g_j \in \mathbb{Z}_p[X]$; since $\pi(g)$ is squarefree, we have

$$\gcd(g_j, g_i) = 1, \ i \neq j.$$

For a suitable $n \in \mathbb{N}$, let us set $q := p^n$; we recall that there are monic polynomials $G_1, \ldots, G_r \in \mathbb{Z}[X]$ such that

for all j, $\pi(G_j) = g_j$;
$\pi_n(g) = \pi_n(lc(g)) \prod_{j=1}^{r} \pi_n(G_j)$;

---

[4] In this presentation I limit myself to giving just a sketch of the L^3 algorithm and I direct the interested reader to the original paper.

A.K. Lenstra, H.W. Lenstra, L. Lovász, Factoring Polynomials with Rational Coefficients, *Math. Ann.* **261** (1982) 515–534.

for all j, $\pi_n(G_j)$ is irreducible;
$\gcd(G_i, G_j) = 1$, $i \neq j$.

Moreover, for each polynomial $h(X) := \sum_{i=0}^{\delta} a_i X^i \in \mathbb{Z}[X]$ we will denote

$$\|h\| := \sqrt{\sum_{i=0}^{\delta} |a_i|^2}.$$

With this notation, it is clear that

for all J, $1 \leq J \leq r$, there exists I, $1 \leq I \leq s : \pi(G_J) = g_J$ divides $\pi(f_I)$;

moreover

Proposition 18.7.1. *Let $h(X) \in \mathbb{Z}[X]$ be a factor of g. Then the following conditions are equivalent:*

(1) $\pi(G_J)$ divides $\pi(h)$ in $\mathbb{Z}_p[X]$,
(2) $\pi_n(G_J)$ divides $\pi_n(h)$ in $\mathbb{Z}_q[X]$,
(3) f_I divides h in $\mathbb{Z}[X]$.

In particular $\pi_n(G_J)$ divides $\pi_n(f_I)$ in $\mathbb{Z}_q[X]$.

Proof

(3) $\Longrightarrow$ (2) : obvious.
(2) $\Longrightarrow$ (1) : obvious.
(1) $\Longrightarrow$ (3) : since $\pi(g)$ is squarefree, $\pi(G_J)$ does not divide $\pi(\frac{g}{h})$ in $\mathbb{Z}_p[X]$; therefore f_I does not divide $\frac{g}{h}$ in $\mathbb{Z}[X]$, so that it divides h.

The last statement follows taking $h = f_I$. ♃

The Lenstra–Lenstra–Lovász paper gives a proof of the following

Fact 18.7.2. *With the notation above, let G_J be a factor of $\pi_n(g)$ and let $l := \deg(G_J)$. For each integer $m \geq l$ satisfying the inequality*

$$p^{nl} > 2^{\frac{dm}{2}} \binom{2m}{m}^{\frac{d}{2}} \|g\|^{m+d} \tag{18.2}$$

it is possible – with polynomial complexity in m, and $\log(p)n$ – to decide whether $\deg(f_I) \leq m$, in which case it is also possible to determine f_I.

Fig. 18.4. Lenstra–Lenstra–Lovász Algorithm

$[f_1, \ldots, f_s] := \mathbf{L}^3 - \mathbf{Factorization}(f)$
where
    $f(X) \in \mathbb{Q}[X]$ is a squarefree polynomial
    $f_i(X) \in \mathbb{Q}[X]$ are monic irreducible polynomials
    $f = f_1, \ldots, f_r$
List := []
$g := \mathrm{Prim}(f)$
Repeat
    chooserandom $p \in \mathbb{Z}$ prime
until p does not divide $\mathrm{lc}(g)$ and $(g \bmod p)$ is squarefree
$[g_1, \ldots, g_r] := \mathbf{Factorization}(g, \mathbb{Z}_p)$
$\mathcal{I} := \{i : 1 \leq i \leq r\}$
While card$(\mathcal{I}) > 1$ **do**
    Choose $J \in \mathcal{I}$
    $\mathcal{I} := \mathcal{I} \setminus \{J\}$
    $l := \deg(g_J), d := \deg(g), m := d$
    Compute n such that $q := p^n$ satisfies Equation 18.2
    $(G_J, H_J) := \mathbf{HenselLifting}(g, \rho(g_J), \prod_{i \in \mathcal{I}} \rho(g_i), p, n)$
    $m := \deg(g_J)$
    Repeat
        (**bool**, h) := **Query**(g, G_J, m)
        $m := m + 1$
    until bool = true
    List := List $\cup\ [h], g := \frac{g}{h}$
    $\mathcal{I} := \{i \in \mathcal{I} : g_i \text{ does not divide } (h \bmod p)\}$
List $\cup\ [g]$

Algorithm 18.7.3 (Lenstra–Lenstra–Lovász). Thanks to Fact 18.7.2, the L^3 algorithm (Figure 18.4) allows us to factorize a polynomial $f(X) \in \mathbb{Q}[X]$; the complexity of this algorithm can be proved to be

$$\mathcal{O}\big(\deg(f)^{12} + \deg(f)^9 \log^3(\|f\|)\big).$$

In my presentation I denote by (**bool**, h) := **Query**(g, G_J, m) the algorithm whose existence is implied by Fact 18.7.2, and whose output is

$$(\mathbf{bool}, h) := \begin{cases} (\mathbf{true}, f_I) & \text{if } \deg(f_I) \leq m, \\ (\mathbf{false}, G_J) & \text{otherwise.} \end{cases}$$

The proof of Fact 18.7.2 is based on study of the *lattice*

$$L(m) := \{h(X) \in \mathbb{Z}[X] : \deg(h) \leq m, \pi_n(h) = \pi_n(G_J)\}.$$

Definition 18.7.4. *For two vectors* $v := (v_1, \ldots, v_k)$, $w := (w_1, \ldots, w_k)$, *in* $\mathbb{R}^k$,

$$(v, w) := \sum_{i=1}^{k} v_i w_i$$

denotes the usual scalar product in $\mathbb{R}^k$ and

$$\|v\| := \sqrt{\sum_{i=0}^{d} v_i^2}$$

the ordinary euclidean length, so that $\|v\|^2 = (v, v)$.

A subset $L \subset \mathbb{R}^k$ is called a **lattice** if there is a basis $\mathbf{B} := \{b_1, \ldots, b_k\}$ of $\mathbb{R}^k$ such that

$$L = \left\{ \sum_{i=1}^{k} a_i b_i : a_i \in \mathbb{Z} \right\}.$$

If $\mathbf{B} := \{b_1, \ldots, b_k\}$ is a basis of a lattice $L \subset \mathbb{R}^k$ with $b_i := (b_{i1}, \ldots, b_{ik})$, for all i, the **discriminant** of L is

$$d(L) := |\det(b_{ij})|$$

which can be proved to be independent of the choice of the basis $\mathbf{B}$ and to be the volume of the parallelepiped whose sides are the b_is.

Given a basis $\mathbf{B} := \{b_1, \ldots, b_k\}$ of $\mathbb{R}^k$, the *Gram–Schmidt process* allows us to compute an orthogonal basis $\mathbf{B}^* := \{b_1^*, \ldots, b_k^*\}$ and real numbers μ_{ij}, $1 \leq j < i \leq k$ by the inductive formula

$$\mu_{ij} := \frac{(b_i, b_j^*)}{(b_j^*, b_j^*)}, \qquad b_i^* := b_i - \sum_{j=1}^{i-1} \mu_{ij} b_j^*.$$

A basis $\mathbf{B} := \{b_1, \ldots, b_k\}$ of a lattice $L \subset \mathbb{R}^k$ is called **reduced** if

$|\mu_{ij}| \leq \frac{1}{2}$, for all $i, j, 1 \leq j < i \leq k$,
$\|b_i^* + \mu_{ii-1} b_{i-1}^*\|^2 \geq \frac{3}{4} \|b_{i-1}^*\|^2$, for all $i, 1 < i \leq k$.

Lemma 18.7.5. *Let $\mathbf{B} := \{b_1, \ldots, b_k\}$ be a reduced basis of a lattice $L \subset \mathbb{R}^k$, and let $\mathbf{B}^* := \{b_1^*, \ldots, b_k^*\}$ be the orthogonal basis produced by the Gram–Schmidt process. Then:*

(1) $d(L) = \prod_i \|b_i^*\|$.

(2) $\|b_i\| \geq \|b_i^*\|$, *for all i.*

(3) $d(L) \leq \prod_i \|b_i\|$ *(Hadamard's inequality).*

(4) $\|b_j\|^2 \leq 2^{i-1} \|b_i^*\|^2$, *for all $i, j, 1 \leq j \leq i \leq k$.*

(5) $\|b_1\| \leq 2^{\frac{k-1}{2}} \|\beta\|$, *for all $\beta \in L, \beta \neq 0$.*

(6) *If $\{\beta_1, \ldots, \beta_t\} \subset L$ is a linearly independent set, then*

$$\|b_j\|^2 \leq 2^{k-1} \max\{\|\beta_i\|^2 : 1 \leq i \leq t\}, \text{ for all } j, 1 \leq j \leq t.$$

Proof

(1) Since $\mathbf{B}^*$ consists of an orthogonal basis.
(2) $\|b_i\|^2 = \|b_i^* + \sum_{j=1}^{i-1} \mu_{ij} b_j^*\|^2 = \|b_i^*\|^2 + \sum_{j=1}^{i-1} \mu_{ij}^2 \|b_j^*\|^2 \geq \|b_i^*\|^2$.
(3) The claim follows since for all i, $\|b_i\| \geq \|b_i^*\|$.
(4) We have

$$\|b_i^*\|^2 \geq \left(\frac{3}{4} - \mu_{ii-1}^2\right) \|b_{i-1}^*\|^2 \geq \frac{1}{2} \|b_{i-1}^*\|^2, \text{ for all } i,$$

so that

$$\|b_j^*\|^2 \leq 2^{i-j} \|b_i^*\|^2, \text{ for all } i, j, \ j \leq i.$$

Therefore

$$\begin{aligned}
\|b_i\|^2 &= \left\| b_i^* + \sum_{j=1}^{i-1} \mu_{ij} b_j^* \right\|^2 \\
&= \|b_i^*\|^2 + \sum_{j=1}^{i-1} \mu_{ij}^2 \|b_j^*\|^2 \\
&\leq \|b_i^*\|^2 + \sum_{j=1}^{i-1} \frac{1}{4} 2^{i-j} \|b_i^*\|^2 \\
&= \left(1 + \frac{1}{4}(2^i - 2)\right) \|b_i^*\|^2 \\
&\leq 2^{i-1} \|b_i^*\|^2.
\end{aligned}$$

As a consequence

$$\|b_j\|^2 \leq 2^{j-1} \|b_j^*\|^2 \leq 2^{i-1} \|b_i^*\|^2 \text{ for all } i, j, \ j \leq i.$$

(5) Let $a_i \in \mathbb{Z}$ and $\alpha_i \in \mathbb{R}$ be such that $\beta = \sum_i a_i b_i = \sum_i \alpha_i b_i^*$. Let $I := \max\{i : a_i \neq 0\}$, so that $0 \neq \alpha_I = a_I$. Therefore

$$\|\beta\|^2 \geq \alpha_I^2 \|b_I^*\|^2 \geq \|b_I^*\|^2,$$

and

$$2^{k-1} \|\beta\|^2 \geq 2^{k-1} \|b_I^*\|^2 \geq 2^{I-1} \|b_I^*\|^2 \geq \|b_1\|^2.$$

(6) With an argument analogous to the above proof, we can deduce that for all j, $1 \leq j \leq t$, there is $i(j)$, $1 \leq i(j) \leq k$ such that

$\|\beta_j\|^2 \geq \|b_{i(j)}\|^2$,
β_j is contained in the $\mathbb{R}$-vector space spanned by $\{b_1, \ldots, b_{i(j)}\}$.

Therefore, after renumbering the β_js so that $i(1) \leq i(2) \leq \cdots \leq i(t)$, we can deduce that $j \leq i(j)$: in fact, we would otherwise get the contradiction that the linearly independent set $\{\beta_1, \ldots, \beta_j\}$ is contained in the $\mathbb{R}$-vector space spanned by $\{b_1, \ldots, b_{j-1}\}$.

Therefore

$$\|b_j\|^2 \leq 2^{i(j)-1} \|b^*_{i(j)}\|^2 \leq 2^{k-1} \max\{\|\beta_i\|^2\}.$$

□

Fact 18.7.6. *Let* $\mathbf{B} := \{b_1, \ldots, b_k\}$ *be a basis of a lattice* $L \subset \mathbb{R}^k$. *Then, there is a basis* $\mathbf{B'} := \{b'_1, \ldots, b'_k\}$ *with* $b'_i := (b'_{i1}, \ldots, b'_{ik})$, *for all i such that* $b'_{ij} = 0$, *for all i, j,* $1 \leq i < j \leq k$.

□

Let us identify the set $\{h(X) \in \mathbb{R}[X] : \deg(h) \leq m\}$ with the $\mathbb{R}$-vector space generated by $\{1, X, \ldots, X^m\}$ and each polynomial $h(X) := \sum_{i=0}^{m} a_i X^i$ with the vector $(a_0, a_1, \ldots, a_m)$.

With this identification it is clear that the lattice

$$L(m) := \{h(X) \in \mathbb{Z}[X] : \deg(h) \leq m, \pi_n(h) = \pi_n(G_J)\}$$

is generated by the basis

$$\{q X^i : 0 \leq i < l\} \cup \{G_J X^i : 0 \leq i \leq m - l\}.$$

Proposition 18.7.7. *Let* $b(X) \in L(m)$ *satisfy*

$$q^l > \|g\|^m \|b\|^d.$$

Then b is divisible by f_I and, in particular, $\gcd(g, b) \neq 1$.

Proof We may assume $b \neq 0$ and let us set $h := \gcd(g, b)$. By Proposition 18.7.1 it is sufficient to prove that $\pi(G_J)$ divides $\pi(h)$ in $\mathbb{Z}_p[X]$.

If this were not the case, then $\gcd(\pi(G_J), \pi(h)) = 1$ and there would be $\lambda', \mu', \nu' \in \mathbb{Z}[X]$ such that

$$\lambda' G_J + \mu' h = 1 - p\nu'. \tag{18.3}$$

In order to derive a contradiction from Equation 18.3, let us set $e := \deg(h)$, $m' := \deg(b)$ so that $0 \leq e \leq m' \leq m$. Then we consider the $\mathbb{Z}$-vector space

$$\{\lambda g + \mu b - \nu : \lambda, \mu, \nu \in \mathbb{Z}[X], \deg(\lambda) < m' - e, \deg(\mu) < d - e, \deg(\nu) < e\},$$

which is a subvector space M of the $\mathbb{Z}$-vector space generated by

$$\{X^i : 0 \leq i \leq d + m' - e - 1\},$$

and we intend to show that M is a lattice whose basis is

$$\{X^i : 0 \leq i < e\} \cup \{X^i g : 0 \leq i < m' - e\} \cup \{X^i b : 0 \leq i < d - e\}.$$

In fact, let us assume that $\lambda g + \mu b - \nu = 0$: then h divides ν while

$$\deg(\nu) < e = \deg(h),$$

so that

$$\lambda g + \mu b = \nu = 0;$$

therefore $\lambda \frac{g}{h} = -\mu \frac{b}{h}$ and $\frac{g}{h}$ divides μ since $\gcd(\frac{g}{h}, \frac{b}{h}) = 1$; however,

$$\deg(\mu) < d - e = \deg\left(\frac{g}{h}\right)$$

so that $\lambda = \mu = \nu = 0$ and the assertion is proved.

From Hadamard's inequality we deduce that

$$d(M) \leq \|g\|^{m'-e} \|b\|^{d-e} \leq \|g\|^m \|b\|^d < q^l. \tag{18.4}$$

Let $\nu \in M$ be such that $\deg(\nu) < e + l$ and let $Q, R \in \mathbb{Z}[X]$ be such that

$\nu = Qh + R,$
$\deg(R) < e.$

Multiplying Equation 18.3 by $Q \sum_{l=0}^{n-1} p^i \nu'^i$ we obtain $\lambda'', \mu'', \nu'' \in \mathbb{Z}[X]$ such that

$$\lambda'' G_J + \mu'' Qh = Q + p^n \nu''. \tag{18.5}$$

Now $\pi_n(G_J)$ divides both $\pi_n(b)$ – since, by assumption, $b \in L(m)$ –, and $\pi_n(g)$; therefore $\pi_n(G_J)$ divides $\pi_n(h)$ and, consequently, $\pi_n(Q)$; however, since G_J is monic,

$$\deg(\pi_n(G_J)) = l > \deg((\pi_n(Q)),$$

so that $\pi_n(Q) = 0$.

This implies that

$$\text{for all } \nu \in M, \deg(\nu) < e + l \implies \deg(\pi_n(\nu)) < e.$$

Since we can choose a basis $\mathbf{B} := \{b_i : 0 \leq i \leq d + m' - e - 1\}$ of M such that $\deg(b_i) = i$, for all i and, by the result above,

$$\text{lc}(b_i) \equiv 0 \pmod{q}, \text{ for all } i, e \leq i < e + l,$$

we deduce that $d(M) \geq q^l$ which gives the required contradiction with Equation 18.4. ♃

Proposition 18.7.8. *Suppose that* $\mathbf{B} := \{b_i : 1 \leq i \leq m+1\}$ *is a reduced basis of* $L(m)$ *and that Equation 18.2 holds. Then*

$$\deg(f_I) \leq m \iff \|b_1\| < \sqrt[d]{p^{nl}/\|g\|^m}.$$

Proof One implication is a corollary of Propoistion 18.7.7 since

$$\deg(f_I) \leq \deg(b_1) \leq m.$$

To prove the converse, let us assume $\deg(f_I) \leq m$ so that $f_I \in L(m)$ by Proposition 18.7.1; since (by Proposition 18.4.6)

$$\|f_I\| \leq \binom{2m}{m}^{1/2} \|g\|,$$

applying Lemma 18.7.5(5) with $\beta = f_I$ we deduce

$$\|b_1\| \leq 2^{m/2} \|f_I\| \leq 2^{m/2} \binom{2m}{m}^{1/2} \|g\|,$$

from which the claim follows, applying Equation 18.2. □

Proposition 18.7.9. *Suppose that* $\mathbf{B} := \{b_i : 1 \leq i \leq m+1\}$ *is a reduced basis of* $L(m)$, *that Equation 18.2 holds, and that there is a* $j, 1 \leq j \leq m+1$ *such that*

$$\|b_j\| < \sqrt[d]{p^{nl}/\|g\|^m} \tag{18.6}$$

and let t be the largest such j. Then

$\deg(f_I) = m + 1 - t$;
$f_I = \gcd(b_1, \ldots, b_t)$;
Equation 18.6 holds for each $j \leq t$.

Proof Let

$$\mathcal{I} := \{j : 1 \leq j \leq m+1, \text{ Equation 18.6 holds for } j\}$$

and $\mathbf{f} := \gcd(b_j : j \in \mathcal{I})$.
By Proposition 18.7.7 we know that f_I divides b_j, for all $j \in \mathcal{I}$, and so f_I divides $\mathbf{f}$.

18.7 L^3 Algorithm

For each $j \in \mathcal{I}$, $\deg(b_j) \leq m$ and b_j is in the $\mathbb{Z}$-vector space spanned by

$$\{fX^i, 0 \leq i \leq m - \deg(f)\}.$$

Since the b_js are linearly independent, this implies

$$\operatorname{card}(\mathcal{I}) \leq m + 1 - \deg(f).$$

By Proposition 18.4.6

$$\|f_I X^i\| = \|f_I\| \leq \binom{2m}{m}^{1/2} \|g\|, \text{ for all } i, 0 \leq i \leq m - \deg(f_I);$$

therefore, applying Lemma 18.7.5(6) with $\beta_i = f_I X^i$, we deduce

$$\|b_j\| \leq 2^{m/2}\|f_I\| \leq 2^{m/2}\binom{2m}{m}^{1/2} \|g\|, \text{ for all } j, 1 \leq j \leq m + 1 - \deg(f_I).$$

Therefore, by Equation 18.2, we can conclude that $\{j : 1 \leq j \leq m + 1 - \deg(f_I)\} \subset \mathcal{I}$.

Since f_I divides f, we can conclude that

$\deg(f_I) = \deg(f)$,
$\{j : 1 \leq j \leq m + 1 - \deg(f_I)\} = \mathcal{I}$;
$t = m + 1 - \deg(f_I)$.

In order to conclude the proof by showing that f_I and f are associated, we just need to show that f is primitive: since $\operatorname{Prim}(b_1)$ is divisible by f_I, it is an element of $L(m)$; since $b_1 \in \mathbf{B}$, then $\operatorname{Cont}(b_1) = 1$ and f is primitive, being a factor of b_1. ♃

Theorem 18.7.10 (Lenstra–Lenstra–Lovász). *Given a basis*

$$\mathbf{B} := \{b_1, \ldots, b_k\}$$

of a lattice $L \subset \mathbb{R}^k$ there is an algorithm which computes a reduced basis of L with complexity $\mathcal{O}(k^4 \log(B))$ where $B \in \mathbb{R}$ is such that

$$B \geq \max\left(2, \|b_1\|^2, \ldots, \|b_k\|\right).$$

♃

Algorithm 18.7.11. We can now prove Fact 18.7.2, except the complexity claim, by describing the algorithm $(\mathbf{bool}, h) := \mathbf{Query}(g, G_J, m)$.

Let us compute a reduced basis $\{b_1, \ldots, b_{m+1}\}$ of $L(m)$ from the basis

$$\mathbf{B} := \{qX^i : 0 \leq i < l\} \cup \{G_J X^i : 0 \leq i < m - l\}.$$

Then, we check whether $\|b_1\| < \sqrt[d]{p^{nl}/\|g\|^m}$.

If this is false, by Proposition 18.7.8 we can conclude that $\deg(f_I) > m$ and we return (**false**, G_J);
otherwise, by Proposition 18.7.9 we can conclude that $f_I = \gcd(b_1, \ldots, b_t)$ and we return (**true**, $\gcd(b_1, \ldots, b_t)$).

19
Finale

19.1 Kronecker's Dream

In concluding this part I want to give a résumé of the state of the art of polynomial factorization over $K[X]$, K a field:

The Berlekamp and Cantor–Zassenhaus Algorithms allow factorization if K is a finite field;

The Berlekamp–Hensel–Zassenhaus and Lenstra–Lenstra–Lovász factorization algorithms allows us to lift factorization over $\mathbb{Z}_p$ to one over $\mathbb{Z}$ and, by the Gauss Lemma, to one over $\mathbb{Q}$, so that factorization is available over the prime fields.

Algebraic extensions $K = F(\alpha)$ are dealt with by the Kronecker Algorithm (Section 16.3) if F is infinite, and by Berlekamp otherwise,

while Hensel–Zassenhaus allows us to factorize multivariate polynomials in $F[X_1, \ldots, X_n]$ if factorization over F is available, so that

the Gauss Lemma allows us to deal with transcendental extensions $K = F(X)$

so that factorization is available over every field explicitly given in Kronecker's Model.

19.2 Van der Waerden's Example

Within the development of computational techniques for polynomial ideal theory, started by the benchmark work by G. Herrmann[1], van der Waerden pointed to a fascinating limitation of the ability to build fields within Kronecker's

---

[1] Die Frage der endlichen viele Schritten in der Theorie der Polynomideal, *Math. Ann.* **95** (1926) 736–788.

Model. He proved:

Proposition 19.2.1 (van der Waerden (1929)). *The following are equivalent:*

*the existence of a general method for solving any problem of the kind
'does an integer n exit satisfying the property $E(n)$?',
where E is any property of the integers whose validity for any integer n is
solvable* in endlichvielen Schritten;
*the existence of a general method for factorizing any polynomial with coefficients in any explicitly given field $\mathbb{Q}(\alpha_1, \ldots, \alpha_n, \ldots)$ which is obtained
from $\mathbb{Q}$ by the adjunction of a (countable) infinite set of elements.*

Proof Assume there is a property E of the integers for which

for all $n \in \mathbb{N}$, it is possible to decide whether n satisfies $E(n)$;
it is impossible to decide whether there exists $n \in \mathbb{N}$ satisfying $E(n)$.

Then for all $n \in \mathbb{N}$, define

$$\alpha_n := \begin{cases} i & \text{if } n > 0 \text{ is the least integer which satisfies } E \\ \sqrt{p_n} & \text{otherwise,} \end{cases}$$

where p_n is the nth prime.
Setting $\mathbb{K} := \mathbb{Q}(\alpha_1, \ldots, \alpha_n, \ldots)$ and $f := X^2 + 1$, the ability to decide, in $\mathbb{K}[X]$, whether f is irreducible or factorizes as

$$f = (X - i)(X + i),$$

is equivalent to the ability to decide whether $i \in \mathbb{K}$ and so to decide whether there exists $n \in \mathbb{N}$ satisfying $E(n)$.
Conversely, assume that for any general property E of the integers whose validity is solvable *in endlichvielen Schritten* for any integer n, there is a method for solving any problem of the kind *'does an integer n exits satisfying the property $E(n)$?'*.
Let then $\mathbb{K} := \mathbb{Q}(\alpha_1, \ldots, \alpha_n, \ldots)$ and $f[X]$ be any polynomial in $\mathbb{K}[X]$. Let us just consider for the property $E(n)$ that f is reducible in $\mathbb{Q}(\alpha_1, \ldots, \alpha_n)$. It is clear that f is reducible in $\mathbb{K}[X]$, if there is an n such that f is reducible in

$$\mathbb{Q}(\alpha_1, \ldots, \alpha_n)[X],$$

i.e. $E(n)$ holds.

19.2 Van der Waerden's Example

Since for all $n \in \mathbb{N}$ it is possible to decide *in endlichvielen Schritten* whether f is reducible in $\mathbb{Q}(\alpha_1, \ldots, \alpha_n)[X]$, i.e. whether $E(n)$ holds, then the assumption guarantees the ability to decide the reducibility of f in $\mathbb{K}[X]$. ♃

Corollary 19.2.2. *Assume there is a property E of $\mathbb{N}$ for which*

for all $n \in \mathbb{N}$, it is possible to decide whether n satisfies $E(n)$;
it is impossible to decide whether there exists $n \in \mathbb{N}$ satisfying $E(n)$.

Then there is a field $\mathbb{K}$, explicitly given in Kronecker's Model and a polynomial $f(X) \in \mathbb{K}[X]$ for which factorization is not computable. ♃

Historical Remark 19.2.3. About the existence of such a property E, van der Waerden remarked:

Bei der Bildung des Beispiels, das den wesentlichen Inhart dieses Bewieses ausmacht, hätte ich mich naturlich auch auf eine bestimmte Eigenschaft $E(n)$, etwa ein bestimmtes bis jetzt noch nicht beobachtetes Vorkommnis in der Dezimalbruchentwicklung von π, stützen können. Ich habe das vermieden, weil die Voraussetzung der Unentscheidbarkeit eines solchen Existenzproblems nicht nur völlig unberechtigt, sondern auch für den Beweis zu einem gewissen Grade unwesentlich ist. Wesentlich ist nur die Voraussetzung einer Unentscheidbarkeit überhaupt, eines 'Ignorabimus' in bezug auf Existenzprobleme der genannten Art.
In the construction of the example that establishes the essential content of this proof, I could naturally also have supported my proof on a determined property, perhaps a determined event not yet observed in the decimal expansion of π. I have avoided this, since the assumption of the undecidability of such problems existence is not only completely unjustified, but also the proving of it is to a certain degree of minor importance. It is especially essential only to the hypothesis of undecidability, an 'Ignorabimus' in respect of existence problems of this kind.

To give a flavour of the kind of property which van der Waerden was thinking of, let me suggest the following:

$E(n)$ is the property that
    there are n 9s in the first $2n$ digits of the expansion of π:
    for any n, it is possible to compute the first $2n$ digits of the expansion of π and so decide whether $E(n)$ holds. However, any finite expansion of π allows us to get, say, $2m$ digits and so to decide on the validity of $E(n)$ for $n \leq m$, but, when $E(n)$ is false for all $n \leq m$, we still do not know whether there exists any $n > m$ such that $E(n)$ holds.
$E(n)$ is the property that
    $2n$ cannot be represented as a sum of 2 prime numbers.

Clearly it is possible to check in a finite number of computations whether $E(n)$ holds for any n. Deciding whether there is $n \in \mathbb{N}$ satisfying $E(n)$ is the same as disproving the Goldbach Conjecture.

A more deeply puzzling example, which I used in my lessons until 1994 is the following: $E(n)$ is the property that

there are non-trivial solutions of the diophantine equation $x^n + y^n = z^n$.

In 1994 Fermat's Theorem was proved, showing that there is no $n \in \mathbb{N}$ satisfying $E(n)$.

The kind of property E which van der Waerden was looking for could be formalized as a function $\mu : \mathbb{N} \mapsto \mathbb{N}$ for which

it is possible to 'compute' $\mu(n)$ for all $n \in \mathbb{N}$;
for any $n \in \mathbb{N}$ it is impossible to decide whether

$$\text{there exists } m \in \mathbb{N} : \mu(m) = n.$$

Functions of this kind are called *semirecursive functions* and their existence was proved in 1936 by Kleene, Church and Turing.

Remark 19.2.4. In the same setting but in a more jocular mode, consider three semirecursive functions μ, ν_1, ν_2 and define

$$\lambda(n) := \begin{cases} 0 & \text{if } \nu_2(n) = 0 \text{ and there exists } m \leq n : \mu(m) = 0 \\ 1 & \text{otherwise,} \end{cases}$$

and α_n, β_n, for all $n \in \mathbb{N} \setminus \{0\}$, as follows, where again p_i denotes the ith prime:

$$\alpha_n := \begin{cases} i & \text{if } n \text{ is the least integer such that} \\ & \quad \nu_1(n) = 0 \text{ or } \mu(n) = 0 \\ \sqrt{p_n} & \text{otherwise,} \end{cases}$$

$$\beta_n := \begin{cases} i & \text{if } n \text{ is the least integer such that} \\ & \quad \nu_2(n) = 0 \text{ and there exists } m \leq n : \mu(m) = 0 \\ \sqrt{p_n} & \text{otherwise.} \end{cases}$$

Setting

$$A := \mathbb{Q}(\alpha_1, \ldots, \alpha_n, \ldots), \quad B := \mathbb{Q}(\beta_1, \ldots, \beta_n, \ldots),$$

we have

$$\begin{cases} 0 \in \operatorname{Im} \mu & \implies & i \in A \\ 0 \notin \operatorname{Im} \mu & \implies & \begin{cases} i \notin A & \text{if } 0 \notin \operatorname{Im} \nu_1 \\ i \in A & \text{if } 0 \in \operatorname{Im} \nu_1 \end{cases} \end{cases} \quad \begin{cases} i \notin B & \text{if } 0 \notin \operatorname{Im} \lambda \\ i \in B & \text{if } 0 \in \operatorname{Im} \lambda \end{cases} \quad i \notin B.$$

19.2 Van der Waerden's Example

As a consequence:

if $0 \in \operatorname{Im} \mu$, it is possible to factorize $X^2 + 1$ in A but not in B,
if $0 \notin \operatorname{Im} \mu$, it is possible to factorize $X^2 + 1$ in B but not in A,

and, since it is impossible to decide whether $0 \in \operatorname{Im} \mu$:

there exist two fields A and B, in one of which $X^2 + 1$ is factorizable while in the other it is not, but it is impossible to decide which is which.

This gives a sort of Mathematical Heisenberg Principle of Indetermination.

Bibliography

Arezzo D., *Appunti al corso di Istituzioni di Geometria Superiore* (1981)
Artin M., *Algebra*, Prentice-Hall (1991)
Ben-Or M., Kozen D., Reif J., *The complexity of elementary algebra and geometry*, J. Comput. System Sci. **32** (1986), 251–264
Berlekamp E.R., *Factoring Polynomials over Finite Fields*, Bell System Tech. J. **46**, 1853–1859 (1967)
Berlekamp E.R., *Factoring Polynomials over Large Finite Fields*, Math. Comp. **24**, 713–735 (1970)
Broglia, F. (Ed.), *Lectures in Real Geometry*, De Gruyter (1996)
Brown W.S., *On Euclid's Algorithm and the Computation of Polynomial and Greatest Common Divisors*, J. ACM **18** (1971), 478–504
Buchberger B., Collins G.E., Loos R. (Eds.), *Computer Algebra. Symbolic and Algebraic Computation*, Springer (1982)
Cantor D.G., Zassenhaus H. *On Algorithms for Factoring Polynomials over Finite Fields*, Math. Comp. **36** (1981), 587–592
Char B.W., Geddes K.O., Gonnet G. H., *GCDHEU: Heuristic Polynomial GCD Algorithm Based On Integral GCD Computation*, L. N. Comp. Sci **174** (1984), 285–296, Springer
Collins G.E., *Subresultants and Polynomial Remainder Sequence*, J. ACM **14** (1967), 128–142
Coste M., Roy M.-F., *Thom's lemma, the coding of real algebraic numbers and the topology of semi-algebraic sets*, J. Symb. Comp. **5** (1988), 121–129
Davenport J.H., Siret Y., Tournier E., *Computer Algebra*, Academic Press, (1988)
Davenport J.H., *On the Integration of Algebraic Functions*, L. N. Comp. Sci **102** (1981), Springer
Davenport J.H., Padget J., *HEUGCD: How Elementary Upperbounds Generate Cheaper Data*, L. N. Comp. Sci **204** (1985), 18–28, Springer
Duval D., *Diverses questions relatives au calcul formel avec de nombres algebriques*, These d'Etat, Grenoble (1987)
Ebbinghaus H.-D. et al., *Numbers* Springer (1991)
Edward H.M., *Galois Theory* Springer (1984)
Gauss C.F., *Disquisitiones Arithmeticae*, Braunschweig,1810; republished in *Werke* Band I, Göttingen 1870
Hermann G., *Die Frage der endlich vielen Schritte in der Theorie der Polynomideale*, Math. Ann. **95** (1926) 736–788

Kaltofen E., *Factorization of Polynomials*, in Buchberger *et al.* (1982) 95–114
Kemper G., *The Calculation of Radical Ideals in Positive Characteristic* (2000)
Knuth D.E., *The Art of Computer Programming* Vol. 2 Addison-Wesley (3rd ed, 1997)
Kronecker L., *Grundzüge einer Arithmetischen Theorie der Algebraischen Grössen*, Crelle's Journal, **92** (1882)
Lenstra A.K., Lenstra K.W. Jr., Lovász, L., *Factorization with Rational Coefficients*, Math. Ann. **261** (1982), 515–534
Lidl R., Niederreiter H., *Finite fields*, Cambridge University Press (1983)
Loos R., *Computing in Algebraic Extensions*, in Buchberger *et al.* (1982) 173–188
Macaulay F. S., *The Algebraic Theory of Modular Systems*, Cambridge University Press (1916)
Mignotte M., *An Inequality About Factors of Polynomials*, Math. Comp. **28** (1974), 1153–1157
Mignotte M., *Mathematics for Computer Algebra*, Springer (1992)
Moses J., Yun D.Y.Y., *The EZ GCD Algorithm*, in Proc. of the ACM Annual Conference (1973), 159–166
Remmert R., *The Fundamental Theorem of Algebra*, in Ebbinghaus *et al.* (1991) 97–122
Roy M.-F., *Basis algorithms in real algebraic geometry and their complexity: from Sturm's theorem to the existential theory of reals*, in Broglia (1996) 1–66
Sala M., *Gröbner bases and distance of cyclic codes* (2001)
Toti Rigatelli L., *Evariste Galois*, Birkhäuser (1996)
Trager B. *Integration of Algebraic Functions* Ph.D. Thesis, M.I.T. (1984)
van der Waerden B.L., *Eine Bemerkung über die Unzelegbarkeit von Polynomen*, Math. Ann. **102** (1930) 738–739.
van der Waerden B.L., *Modern Algebra* Vol. I, Ungar (1949)
van der Waerden B.L., *A History of Algebra* Vol. I, Springer, (1985)
van Lint J.H., *Coding Theory*, Springer (1971)
Wang P., *The EZZ-GCD Algorithm*, SIGSAM Bullettin **14** (1980), 50–60
Zariski O., Samuel P. *Commutative Algebra* Vol. I, Van Nostrand (1958)

Index

Abel–Ruffini Theorem 52
Adjunction 82
Admissible Duval sequence 229
Admissible sequence 161
Algebraic extension field 84
Algebraic closure 176
Algebraic dependence 181
Algebraic element 82
Algebraically closed field 176
Associate 10

Bezout's identity 8

Characteristic 54
Characteristic morphism 54
Complete sums 101
Conjugate 192
Constructable by ruler and compass 318
Constructable number 320
Content 92
Cyclotomic field 127
Cyclotomic polynomial 136

Degree of algebraic element 83
Degree of finite extension 84
Degree of inseparability 64
Degree of rational functions 187
Direct sum decomposition 30
Discriminant 110
Distinct degree factorization 126
Distinct power factorization 69
Division theorem 4
Dual space 211
Duval field 229

Effective 5
Eisenstein criterion 318
Elementary symmetric functions 97
Endlichvielen Schritten 5
Euclidan Algorithm 7

Euler totient function 136
Explicitly given 6, 163
Exponent of inseparability 64
Extended Euclidean Algorithm 9

Field extension 82
Finite extension field 84
Finitely generated field extension 82
Formally real fields 265
Frobenius automorphism 57
Fundamental Theorem of Algebra 51
Fundamental Theorem of Galois Theory 303

Galois extension 299
Galois field 121
Galois group 299
Garner Algorithm 29
Generator 130
Gram-Schmidt process 408
Gröbnerian symmetric functions 101

Half-extended Euclidean Algorithm 9

Idempotent 35
Infinitesimal 271
Infinity 271
Inseparable element 64
Inseparable extension field 85
Irreducible 10

Kummer extension 309

Lagrange Chines Remainder Theorem formula 42
Lagrange Interpolation formula 44
Lagrange resolvent 309
Landau–Mignotte Inequality 397
Laplace–Gauss Resolvent 111
Lattice 408
Leading coefficient 4

Index

Lexicographical ordering 97, 165
Linear functionals 211
Little Fermat Theorem 56
Locator polynomial 101

Minimal polynomial 83
Multiplicative order 140
Multiplicity of a root 59
Multiplicity of factors 61

Newton Interpolation Formula 29
Nilpotent 32
Nilpotent ring 32
Norm of a matrix 206
Norm of a polynomial 209
Norm of an element 207
Normal basis 216
Normal extension field 194
Normal subgroup 304

Order of a root of unity 128
Ordered field 232

Perfect field 67
Polynomial remainder sequence 6
Positive cone 232
Pouiseux series 271
Prime field 55
Primitive element 130
Primitive Element 170
Primitive idempotents 37
Primitive polynomial 92
Primitive root of unity 129
Primitive set of idempotents 37
Pure transcendental extension 181
Purely inseparable element 85
Purely inseparable extension 85

Quotient 4

Radical extension 306
Real closed fields 265
Reduced basis 408
Reduced degree 64

Remainder 4
Resultant 112
Root of unity 127
Root tower 306

Section 177
Separable element 64
Separable extension field 85
Separable polynomial 62
Sign 272
Sign pattern 272
Simple field extension 82
Simple root 59
Solvable by radicals 306
Solvable group 312
Splitting field 88
Squarefree 34
Squarefree associate 69
Squarefree decomposition 69
Sturm representation 283
Sum of squares 264
Swinnerton-Dyer Polynomial 402
Sylvester Matrix 112
Symmetric function 97

Thom codification 290
Tower of fields 78
Trace of a matrix 206
Trace of an element 207
Transcendency degree 184
Transcendental basis 181
Transcendental element 82
Transcendental extension field 84
Transcendental set 181
Transitive group 300

Unique factorization domain 10
Universal field 186

Vandermonde determinant 109

Waring functions 101

Zech logarithm 135